METHODS IN MOLECULAR BIOLOGY™

Series Editor
John M. Walker
School of Life Sciences
University of Hertfordshire
Hatfield, Hertfordshire, AL10 9AB, UK

For other titles published in this series, go to
www.springer.com/series/7651

Advanced Protocols in Oxidative Stress II

Edited by

Donald Armstrong

University at Buffalo, Buffalo, NY, USA
University of Florida, Gainesville, FL,USA

☼ Humana Press

Editor
Donald Armstrong
University at Buffalo
Buffalo, NY, USA
University of Florida
Gainesville, FL,USA
donnchris6@gmail.com

ISSN 1064-3745   e-ISSN 1940-6029
ISBN 978-1-60761-410-4   e-ISBN 978-1-60761-411-1
DOI 10.1007/978-1-60761-411-1
Springer New York Dordrecht Heidelberg London

Library of Congress Control Number: 2009930359

Cover illustration: In vivo multiphoton image of the redox ratio (fluorescence intensity of FAD/NADH) from a 50 micron depth in low grade dysplastic tissue in the hamster cheek pouch. The redox ratio highlights metabolic activity in the electron transport chain within cellular mitochondria. Comparisons between redox ratio images can provide relative measurements of metabolic activities in cells, where a decrease in the redox ratio indicates increased oxidative metabolic activity. Pink pixels represent low redox values and yellow pixels represent high redox values. The image is 100 x 100 microns (figure credit: Melissa Skala).

Printed on acid-free paper

Humana Press is part of Springer Science+Business Media (www.springer.com)

Preface

The present volume is the second in the continuation of the new collection of oxidative stress (OS) protocols that focus on novel techniques for detecting ROS/RNS, unique AOX technology and applications, gene expression and biostatistics for evaluating OS-derived experimental data. It presents 30 additional chapters to Volume I. These current methods expand on those in *Advanced Protocols I*, giving researchers the opportunity to have 61 total chapters at their disposal. The number of ROS/RNS and AOX chapters in Volume II is now roughly equal, indicating a trend toward therapy.

Once again, there is broad international participation. This volume includes animal models and numerous studies focusing on mitochondria during hypoxic conditions using advanced methods for pO_2, peroxynitrate, reactive S-nitrosothiols, lipid peroxides, COX and the mitochondrial membrane potential. Cell permeable fluorescent and chemiluminescent probes directly measure the global redox state using flow cytometry, and mutiphoton microscopy is described for metabolic monitoring which requires no sample preparation. Erythrocyte fragility can also be used to analyze the effect of OS relative to change in membrane structure. A PubMed search shows that there were over 40 publications relating to OS and mitochondria in January 2009 alone. It is predicted that studies on mitochondrial OS will increase and represent a major portion of future issues of *Advanced Protocols*.

The format is the same as in Volume I, and chapters in Part I present point-of-care in vivo testing in many diverse clinical settings for use in diagnosis, prognosis, and management. It is envisioned that a "gold standard" using a combination of ROS and AOX biomarkers in body fluids can be identified and confirmed. This will be valuable in giving clinicians a global assessment of redox status in their patients. In fact, many bench tests have already been converted into commercial kits for this purpose which are reproducible, fast, user-friendly, and cost effective. These assays can be applied for real-time monitoring.

Chapters in Part II suggest the use of various ROS and AOX biomarkers in developing an index for summarizing OS status. Comparisons between biomarkers including ORAC, TRAP, FRAP, ABTS, DPPH, TOSC, TEAC, PMRS, and oxHLIA are described, and combinations of several biomarkers are recommended for optimal results. Other new methods are presented using cupric ion assays for hydroxyl radical detection, laser-induced breakdown spectroscopy for trace metals, oxidative hemolysis, AOX hybrids of lipoic acid that increase potency, and pentaerithrityl tetranitrate (PETN) therapy to decrease the OS burden. Bacterial biotransformation can be used as an alternative method to pharmacological synthesis for preparation of estrogenic AOX compounds. Electrochemical detection with a new boron-doped diamond electrode is described for quantifying reduced and oxidized glutathione and for thioesters. A plasma membrane redox system is presented for recycling ascorbate radicals. Turmeric is shown to be effective in reducing lipid peroxides and also has immunomodulating properties.

Chapters in Part III illustrate purification of AOX gene vectors for delivery of large amounts of AOX enzyme DNA, thus increasing new in vivo synthesis in the treatment of metabolic disease. The activation of caveolin-1 promoter by ROS is detected using EMSA and chromatic immunoprecipitation methodology, where the latter technique can actually pinpoint transcription factors.

The last chapter gives a statistical approach using meta-analysis that polls data when there is conflicting or inconclusive results, thus providing more accurate analysis of therapeutic intervention. Taken together with methods published in previous books, i.e., volumes 108, 186, 196, and 477, there are now a total of 169 technologies available to readers on OS in this series. These volumes provide a valuable office and laboratory resource for OS methodology. To emphasize the importance of OS technology, PubMed lists 643 articles linking free radicals with advanced technology and 801 linking AOXs with advanced technology.

Chapters in the *Advanced Protocols* series can be used in graduate education, and the preparation of them is an excellent opportunity for junior researchers to gain experience in writing in a clear and organized manner.

I thank David P. Armstrong again for computer assistance and support, for many author referrals from colleagues, as well as advice from Patrick Marton on developing this publication.

Buffalo, New York, USA   *Donald Armstrong*

Contents

PART I   REACTIVE OXYGEN AND NITROGEN TECHNIQUES

PART III   GENE EXPRESSION

PART IV   BIOSTATISTICS

Contributors

IAN ACWORTH • *ESA Biosciences, Inc, Chelmsford, MA, USA*

MOHAMMAD AFZAL • *Department of Biological Sciences, Faculty of Science, Kuwait University, Safat, Kuwait*

SAMEERA AL-AWADI • *Department of Biological Sciences, Faculty of Science, Kuwait University, Safat, Kuwait*

REŞAT APAK • *Department of Chemistry, Faculty of Engineering, Istanbul University, Avcilar, Istanbul, Turkey*

BRUCE BAILEY • *ESA Biosciences, Inc, Chelmsford, MA, USA*

PETER J. BARNES • *Department of Thoracic Medicine, Imperial College, School of Medicine at the National Heart and Lung Institute, London, UK*

JANINE N. BARTHOLOMEW • *Department of Pharmacology, University of Pittsburgh School of Medicine, Pittsburgh, PA, USA*

BURCU BEKTAŞOĞLU • *Department of Chemistry, Faculty of Engineering, Istanbul University, Avcilar, Istanbul, Turkey*

MUSTAFA BENER • *Department of Chemistry, Faculty of Engineering, Istanbul University, Avcilar, Istanbul, Turkey*

SILVIA BERTUGLIA • *Faculty of Medicine, CNR Institute of Clinical Physiology, University of Pisa, Pisa, Italy*

VIVIANA CAVALCA • *Department of Cardiovascular Sciences, University of Milan and Laboratory of Cellular Biology and Biochemistry of Atherotrombosis, Centro Cardiologico Monzino IRCCS, Milan, Italy*

JIANBIN CHEN • *Food Function and Labeling Program, Incorporated Administrative Agency, National Institute of Health and Nutrition, Tokyo, Japan*

M. IQBAL CHOUDHARY • *International Center for Chemical and Biological Sciences, H.E.J. Research Institute of Chemistry, University of Karachi, Karachi, Pakistan*

GIOVANNI CIABATTONI • *Department of Drug Sciences, School of Pharmacy, University "G. D'Annunzio", Chieti, Italy*

GIUSEPPE DA CRUZ • *Japan Institute for the Control of Aging (JaICA), Nikken SEIL Co., Ltd., Japan, Shizuoka, Japan*

ANDREAS DAIBER • *II. Medizinische Klinik, Labor für Molekulare Kardiologie, Johannes-Gutenberg-Universität, Mainz, Germany*

ANASTASIA DETSI • *Laboratory of Organic Chemistry, School of Chemical Engineering, National Technical University of Athens, Athens, Greece*

AOIFE M. DUFFY • *Regenerative Medicine Institute, National University of Ireland, Galway, Galway, Ireland*

MARC EGLON • *Regenerative Medicine Institute, National University of Ireland, Galway, Galway, Ireland*

EVGENIY ERUSLANOV • *Shands Cancer Center and Department of Urology, University of Florida, Gainesville, FL, USA*

FERRUCCIO GALBIATI • *Department of Pharmacology, University of Pittsburgh School of Medicine, Pittsburgh, PA, USA*

ANDREW R. GARRETT • *Department of Microbiology and Molecular Biology, Brigham Young University, Provo, UT, USA*

TILMAN GRUNE • *Institute for Biological Chemistry and Nutrition, Biofunctionality and Food Safety, University of Hohenheim, Stuttgart, Germany*

KUBILAY GÜÇLÜ • *Department of Chemistry, Faculty of Engineering, Istanbul University, Avcilar, Istanbul, Turkey*

H. DAVID GUTHRIE • *U.S. Department of Agriculture, Biotechnology and Germplasm Laboratory, Agricultural Research Service, Beltsville, MD, USA*

ALBERTO M. GUZMÁN-GRENFELL • *Laboratorio de Bioquímica y Medicina Ambiental, Instituto Nacional de Enfermedades Respiratorias (INER), México*

GUANGLONG HE • *The Center for Biomedical EPR Spectroscopy and Imaging, Davis Heart and Lung Research Institute and Division of Cardiovascular Medicine, Department of Internal Medicine, The Ohio State University College of Medicine, Columbus, OH, USA*

JUAN J. HICKS • *Laboratorio de Bioquímica y Medicina Ambiental, Instituto Nacional de Enfermedades Respiratorias (INER), México*

ANNIKA HÖHN • *Institute for Biological Chemistry and Nutrition, Biofunctionality and Food Safety, University of Hohenheim, Stuttgart, Germany*

MARCOS INTAGLIETTA • *Department of Bioengineering, University of California, San Diego, La Jolla, CA, USA*

RASHMI JHA • *Department of Biochemistry, University of Allahabad, Allahabad, India*

DOLLY JAISWAL • *Department of Biochemistry, University of Allahabad, Allahabad, India*

TOBIAS JUNG • *Institute for Biological Chemistry and Nutrition, Biofunctionality and Food Safety, University of Hohenheim, Stuttgart, Germany*

SATOKO KINO • *Japan Institute for the Control of Aging (JaICA), Nikken SEIL Co., Ltd., Japan, Shizuoka, Japan*

MARIA KOUFAKI • *National Hellenic Research Foundation, Institute of Organic and Pharmaceutical Chemistry, Athens, Greece*

SERGEI KUSMARTSEV • *Shands Cancer Center and Department of Urology, University of Florida, Gainesville, FL, USA*

XUEBO LIU • *Laboratory of Food and Biodynamics, Graduate School of Bioagricultural Science, Nagoya University, Nagoya, Japan*

BARRY MCGRATH • *Regenerative Medicine Institute, National University of Ireland, Galway, Galway, Ireland*

JILLIAN M. MCMAHON • *National Centre for Biomedical Engineering Science, National University of Ireland, Galway, Galway, Ireland*

RAFAEL MEDINA-NAVARRO • *Laboratorio de Metabolismo Experimental, Centro de Investigación Biomédica de Michoacán (CIBIMI-IMSS), Michoacán, México*

PAOLO MONTUSCHI • *Department of Pharmacology, Faculty of Medicine, Catholic University of the Sacred Heart, Rome, Italy*

THOMAS MÜNZEL • *II. Medizinische Klinik, Labor für Molekulare Kardiologie, Johannes-Gutenberg-Universität, Mainz, Germany*

BYRON K. MURRAY • *Department of Microbiology and Molecular Biology, Brigham Young University, Provo, UT, USA*

MICHAEL MURRAY • *Pharmacogenetics and Drug Development Group, Faculty of Pharmacy, University of Sydney, NSW, Australia*

TAIRIN OCHI • *Japan Institute for the Control of Aging (JaICA), Nikken SEIL Co., Ltd., Japan, Shizuoka, Japan*

IVONNE OLIVARES-CORICHI • *Laboratorio de Bioquímica y Medicina Ambiental, Instituto Nacional de Enfermedades Respiratorias (INER), México Instituto Politécnico Nacional (IPN), México*

TIMOTHY O'BRIEN • *Regenerative Medicine Institute, National University of Ireland, Galway, Galway, Ireland*

KIM L. O'NEILL • *Department of Microbiology and Molecular Biology, Brigham Young University, Provo, UT, USA*

SOSAMMA OOMMEN • *Department of Biological Sciences, Faculty of Science, Kuwait University, Safat, Kuwait*

TOSHIHIKO OSAWA • *Laboratory of Food and Biodynamics, Graduate School of Bioagricultural Science, Nagoya University, Nagoya, Japan*

MUSTAFA ÖZYÜREK • *Department of Chemistry, Faculty of Engineering, Istanbul University, Avcilar, Istanbul, Turkey*

BENIAMINO PALMIERI • *Department of General Surgery and Surgical Specialties, Surgical Clinic, University of Modena and Reggio Emilia Medical School, Modena, Italy*

KANTI BHOOSHAN PANDEY • *Department of Biochemistry, University of Allahabad, Allahabad, India*

NENAD PETROVIC • *School of Pharmacy, Murdoch University, Murdoch, WA, Australia*

ATTA-UR-RAHMAN • *International Center for Chemical and Biological Sciences, H.E.J. Research Institute of Chemistry, University of Karachi, Karachi, Pakistan*

A.K. RAI • *Department of Physics, Laser Spectroscopy Research Laboratory, University of Allahabad, Allahabad, India*

DEVENDRA K. RAI • *Department of Chemistry, Alternative Therapeutics Unit, Drug Discovery & Development Division, Medicinal Research Lab, University of Allahabad, Allahabad, India*

NILESH K. RAI • *Laser Spectroscopy Research Laboratory, Department of Physics, University of Allahabad, Allahabad, India*

PRASHANT KUMAR RAI • *Department of Biochemistry, University of Allahabad, Allahabad, India*

NIRMALA RAMANUJAM • *Department of Biomedical Engineering, Duke University, CIEMAS, Durham, NC, USA*

SYED IBRAHIM RIZVI • *Department of Biochemistry, University of Allahabad, Allahabad, India*

RICHARD A. ROBISON • *Department of Microbiology and Molecular Biology, Brigham Young University, Provo, UT, USA*

LESLIE ROSENTHAL • *Division of Epidemiology, Statistics, and Prevention Research, Eunice Kennedy Shriver National Institute of Child Health and Human Development, Rockville, MD, USA*

KAZUO SAKAI • *Japan Institute for the Control of Aging (JaICA), Nikken SEIL Co., Ltd., Japan, Shizuoka, Japan*

JUAN SASTRE • *Department of Physiology, School of Pharmacy, University of Valencia, Valencia, Spain*

VALERIANA SBLENDORIO • *Department of General Surgery and Surgical Specialties, Surgical Clinic, University of Modena and Reggio Emilia Medical School, Modena, Italy*

ENRIQUE SCHISTERMAN • *Division of Epidemiology, Statistics, and Prevention Research, Eunice Kennedy Shriver National Institute of Child Health and Human Development, Rockville, MD, USA*

GAETANO SERVIDDIO • *Department of Medical and Occupational Medicine, Institute of Internal Medicine, University of Foggia, Foggia, Italy*

BECHAN SHARMA • *Medicinal Research Lab, Drug Discovery & Development Division, Alternative Therapeutics Unit, Department of Chemistry, University of Allahabad, Allahabad, India*

MELISSA SKALA • *Department of Biomedical Engineering, Duke University, CIEMAS, Durham, NC, USA*

AKIHIRO TAI • *Faculty of Life and Environmental Sciences, Prefectural University of Hiroshima, Hiroshima, Japan*

JUN TAKEBAYASHI • *Food Function and Labeling Program, Incorporated Administrative Agency, National Institute of Health and Nutrition, Tokyo, Japan*

MASAO TAKEUCHI • *Japan Institute for the Control of Aging (JaICA), Nikken SEIL Co., Ltd., Japan, Shizuoka, Japan*

ISAO TOMITA • *Japan Institute for the Control of Aging (JaICA), Nikken SEIL Co., Ltd., Japan, Shizuoka, Japan, and Graduate School of Pharmaceutical Sciences, University of Shizuoka, Shizuoka, Japan*

ELENA TREMOLI • *Department of Pharmacological Sciences, University of Milan, and Centro Cardiologico Monzino IRCCS, Milan, Italy*

FABRIZIO VEGLIA • *Unit of Biostatistics, Centro Cardiologico Monzino IRCCS, Milan, Italy*

JOHN WARASKA • *ESA Biosciences, Inc, Chelmsford, MA, USA*

GEETA WATAL • *Department of Biochemistry, University of Allahabad, Allahabad, India*

GLENN R. WELCH • *Biotechnology and Germplasm Laboratory, U.S. Department of Agriculture, Agricultural Research Service, Beltsville, MD, USA*

NARUOMI YAMADA • *Laboratory of Food and Biodynamics, Graduate School of Bioagricultural Science, Nagoya University, Nagoya, Japan*

SAMMER YOUSUF • *International Center for Chemical and Biological Sciences, H.E.J. Research Institute of Chemistry, University of Karachi, Karachi, Pakistan*

Part I

Reactive Oxygen and Nitrogen Techniques

Chapter 1

Current Status of Measuring Oxidative Stress

Beniamino Palmieri and Valeriana Sblendorio

Abstract

Although the healthcare field is increasingly aware of the importance of free radicals and oxidative stress, screening and monitoring has not yet become a routine test, since, dangerously, there are no symptoms of this condition. Therefore, in very few cases is oxidative stress addressed. Paradoxically, patients are often advised supplementation with antioxidants and or diets with increased antioxidant profile, which ranges from vitamins to minerals and acts against oxidative stress states; even more so, no test is advised to assess whether the patient is under attack by free radicals or has a depleted antioxidant capacity.

Oxidative stress i s an imbalance between free radicals (ROS, reactive oxygen species) production and existing antioxidant capacity (AC); living organisms have a complex antioxidant power. A decrease in ROS formation is often due to an increase in antioxidant capacity, while a decrease in the AC may be associated to increased ROS values. But, this is not always apparently so.

Test kits for photometric determinations that are applicable to small laboratories are increasingly available.

Key words: Oxidative stress, Free radicals, Antioxidants, Point of care test, FORT test, FORD test

1. Introduction

Oxidation is a process that occurs naturally in the body when oxygen combines with reduced molecules, such as carbohydrates or fats, and provides energy. When there is decreased oxidation or decreased energy production, the cells can no longer function efficiently and therefore disease results. However, this normal process propagates short-lived intermediates known as free radicals, and some free radicals escape and initiate further oxidation, setting up a chain reaction. So, potentially harmful reactive oxygen species are produced as a consequence of biological metabolism and by exposure to environmental factors. Free radicals are then

D. Armstrong (ed.), *Advanced Protocols in Oxidative Stress II*, Methods in Molecular Biology, vol. 594
DOI 10.1007/978-1-60761-411-1_1, © Humana Press, a part of Springer Science+Business Media, LLC 2010

usually removed or inactivated by a group of natural antioxidants which prevent these reactive species from causing excessive cellular damage.

"Oxidative stress" is the general phenomenon of oxidant exposure and antioxidant depletion or oxidant–antioxidant balance.

Although reactive oxygen/nitrogen species (ROS/RNS) play an important role in immune-mediated defence against invading micro-organisms and serve as cell-signaling molecules, at high concentrations, ROS/RNS are capable of damaging host tissues, i.e., they can modify or damage DNA, lipids, and proteins. As yet mentioned, ROS/RNS levels are controlled by an intricate network of endogenous and exogenous antioxidant molecules that are responsible for scavenging and consumption of specific reactive species. In this regard, intake of dietary antioxidants has received much attention, with the concept being that these molecules can affect disease by modulating the biological reactivity of free radicals.

Over the past four decades, a substantial body of data has accumulated to support the direct or indirect association between free radicals and various human diseases. Given the number of patients world wide suffering from these disorders and the association with free radicals, screening of oxidative stress (OS) and, consequently, lifestyle and dietary changes are fundamental for a preventive approach. The role of OS in aging, neurodegenerative, vascular, and other diseases is more widely accepted; the value of antioxidant strategies may sometimes be controversial although a well-balanced antioxidant diet is undoubtedly important and strongly supported.

In the past years, several laboratory tests have been investigated and produced to assess the whole antioxidant activity of plasma or serum blood (1–5).

Point-of-care diagnostic testing, or testing performed at the patient bedside, allows physicians to diagnose patients more rapidly than traditional laboratory-based testing. Rapid results can enable better patient management decisions, improved patient outcomes, and a reduction in the overall cost of care. These tests are utilized in hospitals, clinics, commercial laboratories, and research institutions for the purpose of diagnosis and monitoring of disease.

Although clinicians may associate point-of-care testing (POCT) with critical care, the reality is that POCT (bedside, decentralized, or near-patient testing) is already being performed in virtually every clinical setting.

POCT began more than 30 years ago, although the phrase came into use only within the past 15 years. The driving force behind this type of testing has always been improvement of patient care through rapid availability of reliable results. The ability to obtain clinical laboratory test results at the site of care in 2 min has immediate medical management benefits as well as resource and time benefits.

The potential benefits of POCT include earlier and more appropriate diagnosis, fewer tests, earlier treatment, and reduction or elimination of unnecessary treatment. An unquantifiable benefit of POCT also offers convenience and decreases the time spent in a department or clinic, which are advantages for providers and patients alike.

By an analytical point of view, the effectiveness of antioxidant plasma barrier can be evaluated by testing its capacity to reduce a specific substrate, i.e. by assessing its capacity to supply oxidized background (e.g. free radicals) with one or more electrons.

For this purpose, different chemical reducing–oxidizing couples are available. For example, transition metals (i.e. iron) exhibit the property of receiving one electron thus shifting from the oxidized state (Fe^{3+}) to reduced state (Fe^{2+}). Such compounds are the reference to assess antioxidant power of biological systems. Indeed, the so-called "plasma antioxidant power" is ultimately a measure of the reducing or "electron-giving" activity of blood plasma.

On the other hand, some molecules share the property to change their absorbance just when they are bound to compounds that are able to switch from the oxidized to reduced state. For example, some thiocyanates are able to reversibly shift from uncolored to colored derivatives, in the presence of ferric or ferrous salts (5). Such "chromagens" can work as excellent "detectors" when coupled with adequate "oxidizing/reducing metres" in the test designed to assess antioxidant activity of biological systems.

Indeed, when a ferric salt is dissolved in an uncolored solution containing a particular thiocyanate derivative, the resulting solution becomes red, as a function of the ferric ions concentration. This process is due to the formation of a complex between ferric salt and thiocyanate. Further adding of a small amount of blood plasma will reduce ferric ions to ferrous ions, thus making the initial red solution uncolored. Such a chromatic change, due to the release of ferrous ions by the colored thiocyanate complex, can be read by means of a photometer, previously set on the wavelength of chromagen.

Therefore, the entity of absorbance change will directly correlate with the antioxidant "potential" of blood plasma against the specific substrate which has been used as oxidant/detector (ferric ions). In other words, the capacity of tested plasma to reduce ferric to ferrous ions will provide a direct measure of the capacity of a sample of such a plasma to give reducing equivalents and then neutralize chemical species lacking electrons, such as ROS, obviously in the reduction–oxidation potential range of chosen oxidant-reducing couple (Fe^{3+}/Fe^{2+}).

Generally, researchers in the free radicals field assert that each assay has its own specific characteristics and therefore the advantages and disadvantages. There are differences in the free radical-generating system, molecular target, reaction type, biological matrix, residence in the lipo or hydrophilic compartments, and physiological relevance. It is, therefore, impossible to identify one assay as a gold standard for measuring total antioxidant status in body fluids.

A combination of biomarkers of OS, i.e. indexes of oxidative damage and the antioxidant profile, provides a global assessment of the oxidant/antioxidant balance of the organism as well as on the nutritional needs of patients and the possible antioxidant strategies.

In the following paragraphs, we will be discussing about FORD[(patent pending)] and FORT assays that bring laboratory testing to the near patient testing fields. Test kits have been developed to provide operators with highly reliable, rapid, and user-friendly methods for the global evaluation of the oxidative status (radical-induced damage index and the total antioxidant capacity) in the body from a single drop of a capillary blood. FORT and FORD tests are completely stored at room temperature and work employing lyophilized chromagens to reduce operator handling and contact with chemical compounds.

Oxidative stress testing is of fundamental importance for preventive medicine and healthcare, disease management, and the control of relevant therapies during pathologies, in a wide range of fields and applications. Some examples are

- Monitoring antioxidant therapies
- Testing pharmacological treatments
- Monitoring lifestyle changes
- Sports sector (e.g., during various phases of training or before and after a performance)
- Preventive medicine
- Anti-aging fields
- Disease management in several fields
- Clinical research

2. Materials

2.1. Equipment

1. Dedicated spectrophotometer instrument (FormPlus, Callegari 1930).
2. Dedicated centrifuge (Callegari 1930).

2.2. Reagents kits
(FORT, 5 or 30 tests)

1. FORT reagent R1 blister.
2. FORT reagent R2 blister.
3. 20 μl capillaries for blood sampling.

2.3. Reagent kits
(FORD, 10 or 30 tests)

1. FORD reagent S1 blister.
2. FORD reagent S2 blister.
3. FORD reagent S3 vial.
4. FORD reagent C1 blister.
5. Pipette tips.
6. 50 μl capillaries for blood sampling.

3. Methods

3.1. The FORT Test

3.1.1. Principle and Standardization

Free oxygen radicals testing (FORT) is a colorimetric test based on the properties of an amine derivative employed as chromagen, $ChNH_2$ (4-Amino-N-ethyl-N-isopropylaniline hydrochloride) to produce a fairly long-lived radical cation (7). When sample is added to a $ChNH_2$ solution, the colored radical cation of the chromagen is formed and the absorbance at 505 nm, which is proportional to the concentration of hydroperoxyl molecules, is associated with the oxidative status of the sample.

1. *Linearity.* The linearity of the FORT test system was tested using two different methods (LOF test and Mandel test). With both of them, the linearity resulted statistically verified. Range of linearity: 148–608 FORT units.

2. *Precision.* Intra-assay coefficient of variation, $CV < 5\%$.

3. *Repeatability.* Intra-assay coefficient of variation, $CV < 5\%$.

4. *Sensitivity.* It is determined by the linearity range of the FORT reaction, that is 160–600 FORT units. $S = 4.0528$, the sensitivity (S) is defined as $S = \Delta Abs/\Delta C$; $Abs = $ absorbance, $C = $ concentration.

5. *Accuracy.* BIAS of the FORT test was determined analyzing a series of H_2O_2 solutions in water. Ten replicates were performed for each level of concentration (C).

   Predicted BIAS was calculated as $((C \text{ expected} - C \text{ obtained})/C \text{ obtained}) \times 100$.

   BIAS $< 4\%$ for 1.43 mM $H_2O_2 \leq C \leq 4.23$ mM H_2O_2

   BIAS $< 7\%$ for $C = 1.214$ mM H_2O_2

   BIAS $< 12\%$ for $C \geq 4.74$ mM H_2O_2

   Sample: 20 μl of whole blood; 10 μl for serum or plasma.

6. *Normal values.* Up to 310 FORT units (corresponding to approximately 2.36 mmol/l H_2O_2) (8).

The higher the FORT result obtained, the higher is the oxidative status of the sample. Although the assay is very reproducible for the same subject during the day and the CVs, both inter and intra-assay are very low; the value of the FORT measured on healthy subjects may be variable. Since the oxidative stress state of an individual depends on the hereditary, dietary, and environmental factors, there is a large heterogeneity in the population that may be related to disease incidence and longevity. For this reason, it is advisable to establish reference value for a patient.

7. Interference factors

The FORT test is based on Fenton's reaction. Fenton chemistry was discovered about 100 years ago, and it has proven to be a cornerstone of free radical biochemistry (9, 10). Fenton's reagent is a mixture of H_2O_2 and ferrous iron, which produces secondary radicals (11, 12), according to the reactions:

$$Fe^{2+} + H_2O_2 \rightarrow Fe^{3+} + OH^{\bullet} + OH^{-}$$

$$RH + OH^{\bullet} \rightarrow H_2O + R^{\bullet} \rightarrow \text{further oxidation}$$

$$R^{\bullet} + O_2 \rightarrow ROO^{\bullet}$$

The ferrous iron (Fe^{2+}) initiates and catalyzes the decomposition of H_2O_2, resulting in the production of hydroxyl radicals ($OH^{\bullet}$). Hydroxyl radicals can oxidize organics (RH) by abstraction of protons producing organic radicals ($R^{\bullet}$), which are highly reactive and can be further oxidized, initiating a radical chain oxidation. During the FORT reaction, overall organic radicals present in the sample are trapped by the FORT chromagen and photometrically measured.

The reactions above suggest that the presence of iron is required in the test reaction, thereby suggesting that the use of any kind of iron-chelating agents (e.g., EDTA, citrate, desferal) and external hydroperoxide and/or antioxidant sources (e.g., H_2O_2, benzoyl peroxide, BHT, ascorbic acid) affects the FORT assay blocking the Fenton's chemistry (*see* Note 1). In fact, solutions of the FORT chromagen and organics lacking the iron showed no specific EPR signal. Therefore, the FORT test cannot be applied when iron-chelating substances are present.

The hydrogen peroxide reacts in the Fenton's reaction, so external sources of H_2O_2, such as some disinfectants, can potentially interfere with the test, resulting in not reliable FORT values. Analogously, the presence of molecules in the sample, such as BHT (3,5-di-tert-butyl-4-hydroxytoluene), having antioxidant action interfere with the correct scheme for the test reaction in accordance with the Fenton's chemistry.

3.1.2. Performance Results

When the test is performed on whole blood, abnormal hematocrit values and hemolyzed samples may affect the results (*see* Note 2). Nevertheless, there is no significant interference when hematocrit is between 38 and 48% (*see* Note 3).

3.1.2.1. Visible Spectra

The visible spectrum of the $ChNH_2$ radical cation, reported in Fig. 1.1, shows two peaks of absorbance at 505 and 550 nm. The overall spectral intensity increased with time.

3.1.2.2. Time Course

As shown in Fig. 1.2, the increase in absorbance in the first 7–10 min is fairly linear, then reaches a plateau after a time interval which depends on temperature (the reaction is completed in approximately 60 min at 37°C).

Hence, a kinetic analysis of the colorimetric reaction at 37°C was selected.

3.1.2.3. EPR Spectra

The $ChNH_2$ solution is also EPR active, and under high magnetic field modulation (m.a. = 1mT), it exhibited a single broad line (Fig. 1.3a, b).

Hydroperoxides (ROOH) are fairly stable molecules under physiological conditions, but their decomposition is catalyzed by transition metals. Both Fe^{2+} and Fe^{3+} are effective catalysts in the reaction of degradation of these compounds resulting in several secondary reactive radical species formation (13). In biological samples, hydroperoxides concentration represents a good index of free radical attack, because it is indicative of intermediate oxidative products of lipids, peptides, and amino acids.

The reactions occurring in the FORT test conditions are based on the capacity of transition metals to catalyze the breakdown

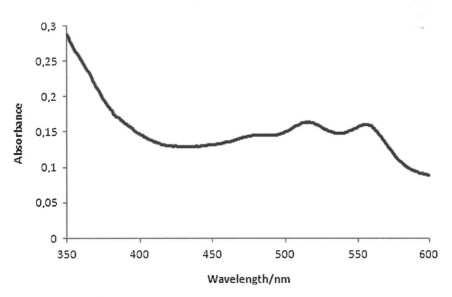

Fig. 1.1. Visible spectrum of the FORT chromagen radical cation.

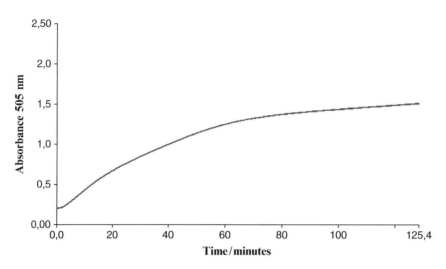

Fig. 1.2. Time course of the FORT chromagen radical cation formation at 37°C.

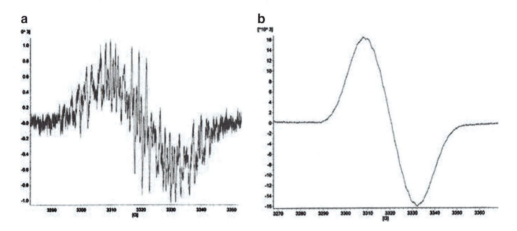

Fig. 1.3. The EPR spectrum of the FORT chromagen radical cation formed in a buffer solution in the presence of ferrous ions and TBH at 0.05 mT of modulation amplitude (**a**) or human plasma (**b**) at 1 mT of modulation amplitude.

of ROOH into derivative radicals, according to Fenton's reaction. Once they are formed, ROOH maintain their chemical reactivity and oxidative capacity to produce proportional amounts of alkoxy (RO$^{\bullet}$) and peroxy (ROO$^{\bullet}$) radicals.

These derivative radicals are then preferentially trapped by a suitably buffered FORT chromagen and develop, in a linear kinetic-based reaction at 37°C, a colored fairly long-lived radical cation photometrically detectable. The intensity of the color correlates directly with the quantity of radical compounds, according to the Lambert–Beer's law, and it can be related to the oxidative status of the sample.

$$R-OOH + Fe^{2+} \rightarrow R-O^{\bullet} + OH^- + Fe^{3+}$$

$$R - OOH + Fe^{3+} \rightarrow R - OO^{\bullet} + H^{+} + Fe^{2+}$$

$$RO^{\bullet} + ROO^{\bullet} + 2CrNH_2 \rightarrow RO^{-} + ROO^{-} + [Cr - NH_2 +^{\bullet}]$$

3.1.3. Definition of Unit

Considering the chemical heterogeneity of the secondary radical species deriving from the iron-dependent breakdown of ROOH during the FORT test, it has been decided to relate the absorbance readings to hydrogen peroxide (H_2O_2) concentration. A reference curve was created and stored in the dedicated instrument (FORM Plus, Callegari Spa, Catellani Group, Parma, Italy), which performs automatically the calculation of equivalent concentrations of H_2O_2.

In order to define a dedicated unit of measure for the FORT test, conventional units called FORT units have been defined. One FORT unit corresponds to approximately 7.6 mmol/l H_2O_2 (equivalent to 0.26 mg/l). Transformations are automatically performed by the dedicated instruments, so that the results are expressed both as concentration of H_2O_2 equivalent and as FORT units. Doing so, the value interpretation results easier for any operators, including lay users.

3.2. The FORD Test

3.2.1. Principle and Standardization

Free oxygen radicals defence (FORD) is a colorimetric test based on the ability of antioxidants present in plasma to reduce a pre-formed radical cation. The principle of the assay is that at an acidic pH (5.2) and in the presence of a suitable oxidant solution ($FeCl_3$), 4-Amino-N,N-diethylaniline, the FORD chromagen, can form a stable and colored radical cation.

Antioxidant molecules (AOH) reduces the FORD chroma-gen radical cation by quenching the color and producing a decol-oration of the solution, which is proportional to their concentration in the sample.

Preliminary experiments showed that the choice of oxidant solution and the ratio between the concentrations of the chroma-gen substance and the oxidative compound are essential for the effectiveness of the method.

$$\text{Chromagen}_{(uncoloured)} + \text{oxidant} \left(Fe^{3+}\right)H^{+} \rightarrow \text{Chromagen}^{\bullet+}_{(purple)}$$

$$\text{Chromagen}^{\bullet+}_{(purple)} + AOH \rightarrow \text{Chromagen}^{+}_{(uncoloured)} + AO$$

3.2.2. Repeatability and Precision

Three different concentrations of Trolox (2.5, 1.25, and 0.25 mM) were assayed 10 times in the same run for the determination of intra-assay coefficient variation (CV). An intra-assay CV < 5% was demonstrated. Additionally, repeatability and precision were established, testing whole human capillary blood. The FORD test was carried out using two different instruments and one level of concentration that is 1.25 mM Trolox equivalent. 20 replicates were performed for each instrument, during the same day.

Repeatability ($N=20$): CV < 5%;
Precision ($N=40$): CV < 5%.

3.2.2.1. Interference

Use of any kind of iron-chelating agents (e.g., EDTA, citrate, desferal); external hydroperoxide and/or antioxidant sources (e.g., H_2O_2, benzoyl peroxide, BHT, BHA, ascorbic acid) (*see* Note 1); abnormal hematocrit values and hemolyzed samples (when the test is performed on whole blood) (*see* Note 2).

Sample. 50 µl of whole blood.

3.2.2.2. Sensitivity

The linearity of the FORD test system was tested using solutions of Trolox as a chemical antioxidant standard. Tests were performed over a wide range of concentrations by subsequent dilutions of a stock solution and measuring the correspondent increment in FORD. The linearity has been statistically verified applying the Mendel's test.

Range of linearity. 0.25–3.0 mmol/l Trolox.

3.2.2.3. Accuracy

Based on a preliminary number of 70 human blood donors (male/female ratio, 37/33; aged 20–70 years, mean age 36 years) and the values cited in the scientific literature (14), at present, the reference values of FORD are estimated to be within the 1.07–1.53 mmol/l Trolox range (mean value = 1.23 mmol/l Trolox), which includes approximately 85% of data.

3.2.3. Performance Results

3.2.3.1. UV Spectrum

The UV-visible spectrum of the FORD chromagen radical cation (Fig. 1.4) shows maximum of absorbance at approximately 330 nm, 510 nm, and 550 nm. Hence, an end-point analysis of the colorimetric reaction at 505 nm and at 37°C was selected.

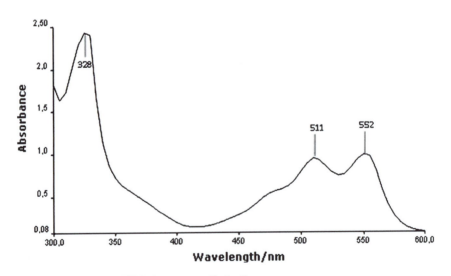

Fig. 1.4. UV-visible spectrum of the FORD chromagen radical cation.

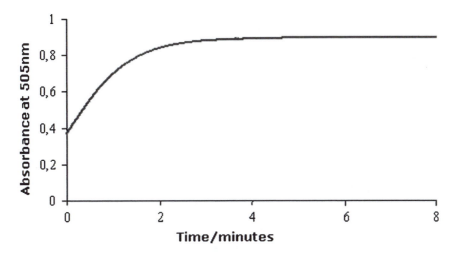

Fig. 1.5. Time course of the FORD chromagen radical formation at 37°C and 505 nm.

3.2.3.2. Time Course

The time course of the FORD chromagen radical formation obtained with an oxidant solution of $FeCl_3$ which gives a stable colored solution is reported in the Fig. 1.5. It is outlined that to have both high sensitivity of the measurements and a sufficient inhibition range, a starting point between 0.80 and 1.00 of absorbance at 505 nm is necessary. These absorbance readings are typically reached after 3–4 min; after that, the optical density remains stable. Hence, a lag time of 4 min was adopted between starting the reaction and reading the measurement of the chromagen radical cation absorbance.

3.2.3.3. Antioxidant Inhibition

The system was tested by using different concentrations of several antioxidant compounds, namely ascorbic acid, albumin, glutathione (GSH), uric acid, and Trolox, the α-tocopherol analog with enhanced water solubility. The dose-response curves obtained (Fig. 1.6) showed that inhibition of the starting absorbance is linear. Moreover, results revealed a relevant participation of FORD from ascorbic acid, Trolox, albumin, and GSH, and no response from uric acid.

Antioxidants tested have comparable kinetics in FORD chromagen radical scavenging, and the absorbance inhibition induced is immediate. Hence, a lag time of 2 min was selected between addition of sample containing antioxidants and measure of the color inhibition.

This provides an assay based on the extent of radical cation reduction at a fixed time point and not on the rate of reduction. This feature rules out complications due to the monitoring of the time course of color inhibitions.

FORD color quenching is determined especially by contribution of antioxidants such as proteins, reduced glutathione, vitamins, etc.

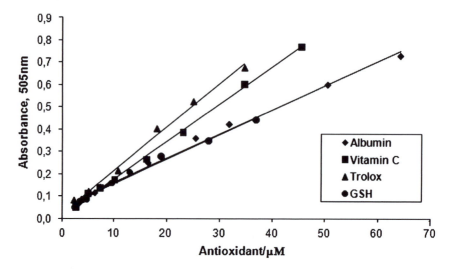

Fig. 1.6. Degree of inhibition of the FORD chromagen absorbance as a function of antioxidant concentration.

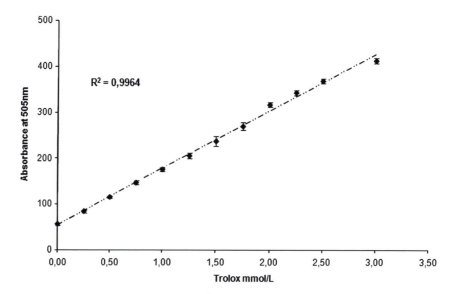

Fig. 1.7. Degree of inhibition of the FORD chromagen absorbance as a function of the Trolox concentration.

These antioxidants (together with uric acid which is not detected by the FORD test but is measured by a different test) are among the most important contributors to antioxidant plasmatic barrier.

3.2.3.4. Trolox Units

Like many other methods (4, 15), FORD results are expressed like Trolox equivalents (mmol/l), using a calibration curve plotted with different amounts of standard Trolox that is stored in the dedicated instrument (FORM Plus, Callegari SpA, Catellani Group, Parma, Italy). An example of dose-response curve obtained by using Trolox is shown in Fig. 1.7. Each datum is the mean of four determinations performed in four different days. The standard deviation is very low and the curve is highly reproducible (Coefficient of Variation, CV < 5%).

3.3. Clinical Results

Sblendorio et al. (16) evaluated oxidative status by FORT and FORD tests in 21 healthy postmenopausal women before and after supplementation of 15 mg of methyltetrahydrofolate for 3 weeks. They obtained a significant decrease in FORT (Fig. 1.8, $p = 0.0335$), increase of FORD (Fig. 1.9, $p = 0.0438$), and antioxidant/free radicals ratio (Fig. 1.10, $p = 0.0389$). There was no significant difference in glucose levels before and after methyltetrahydrofolate administration, but there was a significant decrease in insulin levels (Fig. 1.11, $p = 0.0074$) and resistance, as evaluated by HOMA ($p = 0.0063$). These preliminary results indicate that methyltetrahydrofolate supplementation may reduce oxidative stress and insulin resistance of postmenopausal women.

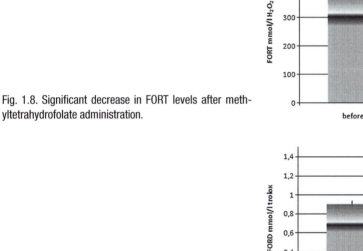

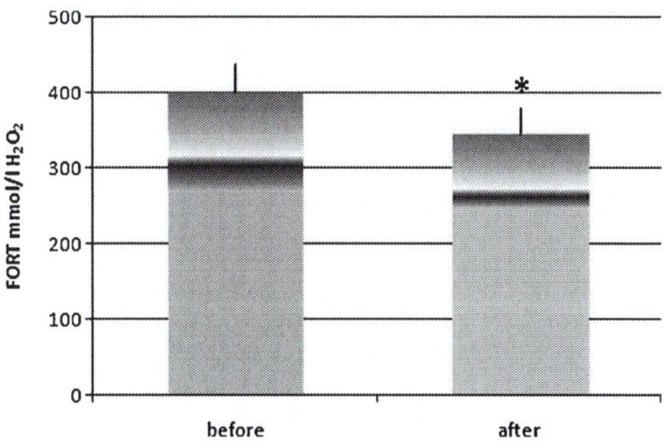

Fig. 1.8. Significant decrease in FORT levels after methyltetrahydrofolate administration.

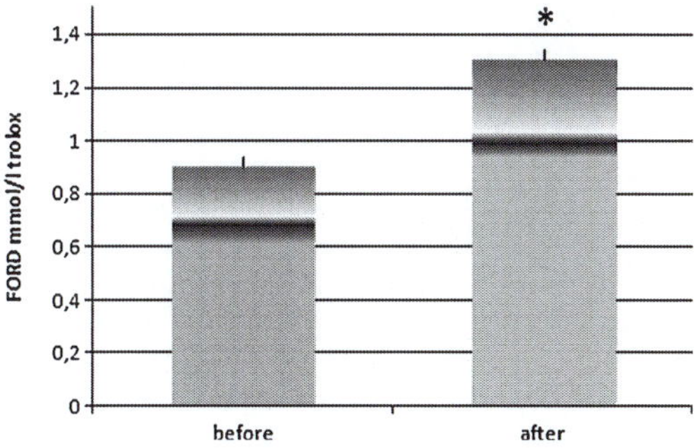

Fig. 1.9. Significant increase in FORD levels after methyltetrahydrofolate administration.

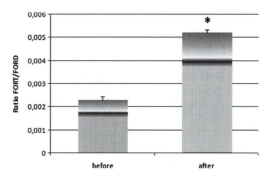

Fig. 1.10. Significant increase in FORD/FORT ratio after methyltetrahydrofolate administration.

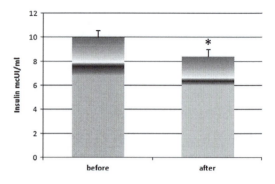

Fig. 1.11. Significant decrease in insulin levels after methyltetrahydrofolate administration.

4. Notes

1. Do not use any kind of iron-chelating agents (e.g., EDTA, citrate, desferal) and external hydroperoxide and/or antioxidant sources (e.g., H_2O_2, benzoyl peroxide, BHT, ascorbic acid) in the FORT assay.

2. Abnormal hematocrit values and hemolyzed samples may affect FORD and FORT tests' results.

3. Do not create bubbles in the silicon microcuvette before spectrophotometric reading.

References

1. Wayner DD, Burton GW, Ingold KU, Locke S (1985) Quantitative measurement of the total, peroxyl radical-trapping antioxidant capacity of human blood plasma by controlled peroxidation. FEBS Lett 187:33–37

2. Glazer AN (1990) Phycoerythrin fluorescence-based assay for reactive oxygen species. Methods Enzymol 186:161–168

3. Cao G, Alessio HM, Cutler RG (1993) Oxygen-radical absorbance capacity assay for antioxidants. Free Rad Biol Med 14:303–311

4. Cao G, Verdon CP, Wu AH, Wang H, Prior RL (1995) Automated assay of oxygen radical absorbance capacity assay using the COBAS FARA II. Clin Chem 41:1738–1744

5. Miller NJ, Rice-Evans C, Davies MJ, Gopinathan V, Milner A (1993) A novel method for measuring antioxidant capacity and its application to monitoring the antioxidant status in premature neonates. Clin Sci 84:407–412

6. Ghiselli A, Serafini M, Maiani G, Azzini E, Ferro-Luzzi A (1995) A fluorescence-based method for measuring total plasma antioxidant capability. Free Rad Biol Med. 18:29–36

7. Biurdon RH (1995) Superoxide and hydrogen peroxide in relation to mammalian cell proliferation. Free Radic Biol Med 18:775–794

8. Koppenol WH (2001) The Haber–Weiss cycle – 70 years later. Redox Rep 6:229–234

9. Dal Negro RW, Visconti M, Micheletto C, Pomari C, Squaranti M, Turati C, Trevisan F, Tornella S (2003) Normal values and reproducibility of the major oxidative stress obtained thanks to FORM system. GIMT; Ital J Chest Dis 57:199–209

10. Neyens E, Baeyens J (2003) A review of classic Fenton's peroxidation as an advanced oxidation technique. J Hazard Mater 98:33–50

11. Yoon J, Lee Y, Kim S (2001) Investigation of the reaction pathway of OH radicals produced by Fenton oxidation in the conditions of wastewater treatment. Water Sci Technol 44:15–21

12. Verde V, Fogliano V, Ritieni A, Maiani G, Morisco F, Caporaso N (2002) Use of N,

N-dimethyl-p-phenylenediamine to evaluate the oxidative status of human plasma. Free Radic Res 36:869–873

13. Miller NJ, Sampson J, Candeias LP, Bramley PM, Rice-Evans CA (1996) Antioxidant activities of carotenes and xanthophylls. FEBS Lett 384:240–242

14. Lu MC, Lin CJ, Liao CH, Ting WP, Huang RY (2001) Influence of pH on the dewatering of activated sludge by Fenton's reagent. Water Sci Technol 44:327–332

15. Kampa M, Nistikaki A, Tsaousis V, Maliaraki N, Notas G, Castanas E (2002) A new automated method for the determination of the Total Antioxidant Capacity (TAC) of human plasma, based on the crocin bleaching assay. BMC Clin Pathol 2:3

16. Sblendorio V, Cannoletta M, Palmieri B, Cagnacci A (2007) Menopause and oxidative stress: effect of methyltetrahydrofolate supplementation. World Conference of Stress. 23–26 August, Budapest, Hungary

pO_2 and ROS/RNS Measurements in the Microcirculation in Hypoxia

Silvia Bertuglia and Marcos Intaglietta

Abstract

We expose methods for in vivo assessment of oxygen, nitric oxide (NO), and reactive oxygen species (ROS)/reactive nitrogen species (RNS), in the microcirculation during normoxia and hypoxia. We provide an example of the related mechanisms of ROS/RNS and oxygen level in the process of regulating capillary perfusion. Namely, we discuss the real time pO_2 measurements in vivo in the microvessels and tissues of the hamster cheek pouch and window chamber preparations during normoxia and hypoxia, as well as the corresponding changes in ROS/RNS in systemic blood during normoxia and hypoxia under conditions where NO availability is maximally reduced.

Key words: Nitric oxide, Oxidative stress, Oxygen, Arterioles, Microcirculation, Blood perfusion

1. Introduction

The assessment of oxygen tension (pO_2) and reactive oxygen species (ROS)/ reactive nitrogen species (RNS) play a crucial role in a multitude of physiological and pathologic conditions. Initially, ROS/RNS were thought to be primarily cytotoxic species that increased tissue injury (1). Conversely, they can be endogenously produced and are essential in the immune response and in many physiological signal transduction pathways (2). There are several ROS that can be generated in the body organs: superoxide anion (O_2^-), hydrogen peroxide (H_2O_2), hydroxyl radicals (OH^-), nitric oxide (NO), peroxynitrite ($ONOO^-$) and others (3–5). NO is the most important in the homeostatic regulation of the immune, cardiovascular, and nervous system (6). The formation of RNS is not an inescapable consequence of synthesizing NO. It is efficiently removed by oxyhemoglobin to form nitrate, which prevents even the highest

D. Armstrong (ed.), *Advanced Protocols in Oxidative Stress II*, Methods in Molecular Biology, vol. 594
DOI 10.1007/978-1-60761-411-1_2, © Humana Press, a part of Springer Science+Business Media, LLC 2010

rates on NO synthesis from directly reacting with oxygen to form a significant amount of nitrogen dioxide. However, the simultaneous activation of superoxide synthesis along with NO will completely transform the biological action of NO by forming peroxynitrite.

There are several mechanisms for NO consumption, including red blood cell reactions, cellular metabolism, and reaction with ROS. Interestingly, most consumptive mechanisms of NO are O_2-dependent according to the metabolic state of the cell. Oxygen is a major substrate that regulates the rate of NO synthesis because change in pO_2 has a profound influence on NO synthase activity. The production of NO from arginine by the synthase requires molecular oxygen. Oxygen dependency is one of the major differences between the NOS isoforms, neuronal (NOS-1), endothelial (NOS-3), and the inducible NOS-2. NO inhibits the mechanism of O_2 consumption in tissues and mitochondrial respiration and therefore, increases in NO will increase O_2 which will, in turn, further increase NOS activity (7). Moreover, cellular NO consumption is O_2 dependent and proportional to oxygen concentration; as the O_2 level increases, so does the rate of NO metabolism (8). Taken together, we see that there is a close relationship between these two gases, which is critical in the regulation of tissue oxygenation and perfusion (9–12). Autoregulatory mechanisms that adjust the diameter of arterioles and blood perfusion to the oxygen demand of tissues have become a cornerstone of vascular pathophysiology. There is substantial evidence to suggest that oxygen reactivity is mediated, at least in part, by the oxygen-dependent production of vasoconstrictors, likely in addition to the production of vasodilators during severe hypoxia (9). The formation of NO-derived S-nitrosothiols (RSNOs) can stimulate guanylate cyclase, and thereby promote vasorelaxation (13). S-nitrosohemoglobin adducts are also involved in oxygen delivery (14).

A major challenge is the development of specific and sensitive methods for measuring pO_2 and quantifying ROS/RNS in cells and tissues. The measurement of pO_2 and transport at the microcirculatory level is technically difficult because of limited accessibility and the necessity of high-resolution techniques to differentiate between arterioles, venules, capillaries, and tissues within the microcirculation (15–19). Blood pO_2 has also been evaluated by measurements of light absorption at different wavelengths of the hemoglobin absorption spectrum (20–24). Mass spectrometry is also invasive and employs a relatively large tissue probe, yielding an "average" rather than a localized measurement (25). The quenching method, based on the relationship between the decay rate of excited phosphorescence from palladium-mesotetra-(4-carboxyphenyl) porphyrin bound to albumin, enables pO_2 measurement in the blood and the tissue as the probe passes to the interstitium according to the exchange of albumin from blood to tissue.

A commonly used approach measures markers of ROS/RNS rather than the actual radical. These markers of oxidative stress are measured using a variety of different assays. Lipid peroxidation has been and remains one of the most widely used indicators of oxidant/free radical formation in vitro and in vivo. Oxidants such as hydroxyl radicals, peroxyl radicals, nitrogen dioxide, peroxynitrite, and higher oxidation states of heme and hemoproteins (ferryl heme) are capable of initiating and propagating peroxidation of polyunsaturated fatty acids, thereby degrading membrane lipids (26).

Unfortunately, many of the methods used to detect lipid peroxidation in urine, blood plasma, or tissue are nonspecific, relying on the detection of thiobarbituric acid (TBA)-reactive substances such as malondialdehyde (MDA) or other reactive aldehydes generated from the in vivo or ex vivo decomposition of lipid peroxidation products (27). However, a variety of different bio-organic substances (e.g., bile acids, carbohydrates, nucleic acids, certain antibiotics, and amino acids) react with TBA to varying degrees, rendering this method sensitive, but not very specific. Elevated MDA levels have been measured in virtually every organ in experimental models of septic shock, hemorrhagic shock, and ischemia/reperfusion injury, where the role of peroxynitrite has been attested by various antiperoxynitrite strategies, such as melatonin (28–31). A particularly useful method is to measure F_2-like prostanoid derivatives of arachidonic acid, termed F_2-isoprostanes (IsoP) (32).

The detection of RSNO has often employed the Saville reaction, which involves the displacement of the nitrosonium ion (NO$^+$) by mercury salts. The resulting nitrite or NO generated from the spontaneous decomposition of NO$^+$ is detected by methods such as chemiluminescence or HPLC. Other techniques for the detection of RSNO employ colorimetric methods such as the Griess reaction to measure the nitrite formed from the treatment of RSNO with mercuric chloride. However, samples that contain large amounts of nitrite can interfere with and limit the detection range of these methods under acidic conditions. To overcome these problems, two methods have been devised to detect RSNO-derived nitrosating species at neutral pH (33). The colorimetric method uses the components of the Griess reaction while the fluorimetric method uses the conversion of DAN to its fluorescent triazole derivative (34). These methods may be conducted at neutral rather than acidic pH, eliminating the interference of contaminating nitrite and allowing the detection of nitrosation mediated by the presence of NO. Lipid peroxidation and the formation of peroxyl radicals and aldehydes damage membranes, membrane-bound enzymes, and receptors. Protein damage may result in their unfolding, fragmentation, and polymerization, while damage to DNA can cause mutations or enzyme activation.

Clinically, the evaluation of this type of ROS generation is limited by the very short half life (e.g., nanoseconds) of these free radicals (5).

In addition, ROS can be measured using electron spin resonance (ESR), also termed electron paramagnetic resonance spectroscopy and the spin trapping method(35). This technique and nuclear magnetic resonance spectroscopy are based on the magnetic properties of unpaired electrons and their molecular environment. These unpaired electrons exist in two orientations, either parallel or antiparallel with respect to an applied magnetic field. The energy differences of these states correspond to the microwave region of the electromagnetic spectrum. Although unpaired electrons of species such as NO, OH, or O_2^- are too low in concentration and too short-lived to be directly detected by ESR in biological systems, this dilemma can be circumvented by the ESR measurement of more stable secondary radical species formed by adding exogenous spin-traps, molecules that react with the primary radical species to give longer-lasting radical adducts with characteristic ESR signatures, which can accumulate to levels permitting detection. Interestingly, this spin-trap approach can also be used to measure tissue oxygen consumption and to non-invasively map spatially localized oxygen concentrations in living tissues. (36). Although instability, tissue metabolism, the sometimes broad reactivity of spin traps, and the cost of ESR spectrometers can be problematic, when combined with parallel strategies for detecting specific reactive species, ESR has proven to be a useful and revealing free radical detection strategy (37). However, these methods involve equipment and expertise not always found in most laboratories.

A major challenge is the development of specific and sensitive methods for measuring pO_2 and quantifying ROS/RNS in cells and tissues. This chapter provides an important example of the related mechanisms of ROS/RNS and oxygen level in the process of regulating capillary perfusion. In what follows, we discuss pO_2 measurements in vivo in microvessels and tissues of the hamster cheek pouch and window chamber preparations (38–44) during normoxia and hypoxia, as well as the corresponding changes in ROS/RNS in systemic blood during normoxia and hypoxia under conditions where NO availability is maximally reduced.

2. Materials

2.1. Equipment

1. Mean arterial blood pressure (MAP)-Heart rate: Viggo-Spectramed P10E2 transducer, Oxnard, CA, USA; Gould Windograf recorder, Mod. 13-6615-10S Gould Inc. Ohio, USA.

2. p$_A$O$_2$, p$_A$CO$_2$, pH: Blood Chemistry Analyzer 248, Bayer, Norwood, MA.

3. Blood Viscosity: Brookfield, Middleboro, MA.

4. *Hematocrit.* Hettick Hct 20 Centrifuge; Tuttlingen, Germany.

5. *Intravital microscopy.* Orthoplan, Leica Microsystem GmbH, Wetzlar, Germany.

6. *Charge-couple device camera.* COHU 4815-2000, San Diego CA.

7. *Videocassette recorder.* Panasonic AG-7355, Tokyo, Japan.

8. *Sony monitor.* PMV-1271Q, Tokyo, Japan.

9. *Red blood cell velocity.* Photo Diode/Velocity Tracker Model 102B, Vista Electronics, San Diego, CA.

10. *Fluorescence decay-curve fitter.* Model 802, Vista Electronics, Ramona, CA.

11. Potentiostat and Keithley Electrometer-Amplifier model 610C, Cleveland, OH.

3. Reagents and Supplies

1. 2-phenyl-4,4,5,5-tetramethylimidazoline-1-oxyl-3-oxide (carboxy-PTIO, CPTIO: Cayman Chemical, Michigan USA).

2. Palladium-meso-tetra(4-carboxyphenyl)porphyrin: Porphyrin Products, Inc., Logan, UT.

3. Analytic method d-ROMs: Diacron s.r.l., Parma Italy.

4. Thiobarbituric acid reactive kit: TBARS, ZeptoMetric, Corp. Buffalo, NY, USA.

5. Aliphatic alcohols Sigma-Aldrich, St Louis, MO.

4. Methods

4.1. Hamster Cheek Pouch Preparation

4.1.1. Systemic Parameters

Male Syrian hamsters (80–100 g, Charles River, Italy) *were anesthetized* by pentobarbital sodium, 50 mg/kg/intraperitoneally injected. Animal handling and care are according to the procedures approved by the Animal Care and Use Committee at the Italian Research Council. The right carotid artery and left femoral vein are cannulated for measurements of blood pressure and arterial blood gases, and the administration of drugs, respectively. Mean arterial blood pressure (MAP) and heart rate (HR) are acquired continuously during the experiment[1]. Arterial and venous blood is sampled from the catheter in the carotid artery and jugular vein into heparinized capillary tubes and analyzed for pO_2, pCO_2, pH[2], hematocrit[3], hemoglobin[4] at 37°C. A decrease of the fraction of inspired oxygen (FIO_2) is induced by normobaric hypoxia (10 O_2, FIO_2 0.1, balance N_2). Between each oxygen level, the animal is allowed 10 min to stabilize before the data acquisition. The groups are subjected to systemic parameters, blood gas analysis, and pO_2 measurements during baseline and after 20 min of exposure to hypoxia. Two groups are infused intravenously with the NO scavenger, the nitronyl nitroxide compound oxidized and reduced form of 2-phenyl-4,4,5,5-tetramethylimidazoline-1-oxyl-3-oxide (carboxy-PTIO, CPTIO, 1.0 mg/Kg, 10 mg/ml solution) or the vehicle (isotonic saline). CPTIO reacts stechiometrically with NO to generate NO_2 and a 2-carboxyphenyl-4,4,5,5-tetramethylimidazoline-1-oxyl (PTI) derivative. CPTIO is water soluble and has been shown to have very strong NO-scavenging properties based on a radical-radical reaction with NO (12). The cheek pouch is spread out over a Plexiglas microscope stage and a region of about 1 cm² in area is prepared as a single layer (9). The cheek pouch is covered with transparent plastic film to prevent both desiccation of the tissue and gas exchange with the atmosphere. Observations are made with an intravital microscope[5]. All selected microvessels and interstitial tissue segments are recorded by a video camera[6] displayed on a monitor[7] and transferred to a video recorder[8]. The hamster's body temperature and cheek pouch temperature are maintained at 37°C with circulating warm water. Baseline characterization (systemic hemodynamics, blood sampling for TBARS and nitrite/nitrate measurements, arteriolar diameter, RBC velocity) is performed after a 30 min stabilization period. Arterioles are chosen for investigation and followed throughout the protocol.

4.2. Dorsal Skin Fold Window Chamber Preparation

4.2.1. Systemic Parameters

Investigations are performed in 55–65 g male Golden Syrian Hamsters fitted with a dorsal window chamber. Animal handling and care follows the NIH Guide for the Care and Use of Laboratory Animals. Catheters are tunneled under the skin, exteriorized at the dorsal side of the neck, and securely attached to the window frame. MAP and HR are recorded continuously[1]. Hematocrit is measured from centrifuged arterial blood samples taken in heparinized capillary tubes[4]. Arterial blood is collected in heparinized glass capillaries (50 µl) and immediately analyzed for p_AO_2, p_ACO_2, base excess (BE), and pH[2]. The unanesthetized animal is placed in a restraining tube with a longitudinal slit from which the window chamber protrudes, then fixed to the microscopic stage for transillumination with the intravital microscope[5]. Animals are given 20 min to adjust to the tube environment before any measurement is carried out. The tissue image is projected onto a charge-coupled device camera[6] connected to a videocassette recorder[8] and viewed on a monitor[7]. Measurements are carried out using a 40× (NA 0.8) water immersion objective. The same sites of study are followed throughout the experiment, so that comparisons to baseline levels could be made directly. A decrease of the fraction of inspired oxygen (FIO$_2$) is induced by normobaric hypoxia (10 O$_2$, FIO$_2$ 0.1, balance N$_2$). Between each oxygen level, the animal is allowed 10 min to stabilize before the data acquisition. The groups are subjected to systemic parameters, blood gas analysis, and pO$_2$ measurements during baseline and after 20 min of exposure to hypoxia. Two groups are infused intravenously with CPTIO (1.0 mg/kg, 10 mg/ml solution) or the vehicle (isotonic saline) (12)

4.3. Measurement of Microvascular Parameters

4.3.1. Microhemodynamics

Arteriolar and blood flow velocities are measured on-line by using the photodiode cross-correlation method. The measured center-line velocity (V) is corrected according to vessel size to obtain the mean RBC velocity[9]. A video image-shearing method is used to measure the vessel diameter (D). Blood flow (Q) is calculated from the measured values as $Q = \pi \times V(D/2)^2$. Changes in arteriolar diameter from the baseline are used as indicators of a change in the vascular tone (40, 41).

4.3.2. Functional Capillary Density

Functional capillaries, defined as those capillary segments that have an RBC transit of at least a single RBC in a 45 s period in ten successive microscopic fields, were assessed in a region totaling 0.46 mm^2. Each field had between two and five capillary segments with RBC flow. Functional capillary density (FCD) (cm^{-1}), i.e., the total length of the RBC perfused capillaries divided by the area of the microscopic field of view, was evaluated by measuring and adding the length of the capillaries that had an RBC transit in the field of view. The relative change in FCD from the baseline levels after each intervention is indicative of the extent of capillary perfusion (41, 42)

4.3.3. Measurement of Microvascular pO$_2$

Phosphorescence quenching microscopy is based on the oxygen-dependent quenching of phosphorescence emitted by an albumin-bound metalloporphyrin complex after pulsed light excitation (*see* Note 1). The phosphorescence quenching microscopy is independent of the dye concentration within the tissue and is well suited for detecting hypoxia because its decay time is inversely proportional to the pO$_2$ level, causing the method to be more precise at low pO$_2$. The phosphorescence decay curves are converted to oxygen tension using a fluorescence decay-curve fitter. This technique has been used in this animal and others for both intravascular and extravascular oxygen tension measurements. Tissue pO$_2$ measurements are possible in this preparation because the albumin-bound dye equilibrates between the plasma and tissue compartments as a consequence of the increased permeability of the subcutaneous connective and adipose tissue to albumin. The animals receive a slow intravenous injection of 15 mg/kg bw at a concentration of 10.1 mg/ml of a palladium-meso-tetra(4-carboxyphenyl) porphyrin. The dye is allowed to circulate for 20 min before pO_2 measurements are carried out(12, 15). The phosphorescence is excited by pulsed light (30 Hz, 4 μs duration) for a period of <5 s, and intravascular measurements are made by placing an optical rectangular window (5×15 μm) within the vessel of interest with the longest side of the rectangular slit positioned parallel to the vessel wall. Tissue pO_2 is measured in regions void of large vessels within intracapillary spaces (10×10 μm) (24). In the present configuration of the oxygen measuring system, tissue pO_2 values are obtained with a repeatability of 1–3 Torr capturing the emission from a tissue area of 75–100 μm^2 (43). The phosphorescence decay curves are converted to oxygen tensions using a fluorescence decay curve fitter. Measurements in regions with large tissue gradients in the vicinity of arterioles are made using a shaping list in a rectangular format, which is placed along the outside of the vessel wall, and by varying the length at which the acceptable signal-to-noise ratio was obtained (44). Perivascular pO_2 measurements are made by placing the centerline of the measuring slit at a distance of one-tenth of the diameter of the inner vessel that stops at the blood tissue interface (*see* Notes 2–3).

4.3.4. Measurement of Lipid Peroxides (ROS)

4.3.4.1. Analytical

This section describes the analytic method d-ROMs (39, 40) to measure plasma hydroperoxides, based on Fenton's reaction or on radical formation during lipid peroxidation. The oxyradical species produced, which is directly proportional to the amount of plasma peroxides, is trapped by alchylamine, a phenolic compound that forms a coloured stable radical detectable spectrophotometrically at 505 nm. The concentration of the coloured complex is directly correlated to the concentration of the hydroperoxides. Ten μl of a chromogenic substance and 1 ml of the kit buffer are mixed with 10 μl of blood for 1 min at 37°C. The results are expressed in arbitrary units (a.u.; 1 a.u. = 0.08 mg/100 ml H_2O_2). Blood samples are taken at the baseline and after hypoxia from the cannulated carotid artery.

4.3.4.2. TBARS

Lipid peroxidation is also measured in plasma using an assay to estimate the levels of thiobarbituric acid reactive substances. TBARS, including mainly lipid peroxides and malondialdehyde, generated during the peroxidative process are determined using 100 μl of plasma and 60 min incubation under acidic conditions at 96°C, via spectrophotometric measurements at 532 nm. TBARS values are expressed in terms of malondialdehyde equivalents as nmol/ml.

4.3.5. NO Measurements

4.3.5.1. Electrodes

Perivascular NO levels were measured using amperiometric bi-polymer coated (Nafion and *o*-phenylenediamine) carbon fiber microelectrodes. The electrodes were fabricated by sequential dipping and drying in Nafion (5% in aliphatic alcohols). They were additionally coated with 5 mM *o*-phenylenediamine dihydrocholoride (1, 2-benzenediame solution), which selectively repels ascorbic acid and dopamine (45). The current generated was measured with a potentiostat and an electrometer-amplifier[11]. The sensitivity of each electrode to ascorbic acid was evaluated by measuring its response to a 30 mM solution equilibrated with 100% Argon and compared to the magnitude of the response to 1 nM of NO. The electrodes selected for use in these studies had less than a 2% response to ascorbic acid. NO sensitivity was of 7 nM. The stability of the electrodes was re-assessed after the measurements by repeating the calibration procedure. Data from the electrodes outside the defined characteristics or presenting calibration changes greater than 5% were discarded from the study. Measurements were made in the perivascular space, in the abluminal side of the microvascular walls to obtain an estimate of the NO level in the vascular smooth muscle. This required removing the window chamber cover glass and superfusing the tissue (~5 ml/min) with heated physiological Krebs salt solution equilibrated with 95% N$_2$ and 5% CO$_2$, which maintained the suffusate at pH 7.4. The fluid was dripped onto the tissue to maintain a 35–36°C temperature and minimize oxygen delivery to the tissue. NO measurements began 20 min after the removal of the window to allow the tissue to stabilize and the microvascular hemodynamics to return to baseline levels (46). Measurements were made by penetrating the tissue with the NO electrode maneuvered with a micromanipulator and the tip positioned as close as possible to the microvessel without touching the wall, as this stimulation causes transient and sustained changes in NO concentration (47–49).

4.3.5.2. Griess Reaction

One method for the indirect determination of NO involves the spectrophotometric measurement of its stable decomposition products NO$_3^-$ and NO$_2^-$. This method requires that NO$_3^-$ first be reduced to NO$_2^-$ and then NO$_2^-$ determined by the Griess reaction (50). This reaction is a two step-diazotization reaction in which the NO derived nitrosating agent (e.g. N$_2$O$_3$) generated from the acid-catalyzed formation of nitrous acid from nitrite (or the interaction of NO with oxygen), reacts with sulfanilic acid to produce a diazonium ion that is then coupled to *N*-(1-napthyl) ethylediamine to form a chromophoric azo product that absorbs strongly at 543 nm. Briefly, blood samples are centrifuged, separated from RBCs, and the plasma is stored at –70°C. After ultra-filtration of the plasma through a 10-kDa membrane, the samples are incubated with nitrate reductase and enzyme cofactors for 3 h for conversion of nitrate to nitrite. Absorbance is read at 543 nm using a plate reader after addition of the Griess reagents that convert nitrite into a deep purple azo compound.

All reported values are means ± (SD). (GraphPad Software, Inc., San Diego California USA) was used to analyze statistical differences. The data were analyzed by the Mann Witney test to examine the differences between the groups. The Friedman two way analysis of variance by ranks test was used to determine the differences between the groups at different times. Where significant differences were indicated by this test, a further comparative analysis was undertaken using the Dunn's test. Data were presented as absolute values and ratios, as relative to the baseline values. A ratio of 1.0 would signify no change from the baseline, whereas lower and higher ratios are indicative of changes proportionally lower and higher than the baseline (i.e. 1.5 would mean a 50% increase from the baseline level). All measurements were compared with the baseline levels obtained before the experimental procedure. Differences are considered significant at $P < 0.05$.

4.3.6. Results

4.3.6.1. Hamster Cheek Pouch

4.3.6.1.1. Systemic Parameters and Microhemodynamics

Twenty animals were used for the study; ten received a continuous infusion of the NO scavenger and ten, a continuous infusion of the vehicle. Table 2.1 summarizes the changes in the systemic parameter response to hypoxia that caused significant changes in blood pO_2, pCO_2, and pH. These changes are the consequence of hyperventilation, a normal response to hypoxia. The microvascular diameter and blood flow responses to hypoxia with and without CPTIO are shown in Fig. 2.1a. CPTIO reduced the diameter of the arterioles in normoxia. Hypoxia resulted in a statistically significant vasodilation. CPTIO and hypoxia maintained a significant vasoconstriction. The continuous infusion of the NO scavenger significantly reduced FCD to 0.95 (SD 0.07) of the normoxic baseline [1.35 (SD 0.08), $P < 0.05$]. Hypoxia resulted in a decrease in FCD [0.90 (SD 0.05)] for the vehicle. CPTIO and hypoxia further decreased FCD [0.68 (SD 0.07), $P < 0.05$], which was significantly lower than normoxia and hypoxia without the NO scavenger.

4.3.6.1.2. Microvascular Oxygen Distribution

Figure 2.2a shows the distribution of pO_2 in the arterioles and in the interstitial space for normoxia and hypoxia with and without the NO scavenger. During normoxia, CPTIO decreased the intravascular pO_2. Tissue pO_2 was significantly reduced with CPTIO when compared with normoxic animals. With CPTIO and hypoxia, the intravascular pO_2 was similar and lower than during hypoxia in the arterioles.

4.3.6.1.3. ROS/RNS Measurements

CPTIO significantly increased ROS formation vs. normoxic animals (Fig. 2.3) Hypoxia significantly increased ROS formation compared with normoxia. Nitrate/nitrite concentrations in the plasma were significantly increased after hypoxia and reduced after CPTIO (Fig. 2.4).

Table 2.1

Systemic parameters during normoxia and hypoxia with and without the treatment with NO scavenger in hamster skin fold and in hamster cheek pouch

	Cheek pouch normoxia	Cheek pouch hypoxia	Skin fold window normoxia	Skin fold window hypoxia
	(n = 10)	(n = 10)	(n = 6)	(n = 6)
MAP (mmHg)				
Vehicle	88.3 ± 5	86.5 ± 7	103 ± 5	105 ± 7
Treatment	98.4 ± 4*°	92.5 ± 6*	149 ± 8*	128 ± 7*
HR (beats/min)				
Vehicle	320 ± 12	330 ± 10	418 ± 23	468 ± 41
Treatment	350 ± 10	360 ± 10	470 ± 39	489 ± 27
paO$_2$, (Torr)				
Vehicle	70.8 ± 5.4	29.3 ± 2.7*	60.4 ± 7.2	27.4 ± 3.2*
Treatment	68.3 ± 7.2	30.5 ± 3.5*	67.8 ± 6.2	30.5 ± 8.8*
paCO$_2$, (Torr)				
vehicle	59.4 ± 4.2	36.3 ± 2.4*	53.7 ± 6.4	33.4 ± 3.2*
Treatment	53.3 ± 3.5	32.5 ± 3.2*	44.6 ± 5.4*	30.2 ± 5.4*
Arterial pH				
Vehicle	7.36 ± 0.02	7.48 ± 0.03*	7.35 ± 0.02	7.47 ± 0.02
Treatment	7.38 ± 0.03	7.49 ± 0.02*	7.38 ± 0.03	7.48 ± 0.04

The mean blood pressure (MAP), heart rate (HR), pO$_2$, arterial pO$_2$–pCO$_2$. Values are means ±SE. *n* no of animals
*$P < 0.05$ vs. normoxia

4.3.6.2. Skin Fold Window Preparation

4.3.6.2.1. Systemic Parameters and Microhemodynamics

Twelve animals were used for the study; six received a continuous infusion of CPTIO and six, the continuous infusion of the vehicle. Table 2.1 summarizes the changes in the systemic parameter response to hypoxia. The continuous infusion of the NO scavenger during normoxia induced a significant arteriolar vasoconstriction ratio (Fig. 2.1b). Hypoxia resulted in a statistically significant increase in the diameter of the arterioles. CPTIO and hypoxia maintained significant arteriolar vasoconstriction. Diameter and RBC velocity data were used to compute the microvascular blood flow in each vessel studied (Fig. 2.1b). Continuous infusion of the NO scavenger during normoxia induced a statistically significant reduction of FCD to 0.78 (SD 0.09) of the baseline. Hypoxia resulted in a small decrease in FCD [0.94 (SD 0.08)] for the vehicle.

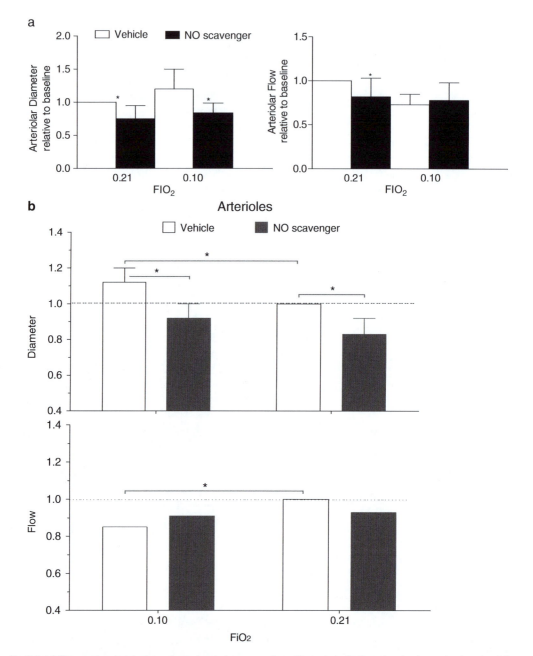

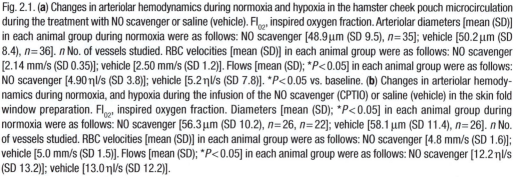

Fig. 2.1. (**a**) Changes in arteriolar hemodynamics during normoxia and hypoxia in the hamster cheek pouch microcirculation during the treatment with NO scavenger or saline (vehicle). FI_{O2}, inspired oxygen fraction. Arteriolar diameters [mean (SD)] in each animal group during normoxia were as follows: NO scavenger [48.9 μm (SD 9.5), $n = 35$]; vehicle [50.2 μm (SD 8.4), $n = 36$]. n No. of vessels studied. RBC velocities [mean (SD)] in each animal group were as follows: NO scavenger [2.14 mm/s (SD 0.35)]; vehicle [2.50 mm/s (SD 1.2)]. Flows [mean (SD); *$P < 0.05$] in each animal group were as follows: NO scavenger [4.90 ηl/s (SD 3.8)]; vehicle [5.2 ηl/s (SD 7.8)]. *$P < 0.05$ vs. baseline. (**b**) Changes in arteriolar hemodynamics during normoxia, and hypoxia during the infusion of the NO scavenger (CPTIO) or saline (vehicle) in the skin fold window preparation. FI_{O2}, inspired oxygen fraction. Diameters [mean (SD); *$P < 0.05$] in each animal group during normoxia were as follows: NO scavenger [56.3 μm (SD 10.2), $n = 26$, $n = 22$]; vehicle [58.1 μm (SD 11.4), $n = 26$]. n No. of vessels studied. RBC velocities [mean (SD)] in each animal group were as follows: NO scavenger [4.8 mm/s (SD 1.6)]; vehicle [5.0 mm/s (SD 1.5)]. Flows [mean (SD); *$P < 0.05$] in each animal group were as follows: NO scavenger [12.2 ηl/s (SD 13.2)]; vehicle [13.0 ηl/s (SD 12.2)].

The NO scavenger and hypoxia further reduced FCD [0.71 (0.09)] to a statistically significant low when compared with normoxia and hypoxia without the NO scavenger

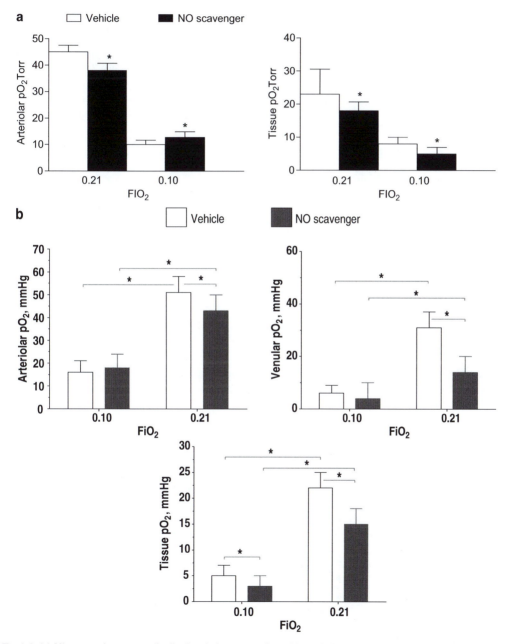

Fig. 2.2. (a) Microvascular oxygen distribution during normoxia and hypoxia in hamster cheek pouch with and without (CPTIO) or saline (vehicle). Values are means (SD). Each point represents an average of at least 36 measurements per group. *P < 0.05. (b) Microvascular oxygen distribution during normoxia and hypoxia in skin window preparation with and without NO scavenger. Values are means (SD). Each point represents an average of at least 36 measurements per group. *P < 0.05.

c

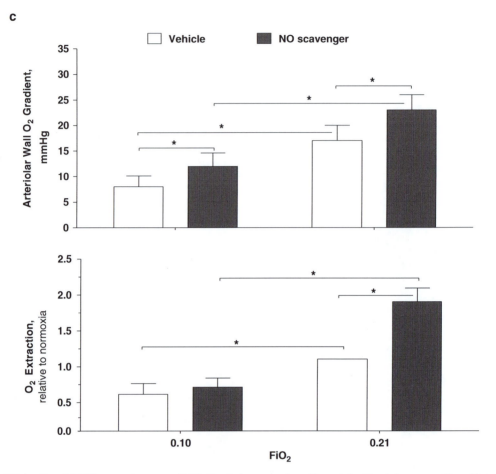

Fig. 2.2. (continued) (c) Microvascular oxygen distribution during normoxia and hypoxia in skin window preparation with and without NO scavenger. Values are means (SD); each point represents an average of at least 36 measurements per group. *$P < 0.05$ vs. baseline.

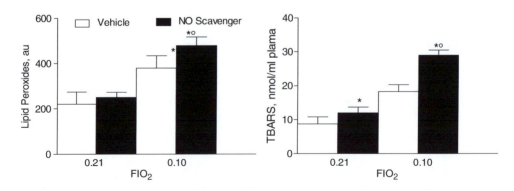

Fig. 2.3. Measurements of lipid peroxides with the two different methods, D-ROMS (*first panel*) and TBARS (*second panel*) in the systemic blood of hamsters in normoxia and hypoxia. Values are expressed as mean ($\pm$SD); $n = 7$ animals in each group. *$P < 0.05$ vs. baseline.

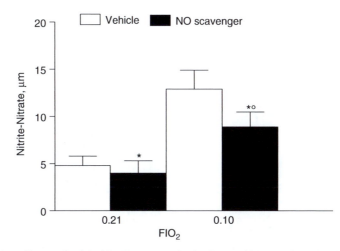

Fig. 2.4. Changes in nitrite/nitrate concentration in plasma of hamsters in normoxia and hypoxia with and without the NO scavenger. Values are means (SD); $n=7$ animals in each group. *$P<0.05$ vs. baseline.

4.3.6.2.2. Microvascular Oxygen and NO Measurement

Figure 2.2b shows the distribution of pO$_2$ in the microvascular arterioles and in the interstitial space (tissue) for normoxia and hypoxia, with and without the NO scavenger. Arteriolar wall gradients were determined from the difference between the intravascular and perivascular pO$_2$ measurements taken across the vessel wall (Fig. 2.2c). This parameter has been shown to be directly related to the rate of oxygen consumption of the vessel wall (20, 23). The vessel wall gradient in the arterioles in normoxia conditions was 16 mmHg (SD 4) [$n=10$; diameter 56 µm (SD 8)]. During the continuous infusion of the NO scavenger in normoxia, the vessel wall gradient in the arterioles was significantly increased to 23 mmHg (SD 3) [$n=12$; diameter 54 µm (SD 9)]. Hypoxia decreased the wall oxygen gradient with the NO scavenger to 11 mmHg (SD 3) [$n=8$; diameter 57 µm (SD 8)] and without the scavenger to 9 mmHg (SD 3) [$n=8$; diameter 55 µm (SD 7)]. NO concentrations were significantly increased after hypoxia and reduced after CPTIO (Fig. 2.5).

4.3.6.3. Critical Analysis of Methods

4.3.6.3.1. Oxygen Measurements

Surgical intervention, exposure to the environment, or at least, inflammation of the tissue can, in principle, increase the production of ROS/RNS as it must be assumed that these interventions and procedures should change the redox state of the membranes. However, the experimental evidence shows that the level of ROS/RNS in basal experimental conditions were low, and became significantly increased following exposure to low oxygen levels. Consequently, results from the experimental models should be representative and a consequence of the hypoxic stimulus. The analysis and measurement of oxygen distribution in the microcirculation and the tissues remains a controversial issue, primarily because there is no universally accepted method based on a common instru-

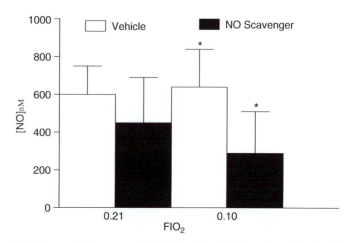

Fig. 2.5. NO concentration measured by electrode in arterioles of hamster skin fold window preparation in normoxia and hypoxia with and without the NO scavenger. Values are means (SD); $n=7$ animals in each group. *$P<0.05$ vs. normoxia.

mentation and methodology. The recent development of the phosphoresce quenching technique, although technically simple to implement, being optically based rather than using the arduous and skilful manipulation of oxygen microelectrodes, has led to both solutions and new problems. The principal new problem is that implementation with approximately similar method s in several laboratories, leads to the discovery and definition of the relative importance of mechanisms of oxygen transfer in the tissues that could not be previously evidenced. The principal problem issues on which there is no clear agreement are (1) the level of oxygen consumption by the method (2) the attainable geometrical resolution (3) the extent to which the results are affected by the toxicity of the metallo-proteins injected to obtain measurements. These features are all notably interrelated, finally leading to a significantly different understanding of the way tissue is oxygenated. There appears to be a general agreement on the validity and congruence of the measurements of intra microvascular pO_2, the concept of the longitudinal pO_2 oxygen gradient in the microcirculation having been validated by all laboratories. It is well established and functionally reasonable that the endothelium regulates the release of vasoactive substances in response to O_2 and NO availability. However, the remaining critical issues depend on the importance of O_2 consumption by the method and the magnitude of the oxygen gradients at the vessel wall. Golub et al. (51) have provided experimental evidence on the lack of significant gradients across the arteriolar wall, thus concluding that this tissue compartment is essentially transparent to the passage of oxygen, providing limited diffusional resistance and oxygen consumption. These authors indicate that the oxygen gradient at the arteriolar vascular wall is not statistically significant, and could be at most 0.5 mmHg/μ (52). A different approach has been taken by Intaglietta et al. and Shibata et al. (15, 53), who minimized oxygen consumption by the technique, and established oxygen mass balances for the measurements, i.e., the correspondence between the longitudinal oxygen gradients correspondence with the radial rate of oxygen exit. This approach, combined with a demonstrated decreased rate of oxygen consumption show that the arteriolar wall is a significant consumer of oxygen, as evidenced by the larger that diffusion limited oxygen gradient in this district. It should be apparent that the concept of the longitudinal pO$_2$ gradient, or a significant oxygen exit along the microvasculature, is not compatible with the lack of a significant radial gradient in the arteriolar circulation that extracts oxygen from the blood column. It is a substantially acceptable concept to assume that the vessel wall consumes a variable but significant percentage of the available oxygen in the context of blood flow regulation.

4.3.6.3.2. ROS/RNS

Measurement of the production of ROS is difficult for several reasons. For example, in many tissues, low intracellular steady-state concentrations of superoxide occur as a result of the balance between the basal rates of partial reduction of oxygen to superoxide and the diffusion-limited scavenging of superoxide by both cytoplasmic and mitochondrial SODs, resulting in intracellular superoxide concentrations estimated to rarely exceed 1 nM (54). The extracellular release of small proportions of intracellularly formed superoxide may occur via diffusion through anion channels, and superoxide formed from plasma membrane-bound oxidases remains at relatively low levels because of the serum and extracellular fluid components, including the low molecular weight oxidant scavengers and the heparin-binding extracellular-SOD (55–57). Thus, the relatively short half-life (seconds) of ROS and the efficient and redundant systems that have evolved to scavenge them require that any detection technique must be sensitive enough to effectively compete with these intra- and extracellular antioxidant components for reaction with the substance in question. Additionally, the methods for the analysis of ROS must have adequate intracellular access to faithfully reflect intracellular conditions. Finally, the often overlapping reactivity of ROS with the detection systems may hamper unequivocal identification and quantification of the responsible substance.

4.3.6.3.3. NO Measurements

Hypoxia regulates oxygen delivery to the tissue, reducing vascular resistance in hypoxia to maintain the delivery at a constant level via signals that control the vessel tone, allowing the blood flow to vary to meet the metabolic needs (58). Hypoxia inhibits endothelial respiration in the presence of NO, rendering the endothelial cell oxygen consumption dependent on the oxygen concentration (59). The oxygen consumption by the constituents of the wall is evidenced by the difference in the pO$_2$ across the arteriolar wall, or the vessel wall oxygen gradient (23). The oxygen consumption of the vessel wall and its pO_2 gradient increases because of the effect of vasoconstriction such as arginine vasopressin (42) and decreases in response to vasodilators, such as verapamil (60). The vasodilator effect of hypoxia also lowers the vessel wall oxygen consumption and the vessel wall gradient. Considering that NO synthesis has a stoichiometry of 1/1 with regard to the consumed O_2, the basal production of NO by eNOS in a rat aorta single endothelial cell requires the consumption of about 6.2×10^{-16} mol $O_2 \times s^{-1} \times cell^{-1}$ (61). Moreover, eNOS has been shown to be activated by endothelial cell stimulation with acetylcholine, leading to the activation of NO release. In the endothelium from rabbit aorta, Ach increases the NO extracellular level, and therefore, the production by ≈ 10 times and probably more, if the intracellular NO degradation is taken into account. Therefore, the corresponding quantity of O_2 transiently consumed by eNOS can be evaluated in the range of 10 times the basal consumption, i.e., 6.2×10^{-15} mol $O_2 \times s^{-1} \times cell^{-1}$ (62, 63). In hypoxia, the O_2 diffusion is insufficient to continuously provide O_2 as well as the enzymes, such as eNOS that consumes high levels of O_2, whereas NO modulates oxygen consumption by the tissues10, 12).

Thus, delivery of tissue oxygen is increased with NO availability because oxygen demand is reduced and vice versa. When the level of pO_2 is lower, as in hypoxia, the concentration of NO determines the tissue pO_2 and capillary perfusion.

4.4. Conclusions

In summary, our results show that reducing NO synthesis decreases the dilator response to hypoxia and tissue pO_2. Therefore, the increased NO formation reduces the O_2 consumption and decreases ROS/RNS production. Endothelial NOS while consuming oxygen, induces a significant decrease in pO_2; also, the amplitude of the consumption of oxygen by the vessel wall can be related to endothelial NOS activity. These findings may have an important implication in conditions where the NO and oxygen level decrease at the same time. Thus, the NO level in combination with decreased oxygen availability present a significant challenge to the maintenance of tissue metabolism, aggravated by the negative effects on capillary perfusion.

ROS/RNS are not simple signaling molecules but they have numerous layers of regulation to consider when assessing the outcome from possible exposure to various pO_2 tensions. NO synthesis, unlike respiration, influences intracellular oxygen tension. A complex relationship exists between the fundamental chemistry of NO and the important influences of the cellular redox state. Understanding the physiology and pathology of ROS/RNS challenges the paradigms of biological thinking and pushes the lower sensitivity limits of analytical chemistry. We conclude that it is possible to monitor real time-NO and pO$_2$ dynamics under different conditions through the use of electrodes selective for NO and oxygen molecules. The importance of finding methods to measure the real in vivo concentration of NO, ROS/RNS, and oxygen is crucial because their concentrations play an important and varied role in modulating vascular injury as well as proinflammatory responses.

5. Notes

1. The phosphorescence method is based on the relationship between the decay rate of excited phosphorescence from palladium_mesotetra_(4_carboxyphenyl)porphyrin bound to albumin and the partial pressure of oxygen, according to the Stern_Volmer equation . In this method, animals receive a slow intravenous injection of the porphyrin dye (15 mg/kg body wt) at a concentration of 10 mg/ml approximately 10 min before pO$_2$ measurements. The dye is made to phosphoresce by excitation with light flashes, and the oxygen concentration in an adjustable optical window that delineates the area where the measurement is to be made, is deduced from the rate of decay of the phosphorescence, which depends on the amount of oxygen that surrounds the dye. Phosphorescence is the emission of photons due to the electronic transition in molecules that are excited into a triplet state by absorbing light and then the transition from this state to a singlet ground state. Molecules such as Pd-porphyrin either release the absorbed energy as light or transfer this energy to oxygen, which prevents light emission, thus "quenching" the phosphorescence. The intensity of light emission $I(t)$ from many molecules is described by an exponential decay of the form: $I(t) = I_0 \exp(-t/a)$ where I_0 is the maximum intensity at $t = 0$ and a is the decay time constant. When light emission is quenched, fewer photons are emitted, which translates into a shorter time constant a. For quenching to occur, it is necessary that the phosphorescent molecule and the quenching agent (oxygen) collide before the occurrence of the triplet to ground state transition, a process that is, in part, dependent on the abundance (or concentration) of quenchers. The relationship between the decay constant a and the oxygen concentration (given by the product of oxygen solubility in the medium, $\forall$, and the local partial pressure pO$_2$ of oxygen) is given by the Stern–Volmer equation $J_0/a = 1 + K_Q J_0 \forall$ pO$_2$, where J_0 is the phosphorescence decay constant in the absence of oxygen, and K_Q is the quenching constant. An advantage of this method is that mixing Pd-porphyrin with excess albumin leads to the formation of a probe whose sensitivity to oxygen

quenching is independent of the probe concentration. In other words, the decay constant becomes independent of the concentration of the albumin/porphyrin complex.

2. The phosphorescence generated by the light excitation of the porphyrin probe consumes O_2. The amount consumed depends upon the concentration of the dye and the total energy delivered by the light source. With a very intense illumination, it is possible to make a determination of the oxygen level with a single flash. In this implementation, the flash lamp employed had a 25 s decay constant, which precluded the acquisition of phosphorescence decay data prior to about 80 s after flash extinction, and therefore, prevented reliable measurements of pO_2 above about 50 mmHg, which corresponds to a phosphorescence decay time of similar duration. The emission obtained with this technique is intense, and the phosphorescence decay curve may be the summation of signals from adjoining areas that would not normally have sufficient intensity to affect the principal component present, particularly if the measurement is made in the neighborhood of microvessels where the oxygen field is not uniform. In addition, oxygen is consumed as the phosphorescence decays, further distorting the decay signal. These problems have been reported to be amenable to solution by deconvolving the decay signal using mathematical techniques, an approach that may be useful if the signal noise is very small.

3. All implementations use a probe injection of 20–30 mg porphyrin/kg bw, leading to a blood concentration of 0.3 mg/ml, and a tissue fluid concentration smaller than 0.1 mg/ml. The multi-flash system of Torres Filho and Intaglietta with a flash decay constant of 10 s requires 100 flashes to obtain an interpretable signal in blood, each flash consuming oxygen, causing the concentration of oxygen in stationary plasma to decrease by 0.01 mmHg/flash at steady state. Thus, a single flash system that gives an interpretable signal will introduce an error <1 mmHg when used in stationary tissue fluid. This error may be lower in tissue because the amount of probe present is about 1/3 that of plasma; but this causes a proportional decrease in phosphorescence signal, requiring a significant increase of flash intensity and a corresponding increase in the consumption of oxygen, which may explain why not all systems are able to obtain tissue measurements. Comparison of in vivo pO_2 measurements at the same site with the multi-flash phosphorescence method and the microelectrode technique were obtained by two groups of investigators using avascular tissue areas of the hamster skin fold preparation and rat skeletal muscle. In the hamster preparation superfused with Krebs solution bubbled with 100% N_2, there was a maximum divergence of 2% between the methods over the tissue pO_2 range of 5–40 mmHg.

References

1. Haliwell B, Gutteridge JMC (1999) Free radical in Biology and Medicine. Oxford University Press, London

2. Dröge W (2001) Free radicals in the physiological control of cell function. Physiol Rev 82:47–95

3. Halliwell B, Gutteridge JMC (1985) The chemistry of oxygen radicals and other oxygen-derived species. In: Free radicals in biology and medicine. Oxford University Press, New York, pp 20–64

4. Barreto JC, Smith GS, Stobel NHP, McQuillin PA, Miller TA (1995) Teraphtalic acid: a dosimeter for detection of hydroxyl radical in vitro. Life Sci 56:89–96

5. Parker L (ed) (1994) Oxygen radicals in biological systems. In: Methods in Enzymology, vol 234, part D. Academic, New York

6. Ignarro LJ (2002) Nitric oxide is an unique signaling molecule in the vascular system: a historical overview. J Physiol Pharmacol 53:503–514

7. Buerk DG (2007) Nitric oxide regulation of microvascular oxygen. Antioxid Redox Signal 9:829–843

8. Abu-Soud HM, Rousseau DL, Stuehr DJ (1966) Nitric oxide binding to the heme of neuronal nitric-oxide synthase links its activity to changes in oxygen tension. J Biol Chem 271:32515–32518

9. Bertuglia S, Giusti A (2005) The role of nitric oxide in capillary perfusion and oxygen delivery regulation during systemic hypoxia. Am J Physiol Heart Circ Physiol 288(2):H525–531

10. Shen W, Xu X, Ochoa M, Zhao G, Wolin MS, Hintze TH (1994) Role of nitric oxide in the regulation of oxygen consumption in conscious dogs. Circ Res 75:1086–1095

11. King CE, Melinyshyin MJ, Mewburn JD, Curtis SE, Winn MJ, Cain SM, Chapler CK (1994) Canine hindlimb flow and O$_2$ uptake after inhibition of EDRF/NO synthesis. J Appl Physiol 76:1166–1171

12. Cabrales P, Tsai AG, Intaglietta M (2006) Nitric oxide regulation of microvascular oxygen exchange during hypoxia and hyperoxia. J Appl Physiol 100:1181–1187

13. Stamler JS (1995) S-nitrosothiols and the bioregulatory actions of nitrogen oxides through reactions with thiol groups. Curr Top Microbiol Immunol 196:19–36

14. Jia L, Bonaventura C, Bonaventura J, Stamler JS (1996) S-nitrosohaemoglobin: a dynamic activity of blood involved in vascular control. Nature 380:221–226

15. Intaglietta M, Johnson PC, Winslow RM (1996) Microvascular and tissue oxygen distribution. Cardiovasc Res 32:632–643

16. Hogan MC (1999) Phosphorescence quenching method for measurement of intracellular pO$_2$ in isolated skeletal muscle fibers. J Appl Physiol 86:720–724

17. Kessler M, Harrison, DK, Hoper J (1986) Tissue oxygen measurement techniques. In: Baker CH, Nastuk WL, Orlando FL (eds) Microcirculatory technology. Academic, New York, pp. 391–425

18. Popel AS, Pittman RN, Ellsworth ML (1989) Rate of oxygen loss from arterioles is an order of magnitude higher than expected. Am J Physiol 256:H921–H924

19. Tsai AG, Johnson PC, Intaglietta M (2003) Oxygen gradients in the microcirculation. Physiol Rev 83:933–963

20. Kerger H, Saltzman DJ, Gonzales A, Tsai AG, van Ackern K, Winslow RM, Intaglietta M (1997) Microvascular oxygen delivery and interstitial oxygenation during sodium pentobarbital anesthesia. Anesthesiology 86:372–386

21. Pawlowski M, Wilson DF (1992) Monitoring of the oxygen pressure in the blood of live animals using the oxygen dependent quenching of phosphorescence. Adv Exp Med Biol 316:179–185

22. Sinaasappel M, van Iterson M, Ince C (1992) Microvascular oxygen pressure in the pig intestine during hemorrhagic shock and resuscitation. J Physiol (Lond) 514:245–253

23. Tsai AG, Friesenecker B, Mazzoni MC, Kerger H, Buerk DG, Johnson PC, Intaglietta M (1988) Microvascular and tissue oxygen gradients in the rat mesentery. Proc Natl Acad Sci USA 95:6590–6595

24. Wilson DF (1993) Measuring oxygen using oxygen dependent quenching of phosphorescence: a status report. Adv Exp Med Biol 333:225–232

25. Seylaz E, Pinard J (1977) Continuous intracerebral PO2 and PCO2 measurements by mass spectrometry: study of the influence of vasoactive drugs. Acta Neurol Scand 64:438–439

26. Grisham MB (1994) Oxidants and free radicals in inflammatory bowel disease. Lancet 344:859–861

27. Pryor WA, Stanley JP, Blair E (1976) Autoxidation of polyunsaturated fatty acids: II. A suggested mechanism for the formation of TBA-reactive materials from prostaglandin-like endoperoxides. Lipids 11:370–379

28. Bertuglia S, Reiter NJ (2007) Melatonin reduces ventricular arrhythmias and preserves capillary perfusion during ischemia-reperfusion events in cardiomyopathic hamsters. J Pineal Res 42(1):55–63

29. Salvemini D, Mazzon E, Dugo L, Serraino I, De Sarro A, Caputi AP, Cuzzocrea S (2001) Amelioration of joint disage in a rat model of collagen-induced arthritis in M40403, a superoxide dismutase mimetic. Arthritis Rheum 44:2909–2291

30. Szabo A, Hake P, Salzman AL, Szabo C (1999) Beneficial effects of mercaptothylguanidine, an inhibitor of the inducible isoform of NO synthase and a scavenger of peroxinitrite in a porcine model of delayed hemorrhagic shock. Crit Care Med 27:1343–1359

31. Halliwell B, Gutteridge JMC (2000) Detection of free radicals and other reactive species: trapping and finger printing. In: Halliwell B, Gutteridge JMC (eds) Free Radicals in Biology and Medicine. Oxford University Press, Oxford, pp 351–429

32. Morrow JD, Hill KE, Burk RF, Nammour TM, Badr KF, Roberts LJ (1990) A series of prostaglandin F2-like compounds are produced in vivo in humans by a non-cyclooxygenase, free radical-catalyzed mechanism. Proc Natl Acad Sci USA 87:9383–9387

33. Wink DA, Kim S, Coffin D, Cook JC, Vodovotz Y, Chistodoulou D, Jourd'heuil D, Grisham MB (1999) Detection of S-nitrosothiols by fluorometric and colorimetric methods. Methods Enzymol 301:201–211

34. Karlsson J (1997) Introduction to nutraology and radical formation. In: Antioxidants and exercise. Human Kinetics Press, Illinois, pp 1–143

35. Baker JE, Froncisz W, Joseph J, Kalyanaraman B (1997) Spin label oximetry to assess extracellular oxygen during myocardial ischemia. Free Radic Biol Med 22:109–115

36. Velan SS, Spencer RG, Zweier JL, Kuppusamy P (2000) Electron paramagnetic resonance oxygen mapping (EPROM): direct visualization of oxygen concentration in tissue. Magn Reson Med 43:804–809

37. Acworth IN, Bailey B (1997) Reactive oxygen species. In: The handbook of oxidative metabolism. ESA, Northampton, MA, pp 1–4.

38. Bertuglia S, Colantuoni A, Coppini G, Intaglietta M (1991) Hypoxia- or hyperoxia-induced changes in arteriolar vasomotion in skeletal muscle microcirculation. Am J Physiol 260:H362–H372

39. Bertuglia S, Giusti A, Del Soldato P (2004) Antioxidant activity of a nitro derivative of aspirin against ischemia reperfusion in hamster cheek pouch microcirculation. Am. J. Physiol. Gastrointestinal-Liver Physiol. 286(3):G437–G443

40. Bertuglia S, Giusti A (2003) Microvascular oxygenation, oxidative stress, nitric oxide suppression and superoxide dismutase during postischemic reperfusion. Am J Physiol 285:H1064–H1071

41. Cabrales P, Tsai AG, Intaglietta M (2004) Increased tissue PO_2 and decreased O_2 delivery and consumption after 80% exchange transfusion with polymerized hemoglobin. Am J Physiol Heart Circ Physiol 287:H2825–H2833

42. Friesenecker B, Tsai AG, Dunser MW, Mayr AJ, Martini J, Knotzer H, Hasibeder W, Intaglietta M (2004) Oxygen distribution in the microcirculation following arginine vasopressin-induced arteriolar vasoconstriction. Am J Physiol Heart Circ Physiol 287:H1792–H1800

43. Torres Filho IP, Intaglietta M (1993) Microvessel PO_2 measurement by phosphorence decay method. Am J Physiol Heart Circ Physiol 265:H1537–H1545

44. Kerger H, Groth G, Kalenka A, Vajkoczy P, Tsai AG, Intaglietta M (2003) PO_2 measurements by phoroshorence quenching characteristics and applications of an automated system. Microvasc Res 15:93–101

45. Friedemann MN, Robinson SW, Gerhardt GA (1996) o-Phenylenediamine-modified carbon fiber electrodes for the detection of nitric oxide. Anal Chem 68:2621–2628

46. Cabrales P, Tsai AG, Intaglietta M (2004) Microvascular pressure and functional capillary density in extreme hemodilution with low and high plasma viscosity expanders. Am J Physiol Heart Circ Physiol 287:H363–H373

47. Tsai AG, Acero C, Nance PR, Frangos JA, Buerk DG, Intaglietta M (2005) Elevated plasma viscosity in extreme hemodilution increases perivascular nitric oxide concentration and microvascular perfusion. Am J Physiol Heart Circ Physiol 288:H1730–H1739

48. Sarelius IH (1968) Cell flow path influences transit time through striated muscle capillaries. Am J Physiol Heart Circ Physiol 250:H899–H907

49. Bohlen HG, Nase GP (2000) Dependence of intestinal arteriolar regulation on flow-mediated nitric oxide formation. Am J Physiol Heart Circ Physiol 279:H2249–H2258

50. Grishman MB, Jonshon GG, Lancaster JR (1966) Quantitation of nitrite and nitrate in extracellular fluids. Methods Enzymol 268:237–246

51. Golub AS, Barker MG, Pittman RN (2007) PO$_2$ profiles near arterioles and tissue oxygen consumption in rat mesentery. Am J Physiol Heart Circ Physiol 293:H1097–H1106

52. Golub AS, Pittman RN (2008) PO$_2$ measurements in the microcirculation using phosphorescence quenching microscopy at high magnification. Am J Physiol Heart Circ Physiol 294:H2095–H2916

53. Shibata M, Ichioka S, Ando J, Kamiya A (2001) Microvascular and interstitial PO$_2$ measurements in rat skeletal muscle by phosphorence quenching. J Appl Physiol 91:321–327

54. Tarpey MM, Wink DA, Grishman MB (2008) Methods for detection of reactive metabolites of oxygen and nitrose: in vitro and in vivo considerations. Am J Physiol Regul Integr Comp Physiol 206:R431–R444

55. Halliwell B (1995) How to characterize an antioxidant: an update. Biochem Soc Symp 61:73–101

56. Rice-Evans CA, Diplock AT (1993) Current status of antioxidant therapy. Free Radic Biol Med 15:77–96

57. Skulachev VP (1997) Membrane-linked systems preventing superoxide formation. Biosci Rep. 17:347–366.

58. Prewitt RL, Johnson PC (1976) The effect of oxygen on arteriolar red cell velocity and capillary density in the rat cremaster muscle. Microvasc Res 12:59–70

59. Edmunds NJ, Marshall JM (2003) The roles of nitric oxide in dilating proximal and terminal arterioles of skeletal muscle during systemic hypoxia. J Vasc Res 40:68–76

60. Hangai-Hoger N, Tsai AG, Friesenecker B, Cabrales P, Intaglietta M (2005) Microvascular oxygen delivery and consumption following treatment with verapamil. Am J Physiol Heart Circ Physiol 288:H1515–H1520

61. Akaike T, Yoshida M, Miyamoto Y (1993) Antagonistic action of imidazolineoxyl N-oxides against endothelium-derived relaxing facto/ NO through a radical reaction. Biochemistry 32:827–832

62. Christie MI, Griffith TM, Lewis MJ (1989) A comparison of basal and agonist-stimulated release of endothelium-derived relaxing factor from different arteries. Br J Pharmacol 98:397–406

63. Coste J, Vial JC, faury G, Deronzier A, Usson Y, Nicoud MR, Verdetti J (2002) NO synthesis, unlike respiration, influences intracellular oxygen tension. Biochem Biophys Res Comm 209:97–104

Chapter 3

Nitrate Reductase Activity of Mitochondrial Aldehyde Dehydrogenase (ALDH-2) as a Redox Sensor for Cardiovascular Oxidative Stress

Andreas Daiber and Thomas Münzel

Abstract

In 2002, mitochondrial aldehyde dehydrogenase (ALDH-2) was identified as an organic nitrate bioactivating enzyme. This so-called nitrate reductase activity denitrates nitroglycerin (glycerol trinitrate) to its 1,2-glycerol dinitrate metabolite and nitrite. This reaction relies on reduced thiols at the active site of the enzyme and on the presence of reduced dithiols as the electron source. During bioconversion of nitroglycerin, and also in the presence of reactive oxygen and nitrogen species, the active site thiols of ALDH-2 are oxidized and the enzyme looses its activity. We, therefore, speculated that ALDH-2 activity could be a useful marker for cardiovascular oxidative stress. Indeed, this hypothesis was supported by a number of studies, indicating that ALDH-2 activity is impaired in experimental animal models of increased oxidative stress and may be used for detection of an imbalance of mitochondrial and cellular redox state.

Key words: Mitochondrial aldehyde dehydrogenase (ALDH-2), Nitrate reductase activity, Thiol oxidation, Reactive oxygen species, Vascular oxidative stress

1. Introduction

The mitochondrial aldehyde dehydrogenase ''(ALDH-2) is an important enzyme for the detoxification of aldehydes to the corresponding carboxylic acid. Besides the well-known aldehyde oxidizing activity, which relies on the cofactor NAD^+, ALDH-2 also shows unspecific esterase activity leading to cleavage of carboxylic acid esters and esters of inorganic acids without requirement of cofactors. In 2002, this enzyme was identified as an organic nitrate bioactivating enzyme (1). This so-called nitrate reductase activity denitrates organic nitrates and forms nitrite. Since organic nitrates

D. Armstrong (ed.), *Advanced Protocols in Oxidative Stress II*, Methods in Molecular Biology, vol. 594
DOI 10.1007/978-1-60761-411-1_3, © Humana Press, a part of Springer Science+Business Media, LLC 2010

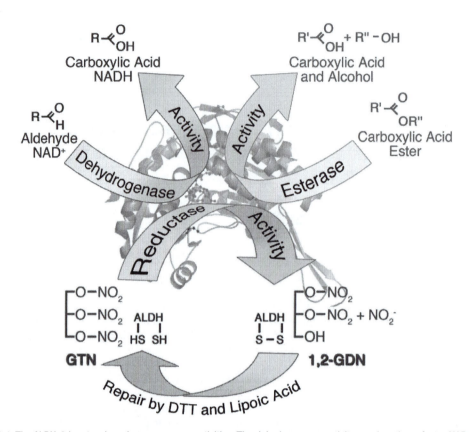

Fig. 3.1. The ALDH-2 has two long-known enzyme activities. The dehydrogenase activity requires the cofactor NAD⁺ and converts aldehydes to the respective carboxylic acid (e.g. acetaldehyde to acetic acid). The esterase activity does not require cofactors and converts carboxylic acid esters (probably also esters from other acids) to the free acid and the respective alcohol. These enzyme activities involve thiol-dependent catalysis that are inhibited by thiol-oxidizing compounds. The third activity is the "reductase activity", which was recently identified by Chen et al., and is responsible for bioactivation of organic nitrates such as GTN yielding nitrite and the dinitrate 1,2-GDN. Reductase activity of this oxidized ALDH-2 can be restored by dithiol compounds such as DTT and dihydrolipoic acid. Adopted from Ref. 3.

such as nitroglycerin are nitric acid esters, this reaction appears to be an ester hydrolysis at first view. However, since the oxidation state of the nitrogen of the denitrated group is changed from +V to +III, the reaction is a two-electron-reduction. The electrons for nitrate conversion do not come from the cofactor NADH since NAD⁺ accelerates the reaction probably due to steric reasons (1). In contrast, the dithiol dithiothreitol and dihydrolipoic acid were found to be essential cofactors and the source of electrons (2–5). These basic findings are summarized in Fig. 3.1.

The exact mechanism of how ALDH-2 bioactivates organic nitrates, how the enzyme is inactivated by the organic nitrate and/or oxidative stress as well as restoration of its enzymatic activity is proposed in Fig. 3.2. Essentially, nitroglycerin and reactive oxygen and nitrogen species (ROS and RNS) will cause formation of a disulfide bridge at the active site, which is reduced by

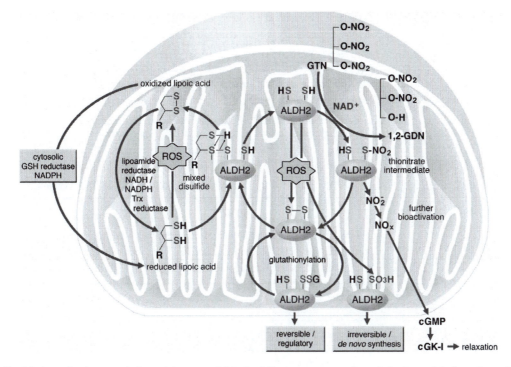

Fig. 3.2. According to our and others data, we postulate the following sequence of events for the metabolism of organic nitrates by ALDH-2: Two adjacent reduced cysteine-thiols are essential for organic nitrate bioactivation. In the first step, one of these thiols will react with the nitric acid ester yielding a thionitrate intermediate and the denitrated metabolite (1,2-GDN). Next, a disulfide and nitrite are formed. The disulfide can also be formed by direct oxidation of the thiols by oxidants such as superoxide and peroxynitrite and even higher oxidation to sulfonic acid ($-SO_3H$) could occur to cause irreversible inhibition of the enzyme. The dithiol dihydrolipoic acid restores enzymatic activity. Oxidized lipoic acid is reduced by reductases in mitochondria or the cytosol. Also, this last step is subject to oxidative impairment in the setting of tolerance. Adopted from Ref. 3.

dithiols such as dithiothreitol and dihydrolipoic acid (3). It should be noted that only dithiols were efficient in restoring ALDH-2 activity upon challenges with nitroglycerin or oxidative stress. Glutathione even further impaired enzymatic activity due to glutathionylation, which might be another physiological regulatory mechanism (3). Nitroglycerin and oxidative stress also caused irreversible inactivation, which was not restored by DTT or DHLA. We speculated that this inhibition was due to formation of sulfonic acid ($-SO_3H$), which requires other reducing factors than thiol-based compounds (6). Since not only ALDH-2 nitrate reductase activity but also aldehyde dehydrogenase and esterase activities rely on thiols and are impaired by hydrogen peroxide as well as thiol-oxidizing reagents such as Ellman's reagent (5), we will also introduce ALDH-2 dehydrogenase activity as a suitable marker for oxidative stress.

We were the first to show that in in vivo nitroglycerin treatment, induction of oxidative stress and clinical nitrate tolerance are associated with inhibition of ALDH-2 activity (2). Meanwhile, we could repeatedly show that ALDH-2 activity in aorta or cardiac mitochondria nicely correlates with the levels of oxidative stress in experimental animal models and also in human studies. Nitroglycerin infusion increases mitochondrial and vascular oxidative stress significantly and decreases ALDH-2 activity to a similar extent in animals (7) and also in human blood vessels from bypass surgery (8). ALDH-2 activity in vessels and mitochondria from rats under chronic nitroglycerin treatment was normalized by cotreatment with the efficient antioxidant hydralazine (7) and also by induction of intrinsic antioxidative systems such as the heme oxygenase-1/ferritin system (9). This concept was further supported at a molecular basis by investigations in an experimental model of increased mitochondrial oxidative stress, using mice with partial (heterozygous) deficiency in manganese superoxide dismutase ($MnSOD^{+/-}$), which is the mitochondrial isoform of this enzyme family (10). Indeed, these mice showed more pronounced tolerance and ALDH-2 inactivation in response to low dose of nitroglycerin, clearly identifying mitochondrial oxidative stress (or more precisely superoxide) as a major determinant for ALDH-2 activity state.

There were also differences in the efficacy of ROS and RNS to inhibit ALDH-2 in cardiac mitochondria. Superoxide, peroxynitrite, and nitroglycerin itself were highly efficient in inhibiting mitochondrial aldehyde dehydrogenase activity (Fig. 3.3), which was restored by dithiol compounds such as dihydrolipoic acid and dithiothreitol, whereas hydrogen peroxide and nitric oxide were rather insensitive inhibitors (3). Our observations indicate that mitochondrial oxidative stress (especially superoxide and peroxynitrite) in response to organic nitrate treatment may inactivate ALDH-2 thereby leading to nitrate tolerance.

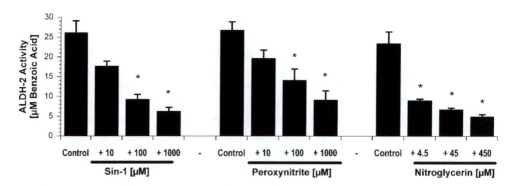

Fig. 3.3. Effect of oxidants on ALDH-2 dehydrogenase activity in isolated rat heart mitochondria. The ALDH-2 activity was determined by an HPLC-based assay measuring the conversion of the substrate benzaldehyde to the product benzoic acid. The effects of authentic peroxynitrite (5 min, room temperature), the peroxynitrite generator Sin-1 or GTN (90 min, 37°C) on ALDH-2 dehydrogenase activity were tested in sonicated mitochondrial suspensions (0.5 mg/ml total protein). Data are mean ± SEM of three independent experiments. *$p < 0.05$ vs. control. Adopted from Ref. 3.

Recently, ALDH-2 deficiency was shown to render mice more susceptible to exogenously triggered oxidative stress. ALDH-2$^{-/-}$ mice were more susceptible to nitroglycerin, acetaldehyde, and doxorubicin-induced cardiovascular damage (11). Cultured cells displayed increased oxidative damage in the absence of ALDH-2 (12). Therefore, ALDH-2 seems to be a marker for oxidative stress but its absence or deficiency in ALDH-2 obviously also increases oxidative stress.

2. Materials

2.1. Synthesis of 2-Hydroxy-3-Nitrobenzoic Acid

1. Yeast aldehyde dehydrogenase (lot 93300620, EC 1.2.1.5) was from Roche Diagnostics (Mannheim, Germany).

2. *250 U in 107 mg solid.* 50 U in 21.4 mg dissolved in 500 µl PBS/water (1:10) yields an ALDH solution of 100 U/ml.

3. Make aliquots with 20 µl (100 U/ml). 1 aliquot diluted in 2 ml PBS yields a final concentration of 1 U/ml.

4. Add 10 µl 2-hydroxy-3-nitrobenzaldehyde (2H3N-BA) from a 10 mM stock in DMSO (yielding a final 2H3N-BA concentration of 50 µM) and 20 µl NAD$^+$ from a 100 mM stock in water (yielding a final NAD$^+$ concentration of 1 mM).

5. Incubate for 5 h at 37°C until 2H3N-BA is completely converted to the 2-hydroxy-3-nitrobenzoic acid product (*see* Note 1), make aliquots with 100 µl and store at –20°C.

2.2. ALDH-2 Aldehyde Dehydrogenase Activity

1. GTN was purchased as an ethanolic stock solution (102 g/l, 450 mM) from UNIKEM (Copenhagen, Denmark).

2. Peroxynitrite (oxoperoxonitrate(1–)) was synthesized according to the quenched-flow method as described previously (13). The concentration was determined at 302 nm by using $\varepsilon_{302} = 1,670$ M^{-1}cm^{-1}. 50 mM stocks were prepared in 0.1 M sodium hydroxyde solution and stored at –80°C (*see* Note 2).

3. Benomyl (methyl-1-(butylcarbamoyl)-2-benzimidazole carbamate) was purchased from Chem Service, Inc. (West Chester, PA). 10 mM stocks were prepared in DMSO and stored at –20°C.

4. Daidzin (7-glucoside of 4-,7-dihydroxy-isoflavone) was purchased from INDOFINE Chemical Company, Inc. (Hillsborough, NJ). 20 mM stocks were prepared in DMSO/water 1:1 and stored at –20°C.

5. 3-morpholino-sydnonimine (Sin-1) was purchased from Calbiochem (San Diego, CA). 100 mM stocks were prepared in 0.1 M hydrochloric acid solution and stored at –80°C (*see* Note 3).

6. *Mitochondria isolation buffer I.* 50 mM HEPES, 70 mM sucrose, 220 mM mannitol, 1 mM EGTA, and 0.033 mM bovine serum albumin.

7. *Mitochondria isolation buffer II.* 10 mM Tris, 340 mM sucrose, 100 mM KCl, and 1 mM EDTA.

8. A 10 mM stock of 2-hydroxy-3-nitrobenzaldehyde was prepared in DMSO. Make 100 µl aliquots and store them at −20°C.

2.3. ALDH-2 Reductase Activity

1. The PETriN-dansyl-labeled compound (KL-61) was synthesized by K. Lange in the laboratory of J. Lehmann (Department of Pharmaceutical Chemistry, University of Jena, Germany). A 10 mM stock of KL-61 (MW: 504.43 Da) was prepared in DMSO and 50 µl aliquots were stored at −20°C.

2. All other required solutions were prepared as described in Subheading 3.2.2.

3   Methods

3.1. ALDH Dehydrogenase and Esterase Activity

3.1.1. Classical Methods

1. Conversion of benazaldehyde or methylbenzoate to benzoic acid is a measure of aldehyde dehydrogenase or esterase activity, respectively (*see* Note 4). This protocol was published in (2, 5).

2. ALDH activity can also be measured photometrically in the supernatant at room temperature by following NADH formation at 340 nm. The assay mixture (1 ml) contained: 100 mM Tris–HCl (pH 8.5), 1 mM NAD$^+$, 1 mM propionaldehyde, and 1 mM 4-methylpyrazole (*see* Note 5). This protocol was published in (1).

3.1.2. Conversion of 2-Hydroxy-3-Nitro-Benzaldehyde to its Benzoic Acid Derivative in Isolated Heart Mitochondria

This section describes how to measure ALDH-2 dehydrogenase activity by an HPLC-based assay using a substrate with high affinity for the ALDH-2 isoform. The isolation of cardiac mitochondria, incubation of mitochondria with different oxidants as well as measurement of activity by HPLC are described in detail. Since ALDH-2 dehydrogenase activity directly relies on reduced thiol groups at the active site of the enzyme, treatment with oxidants will result in loss of activity.

1. Hearts from wild type and ALDH-2$^{-/-}$ mice were homogenized in mitochondria isolation buffer I and centrifuged at 1,500×*g* (10 min at 4°C) and 2,000×*g* for 5 min (the pellets were discarded). The supernatant was then centrifuged at 20,000×*g* for 20 min, and the pellet was resuspended in 1 ml of mitochondria isolation buffer I.

2. The latter step was repeated and the pellet resuspended in 1 ml of mitochondria isolation buffer I. The mitochondrial fraction (total protein approximately 5–10 mg/ml) was kept on ice

and diluted to approximately 1 mg/ml protein in 0.25 ml of PBS (*see* Note 6).

3. To ensure that ALDH-2 activity is measured, we used mitochondria from wild type or ALDH-2$^{-/-}$ mice (Fig. 3.4) or the samples were preincubated for 10 min at room temperature in the presence or absence of the specific ALDH-2 inhibitor daidzin (200 µM) (Fig. 3.5) (*see* Note 7).

4. Benomyl (100 µM) is an unspecific ALDH inhibitor and preincubation with this compound will suppress the dehydrogenase activity almost completely (*see* Note 8).

5. Also oxidants such as peroxynitrite, Sin-1, or GTN (10–1,000 µM) were preincubated 10 min prior to addition of the ALDH-2 substrate (*see* Notes 9–11).

6. For measurement of ALDH-2 dehydrogenase activity, 2-hydroxy-3-nitro-benzaldehyde (100 µM) was added and the samples were incubated for another 30 min at 37°C.

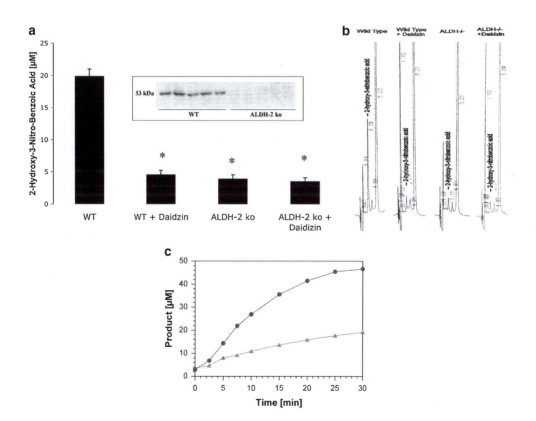

Fig. 3.4. (**a**) Activity of ALDH-2 was assessed by measuring the conversion of 2-hydroxy-3-nitro-benzaldehyde to the benzoic acid product (2H3N-BA) in isolated mitochondria. Insert shows a Western blot for ALDH-2 in extracts of aortic tissue from wild type and ALDH-2-deficient mice. *$P < 0.05$ vs. WT. (**b**) Representative chromatograms for the data presented in panel (**a**). (**c**) Kinetic measurements of the conversion to the benzoic acid product in isolated mitochondria from wild type mice where ALDH-2 is the dominant isoform. Mitochondria (1 mg/ml) was incubated with 50 µM benzaldehyde (triangles) or 2-hydroxy-3-nitro-benzaldehyde (*circles*). The modified substrate showed a much higher rate of conversion, which was almost completely inhibited by daidzin. From Ref. 14

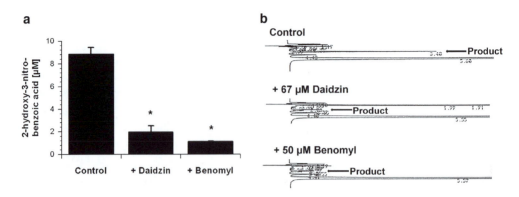

Fig. 3.5. (**a**) The high specificity of the modified substrate (2-hydroxy-3-nitro-benzaldehyde) for ALDH-2 activity could be demonstrated by specific inhibition of the conversion to the benzoic acid product by daidzin, a highly specific inhibitor of the mitochondrial isoform. ALDH-2 activity was measured in cultured smooth muscle cells (10^6 cells per sample). (**b**) Representative chromatograms for three independent experiments. From Ref. 14.

7. Samples were sonicated, centrifuged at $20,000 \times g$ (4°C) for 20 min, and the supernatant was purified by size exclusion centrifugation in a Microcon YM-10 filter device from Millipore (Bedford, USA) (*see* Note 12). Finally, the samples were frozen at –20°C and stored overnight (*see* Note 13).

8. 100 µl of the synthesized 2-hydroxy-3-nitrobenzoic acid standard (50 µM) was injected to the HPLC system to test its accuracy. Next, 100 µl of each sample were subjected to HPLC analysis.

9. The system consisted of a Gynkotek pump and detector and a C_{18}-Nucleosil 125x4 100-3 reversed phase column from Macherey & Nagel (Düren, Germany). The mobile phase contained acetonitrile (35 v/v%) in 50 mM citric acid buffer (65 v/v%) pH 2.2.

10. The substrate and its products were isocratically eluted at a flow rate of 0.8 ml/min, detected at 360 nm, and quantified using internal and external standards (2-hydroxy-3-nitrobenzoic acid and 2-hydroxy-3-nitrobenzaldehyde). The typical retention times were 3.4 and 5.6 min, respectively. *See* Fig. 3.4 for representative chromatograms. This protocol was published in (14).

3.1.3. Conversion of 2-Hydroxy-3-Nitro-Benzaldehyde to its Benzoic Acid Derivative in Isolated Aortic Vessel Segments (Vascular Tissue)

1. For determination of vascular ALDH activity, vessel ring segments (0.5–0.8 cm length) were incubated with 2-hydroxy-3-nitrobenzaldehyde (100 µM) for 30 min at 37°C.

2. The samples were then stored at –20°C over night and the benzoic acid product was measured in the supernatant.

3. The conversion to benzoic acid or to 2-hydroxy-3-nitrobenzoic acid, respectively, was measured in the incubation solution as described above and previously (2, 14).

4. 100 µl of each sample was subjected to high-performance liquid chromatography (HPLC) analysis. This protocol and data from human saphenous veins as well as mammery artery were published in (8).

5. The yield of 2-hydroxy-3-nitrobenzoic acid was normalized to dry weight of the aortic rings (*see* Note 14).

3.1.4. Conversion of 2-Hydroxy-3-Nitro-Benzaldehyde to its Benzoic Acid Derivative in Cultured Cells

1. The medium of smooth muscle cells (10^6 cells per well using four well plates) was replaced by PBS (5 ml) and cells were incubated with the vehicle (0.1% DMSO), 67 µM daidzin or 50 µM benomyl followed by 35 min incubation with 2-hydroxy-3-nitro-benzaldehyde at 37°C.

2. The supernatant can also be removed and stored at –20°C prior to measurement.

3. An aliquot of 100 µl of the supernatant was subjected to HPLC analysis. Since also cytosolic isoforms such as ALDH-1 are active in intact cells, the data clearly demonstrate that most of the conversion of the modified 2-hydroxy-3-nitrobenzaldehyde substrate is due to ALDH-2 activity which can be inhibited by daidzin (Fig. 3.5). This protocol was published in (14).

3.2. Nitrate Reductase Activity

3.2.1. Classical Methods

1. Detection of nitroglycerin and its dinitrate metabolites in cultured RAW 264.7 macrophages by GC–MS analysis was previously reported (5, 15) (*see* Note 15).

2. Detection of ^{14}C-labeled nitroglycerin and its dinitrate metabolites in cell culture and vascular tissue by the TLC and liquid scintillation spectrometry was previously described by Brien et al. (16) (*see* Note 16).

3. Detection of organic nitrates and denitrated metabolites in isolated mitochondria and blood samples by HPLC analysis with chemiluminescence nitrogen detection (CLND), according to a published method (14, 17) (*see* Note 17).

3.2.2. Detection of a Dansyl-Labeled PETriN Compound and its Denitrated Metabolites in Isolated Mitochondria by HPLC Analysis with Fluorescence Detection

The purpose of this new pentaerithrityl trinitrate (PETriN) derivative (KL-61) was to develop an easy HPLC-based assay to assess organic nitrate bioactivation by ALDH-2 by means of optical UV/Vis or fluorescence detection. KL-61 was synthesized in the laboratory of J. Lehmann (Jena, Germany) (for structure *see* Fig. 3.6) which is PETriN linked to a chromophore allowing optical UV/Vis detection at 370 nm (and even fluorescence detection). We observed increased hydrolysis of this compound in mitochondrial preparations that required careful and fast handling of the samples, which could not be injected by an autosampler, but had to be subjected to analysis immediately after thawing.

1. Prior to the assay, liver tissue was removed from freshly killed wild type and ALDH-2–/– mice and a mitochondria preparation

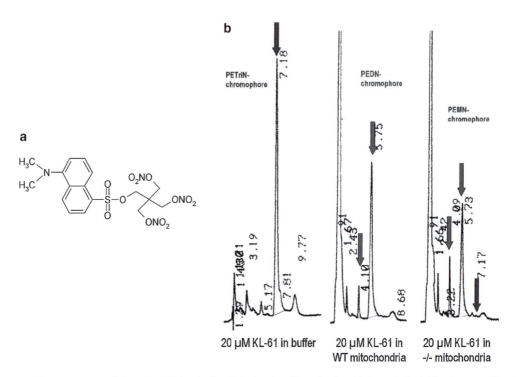

Fig. 3.6. Chromatograms of KL-61 (20 μM) incubations in isolated rat liver mitochondria (5 mg/ml). Mobile phase consisted of 57% acetonitrile/43 % 50 mM citrate pH 2, the column was a Nucleosil C_{18} (125x4) 100-3 from Macherey & Nagel (Düren, Germany) and detection was set at 370 nm. Chromatograms are representative of 2–3 independent measurements. From Ref. 14.

was carried out following the protocol described above (*see* Subheading 3.1.2).

2. The mitochondrial preparation (5 or 15 mg/ml) was incubated with 2.5 or 20 μM of KL-61 for 30 s or 2.5 min and then snap frozen in liquid nitrogen.

3. After thawing, suspensions were centrifuged for 1 min at 20,000×*g*. The supernatant was rapidly mixed with acetonitril (1:1 dilution).

4. Bioactivation of KL-61 by ALDH-2 could be optically detected (*see* Fig. 3.6) and KL-61 (PETriN-dansyl) itself as well as its di- and mononitrate metabolites (PEDN-dansyl and PEMN-dansyl) were identified by LC/MSMS measurements.

5. 100 μl of each sample was subjected to HPLC analysis. The system consisted of a Gynkotek pump and detector and a C_{18}-Nucleosil 125x4 100-3 reversed phase column from Macherey & Nagel (Düren, Germany). The mobile phase consisted of 57% acetonitrile/43% 50 mM citrate pH 2, the column was a Nucleosil C18 (125x4) 100-3 from Macherey & Nagel (Düren, Germany) and detection was set at 370 nm. This protocol was published in (14).

3.4. Notes

1. Conversion was controlled by injection of 50 µl to the HPLC system described under Subheading 3.3.1.

2. Peroxynitrite in alkaline solutions (>0.1 M) can be stored for at least 1 year at −80°C without serious loss of peroxynitrite. However, the peroxynitrite concentration should be determined from time to time before use.

3. Sin-1 in acidic solutions (>0.1 M) can be stored for at least 1 year at −80°C without serious loss of Sin-1.

4. The substrates are quite unspecific and the assay is not very specific for ALDH-2 activity.

5. The advantage is that this assay is easy to establish and very fast. The disadvantage is that the assay is not specific for ALDH-2 and optical detection at 340 nm interferes with disturbed solutions such as mitochondrial suspensions.

6. The protein concentration was determined by a standard DC (Lowry) protein assay from BioRad.

7. Daidzin is highly specific for ALDH-2 isoform but has to be used at rather high concentrations since it is a competitive, reversible inhibitor.

8. Benomyl is an unspecific ALDH inhibitor, but can be used at rather low concentrations since it is an irreversible inhibitor that covalently binds to the thiol groups of ALDH-2. Benomyl probably also inhibits other thiol-dependent enzymes and causes elevation of oxidative stress. Therefore, it is preferably used at concentrations <50 µM.

9. Peroxynitrite ($ONOO^-$) is the reaction product of nitric oxide with superoxide and upon mixing with buffer at pH 7.4 and 37°C will decompose within 2–3 s. Therefore, it is essential to rapidly mix peroxynitrite with the sample (preferably with a Vortex mixer by placing a drop (10 µl) of the peroxynitrite stock solution in the top of the reaction cup, close the top with caution and mix with 1 ml of the sample).

10. Sin-1 is the active metabolite of molsidomine (a nitrovasodilator) and releases NO by an autoxidation process, which will generate superoxide at equimolar fluxes as compared to NO. Therefore, Sin-1 can be regarded as a peroxynitrite donor since NO and superoxide react with a second order rate constant of $6{-}10 \times 10^9\,M^{-1}s^{-1}$ to form peroxynitrite. As soon as Sin-1 is diluted from the acidic stock to buffer at pH 7.4 and 37°C, it will start to decompose with a half-life of approximately

40–45 min in a linear fashion and generate defined fluxes of nitric oxide, superoxide, and accordingly peroxynitrite.

11. Nitroglycerin (GTN) directly inactivates ALDH-2 during the bioactivation process (*see* Fig. 3.2) in a suicide-catalytic fashion. In mitochondria, GTN also causes reactive oxygen and nitrogen species formation (probably also peroxynitrite) further contributing to inactivation of the enzyme.

12. The centrifugation steps should be performed rapidly. In case of highly active samples (e.g. concentrated mitochondrial suspensions), it might be useful to add benomyl ($100\,\mu M$) to the samples to avoid continuation of the ALDH-catalyzed conversion of the aldehyde during the centrifugation and purification steps. After size exclusion centrifugation, all protein is removed from the sample and all dehydrogenase activity is lost. Therefore, after this step the samples can be handled in a more relaxed way. Size exclusion centrifugation is also essential to avoid closing of the HPLC column by mitochondrial particals.

13. Freezing and storage at $-20°C$ overnight seems to be important for disruption of mitochondrial membrane and release of the 2-hydroxy-3-nitrobenzoic acid product to be measured by HPLC.

14. Aortic rings are normally dried within 24 h at room temperature or 4 h at $60°C$.

15. The advantage is the high sensitivity, the disadvantage is that GC–MS method is not available in all laboratories.

16. The advantage is the high sensitivity and that the assay can be used in tissue after chronic or acute treatment with nonlabeled nitroglycerin. The disadvantage is that radioactivity can not be used in all laboratories and that ^{14}C-labeled nitroglycerin is complicated to purchase.

17. The advantage is that the assay can be used for various organic nitrates. The disadvantage is that CLND is not a widely distributed method and that N-containing constituents in body fluids interfere with the detection of organic nitrates.

Acknowledgments

We thank the German Research Foundation for continuous funding of our ongoing research on nitrate tolerance (SFB 553-C17 to T.M. and A.D.) and the University Hospital Mainz for financial support (MAIFOR to A.D.).

References

1. Chen Z, Zhang J, Stamler JS (2002) Identification of the enzymatic mechanism of nitroglycerin bioactivation. Proc Natl Acad Sci U S A 99:8306–11

2. Sydow K, Daiber A, Oelze M, Chen Z, August M, Wendt M, Ullrich V, Mulsch A, Schulz E, Keaney JF Jr, Stamler JS, Munzel T (2004) Central role of mitochondrial aldehyde dehydrogenase and reactive oxygen species in nitroglycerin tolerance and cross-tolerance. J Clin Invest 113:482–9

3. Wenzel P, Hink U, Oelze M, Schuppan S, Schaeuble K, Schildknecht S, Ho KK, Weiner H, Bachschmid M, Munzel T, Daiber A (2007) Role of reduced lipoic acid in the redox regulation of mitochondrial aldehyde dehydrogenase (ALDH-2) activity: Implications for mitochondrial Oxidative stress and nitrate tolerance. J Biol Chem 282:792–9

4. Chen Z, Stamler JS (2006) Bioactivation of nitroglycerin by the mitochondrial aldehyde dehydrogenase. Trends Cardiovasc Med 16:259–65

5. Daiber A, Oelze M, Coldewey M, Bachschmid M, Wenzel P, Sydow K, Wendt M, Kleschyov AL, Stalleicken D, Ullrich V, Mulsch A, Munzel T (2004) Oxidative stress and mitochondrial aldehyde dehydrogenase activity: a comparison of pentaerythritol tetranitrate with other organic nitrates. Mol Pharmacol 66:1372–82

6. Daiber A, Bachschmid M (2007) Enzyme inhibition by peroxynitrite-mediated tyrosine nitration and thiol oxidation. Curr Enzyme Inhib 3:103–17

7. Daiber A, Oelze M, Coldewey M, Kaiser K, Huth C, Schildknecht S, Bachschmid M, Nazirisadeh Y, Ullrich V, Mulsch A, Munzel T, Tsilimingas N (2005) Hydralazine is a powerful inhibitor of peroxynitrite formation as a possible explanation for its beneficial effects on prognosis in patients with congestive heart failure. Biochem Biophys Res Commun 338:1865–74

8. Hink U, Daiber A, Kayhan N, Trischler J, Kraatz C, Oelze M, Mollnau H, Wenzel P, Vahl CF, Ho KK, Weiner H, Munzel T (2007) Oxidative inhibition of the mitochondrial aldehyde dehydrogenase promotes nitroglycerin tolerance in human blood vessels. J Am Coll Cardiol 50:2226–32

9. Wenzel P, Oelze M, Coldewey M, Hortmann M, Seeling A, Hink U, Mollnau H, Stalleicken D, Weiner H, Lehmann J, Li H, Forstermann U, Munzel T, Daiber A (2007) Heme oxygenase-1: a novel key player in the development of tolerance in response to organic nitrates. Arterioscler Thromb Vasc Biol 27:1729–35

10. Daiber A, Oelze M, Sulyok S, Coldewey M, Schulz E, Treiber N, Hink U, Mulsch A, Scharffetter-Kochanek K, Munzel T (2005) Heterozygous deficiency of manganese superoxide dismutase in mice (Mn-SOD+/−): A novel approach to assess the role of oxidative stress for the development of nitrate tolerance. Mol Pharmacol 68:579–88

11. Wenzel P, Muller J, Zurmeyer S, Schuhmacher S, Schulz E, Oelze M, Pautz A, Kawamoto T, Wojnowski L, Kleinert H, Munzel T, Daiber A (2008) ALDH-2 deficiency increases cardiovascular oxidative stress – Evidence for indirect antioxidative properties. Biochem Biophys Res Commun 367:137–143

12. Szocs K, Lassegue B, Wenzel P, Wendt M, Daiber A, Oelze M, Meinertz T, Munzel T, Baldus S (2007) Increased superoxide production in nitrate tolerance is associated with NAD(P)H oxidase and aldehyde dehydrogenase 2 downregulation. J Mol Cell Cardiol 42:1111–8

13. Saha A, Goldstein S, Cabelli D, Czapski G (1998) Determination of optimal conditions for synthesis of peroxynitrite by mixing acidified hydrogen peroxide with nitrite. Free Radic Biol Med 24:653–9

14. Wenzel P, Hink U, Oelze M, Seeling A, Isse T, Bruns K, Steinhoff L, Brandt M, Kleschyov AL, Schulz E, Lange K, Weiner H, Lehmann J, Lackner KJ, Kawamoto T, Munzel T, Daiber A (2007) Number of nitrate groups determines reactivity and potency of organic nitrates: a proof of concept study in ALDH-2−/− mice. Brit J Pharmacol 150:526–33

15. Gerardin A, Gaudry D, Wantiez D (1982) Gas chromatographic mass spectrometric determination of 1, 2, 3-propanetrioltrinitrate (nitroglycerin) in human plasma using the nitrogen-15 labelled compound as internal standard. Biomed Mass Spectrom 9:333–5

16. Brien JF, McLaughlin BE, Breedon TH, Bennett BM, Nakatsu K, Marks GS (1986) Biotransformation of glyceryl trinitrate occurs concurrently with relaxation of rabbit aorta. J Pharmacol Exp Ther 237:608–14

17. Seeling A, Lehmann J (2006) NO-donors, part X (1) : investigations on the stability of pentaerythrityl tetranitrate (PETN) by HPLC-chemoluminescence-N-detection (CLND) versus UV-detection in HPLC. J Pharm Biomed Anal 40:1131–6

Chapter 4

Identification of ROS Using Oxidized DCFDA and Flow-Cytometry

Evgeniy Eruslanov and Sergei Kusmartsev

Abstract

Cells constantly generate reactive oxygen species (ROS) during aerobic metabolism. The ROS generation plays an important protective and functional role in the immune system. The cell is armed with a powerful antioxidant defense system to combat excessive production of ROS. Oxidative stress occurs in cells when the generation of ROS overwhelms the cells' natural antioxidant defenses. ROS and the oxidative damage are thought to play an important role in many human diseases including cancer, atherosclerosis, other neurodegenerative diseases and diabetes. Thus, establishing their precise role requires the ability to measure ROS accurately and the oxidative damage that they cause. There are many methods for measuring free radical production in cells. The most straightforward techniques use cell permeable fluorescent and chemiluminescent probes. $2'$-$7'$-Dichlorodihydrofluorescein diacetate (DCFH-DA) is one of the most widely used techniques for directly measuring the redox state of a cell. It has several advantages over other techniques developed. It is very easy to use, extremely sensitive to changes in the redox state of a cell, inexpensive and can be used to follow changes in ROS over time.

Key words: Reactive oxygen species (ROS), Oxidative stress, $2'$-$7'$-dichlorodihydrofluorescein diacetate (DCFH-DA), $2'$-$7'$-dichlorofluorescein (DCF), Flow-cytometry, Immature myeloid cells (ImC)

1. Introduction

Reactive oxygen species (ROS) are constantly generated under normal conditions as a consequence of aerobic respiration. ROS include free radicals such as the superoxide anion (O_2^-), singlet oxygen ($^1O_2$,) hydroxyl radicals (OH), various peroxides (ROOR'), hydroperoxides (ROOH) and the no radical hydrogen peroxide (H_2O_2). Despite on multiple redox modulation systems, a given proportion of ROS continuously escape from the mitochondrial respiratorial chain inducing a damage cells in various ways including

D. Armstrong (ed.), *Advanced Protocols in Oxidative Stress II*, Methods in Molecular Biology, vol. 594
DOI 10.1007/978-1-60761-411-1_4, © Humana Press, a part of Springer Science + Business Media, LLC 2010

numerous carcinogenic DNA mutations. The cell is equipped with an extensive antioxidant defense system to combat ROS, either directly by interception or indirectly through reversal of oxidative damage. When ROS overcome the defense systems of the cell and redox homeostasis is altered, the result is oxidative stress.

Free radicals and other "reactive oxygen (ROS)/nitrogen/chlorine species" are believed to contribute to the development of several age-related diseases, and perhaps, even to the aging process itself (1, 2) by causing "oxidative stress" and "oxidative damage." For example, many studies have shown increased oxidative damage to all the major classes of biomolecules in the brains of Alzheimer's patients (3–5). Other diseases in which oxidative damage has been implicated include cancer, atherosclerosis, other neurodegenerative diseases and diabetes (6–8). To establish the role of oxidative damage, it is therefore essential to be able to measure it accurately.

1.1. Sources of ROS and Antioxidant Defense Mechanisms

Major sources of ROS production include the mitochondria, endoplasmic reticulum, plasma membrane and cytosol. In normal resting cells, 1–2% of electrons carried by the mitochondrial electron transport chain (ETC) leak from this pathway and form the superoxide free radical O_2^- during respiration. The dismutation of O_2^- by superoxide dismutase (SOD) results in the generation of H_2O_2, which can then react with Fe^{2+} to form hydroxyl radicals via the Fenton reaction: Hydroxyl radicals can also be generated via the metal catalysed Haber–Weiss reaction (9, 10). Other sources of O_2^- include the enzymes xanthine oxidase in the cytosol, NADPH oxidase in the membrane and cytochrome $P450$ in the ER (11–13).

Under normal conditions, antioxidant systems minimize the adverse effects caused by ROS. Antioxidants can be divided into primary or secondary defence mechanisms. Components of the primary antioxidant defense function to prevent oxidative damage directly by intercepting ROS before they can damage intracellular targets. It consists of superoxide dismutase (SOD), glutathione peroxidase (Gpx), catalase and thioredoxin reductase. Four classes of SOD have been identified to date. These are Mn-SOD, Cu, Zn-SOD, Ni-SOD and extracellular SOD. All four SOD enzymes destroy the free radical superoxide by converting it to H_2O_2.

H_2O_2 is one of the major ROS in the cell. While low levels result in apoptosis, high levels can lead to necrosis or caspase-independent apoptosis (14, 15). The primary defence mechanisms against H_2O_2 are catalase (16) and glutathione peroxidase (GPx) through the glutathione (GSH) redox cycle (17). Catalase is one of the most efficient enzymes known and cannot be saturated by H_2O_2 at any concentration (18). It is present, only in the peroxisome fraction whereas the GSH redox cycle exists in the cytosol and mitochondria. Catalase reacts with H_2O_2 to form water and molecular oxygen. Overexpression of catalase in cytosolic or mitochondrial compartments has been demonstrated to protect cells against oxidative injury (19).

The GSH system is probably the most important cellular defence mechanism that exists in the cell. The tripeptide GSH (γ-Glu-Cys-Gly), not only acts as an ROS scavenger but also functions in the regulation of the intracellular redox state. The system consists of GSH, glutathione peroxidase and glutathione reductase. Glutathione peroxidase catalyses the reduction of H_2O_2 and other peroxidases and converts GSH to its oxidized disulphide form (GSSG) as outlined below.

$$ROOH + 2GSH \xrightarrow{\text{GPX ROH}} + GSSG + H_2O$$

GSSG is then reduced back to GSH by glutathione reductase. The ability of the cell to regenerate GSH (either by reduction of GSSG or new synthesis of GSH) is an important factor in the efficiency of that cell in managing oxidative stress. The rate-limiting step for GSH synthesis is catalyzed by the enzyme L-γ-glutamyl-cysteine synthase. Inducers of this enzyme have been reported to prevent glutamate toxicity (20).

Thioredoxin reductase (Trx R)/thioredoxin (Trx) is another powerful system to protect cells against H_2O_2. Thioredoxin reductase (Trx R) utilizes NADPH to catalyze the conversion of oxidized Trx into reduced Trx. Reduced Trx provides reducing equivalents to Trx peroxidase, which breaks down H_2O_2 to water (21).

1.2. Role of Reactive Oxygen Species in Immune System and in Cancer

ROS generation plays an important role in the immune system. Phagocytes, including macrophages and neutrophils, are capable of generating large quantities of RNI and ROS, respectively. These free radicals are important for phagocytic anti-microbial and tumoricidal immune responses. Neutrophils have a short lifespan, usually about 24 h and ROS appear to play an important role in neutrophil survival (22). Activated neutrophils undergo spontaneous apoptosis shortly after producing an ROS burst against invading pathogens. Research suggests that this spontaneous apoptosis is mediated by endogenous ROS production (23). Survival of neutrophils incubated with antioxidants is greatly enhanced over 24 h in culture (24). Generation of ROS by TNF-α is critical for the phagocytic immune response against invading pathogens (25).

The data with knockout animals that lack antioxidant enzymes, supported by data from some animal knockouts of repair enzymes, strongly support the view that ROS contribute to the age-related development of cancer (26, 27). Direct damage to DNA is probably one key event, but it alone is insufficient to produce cancer, suggesting that the ability of RS to suppress apoptosis, and promote proliferation, invasiveness and metastasis (and possibly angiogenesis) are also important (28, 29). The relative contributions of these various mechanisms are unclear. Cancer associated with chronic inflammation may also involve ROS (30, 31). The trend now in cancer treatment is to make therapies based on the gene expression, cell signaling and proteomic profiles of a tumor. Perhaps, we need to do the same for its "oxidative stress status".

It is well established that tumor growth is associated with accumulation of immature myeloid cells (ImC), which in mice are characterized by expression of Gr-1 and CD11b markers. They play an important role in tumor associated immune suppression (32).

ImC from tumor-bearing mice had significantly higher levels of reactive oxygen species (ROS) than ImC obtained from tumor-free mice. Hydrogen peroxide, but not superoxide radical anion was found to be a major part of this increased ROS production. In vitro experiments demonstrated that scavenging of hydrogen peroxide with catalase induced differentiation of ImC from tumor-bearing mice into macrophages. Thus, tumors may prevent differentiation of antigen presenting cells by increasing the level of endogenous hydrogen peroxide in immature myeloid cells (32, 33).

We demonstrated that implantation of human RCC tumor cells into athymic nude mice promotes the appearance of VEGF receptor 1 (VEGFR1)/CD11b double-positive myeloid cells in peripheral blood. Up-regulation of VEGFR1 by myeloid cells could also be achieved in vitro by short-term exposure of naive myeloid cells to oxidative stress. Furthermore, after exposure to oxidative stress, myeloid cells acquire immunosuppressive features and become capable of inhibiting T cell proliferation. Data suggest that tumor-induced oxidative stress may promote both VEGFR1 up-regulation and immunosuppressive function in bone marrow-derived myeloid cells (34).

ImC isolated from tumor-bearing mice but not their control counterparts were able to inhibit antigen-specific response of $CD8^+$ T cells. ImC did not produce nitric oxide, however, ImC obtained from tumor-bearing mice had significantly higher level of reactive oxygen species (ROS) than ImC isolated from tumor-free animals. Accumulation of H_2O_2, but not superoxide was a major contributor to this increased pool of ROS. It appears that arginase activity played an important role in H_2O_2 accumulation in these cells. Inhibition of ROS in ImC completely abrogated the inhibitory effect of these cells on T cells. This indicates that ImC generated in tumor-bearing hosts suppress $CD8^+$ T cell response via release of ROS (34).

1.3. ROS Detection   There are many methods for measuring free radical production in cells: Chemiluminescence of luminol (35, 36) and lucigenin (37), cytochrome c reduction (35), ferrous oxidation of xylenol orange (38) and DCFH-DA (12, 39) have all been used successfully to detect ROS generation. The most straightforward techniques use cell permeable fluorescent and chemiluminescent probes. Flow cytometry or fluorimetry can be used for the detection of ROS with fluorescent probes.

2'-7'-Dichlorodihydrofluorescein diacetate (DCFH-DA) is one of the most widely used techniques for directly measuring the redox state of a cell.

DCFH-DA, a cell permeable, non-fluorescent precursor of DCF can be used as an intracellular probe for oxidative stress. It has many advantages over other techniques developed. It is very easy to use, extremely sensitive to changes in the redox state of a cell, inexpensive and can be used to follow changes in ROS over time.

Intracellular esterases cleave DCFH-DA at the two ester bonds, producing a relatively polar and cell membrane-impermeable product, H_2DCF. This non-fluorescent molecule accumulates intracellularly and subsequent oxidation yields the highly fluorescent product DCF. The redox state of the sample can be monitored by detecting the increase in fluorescence. Accumulation of DCF in cells may be measured by an increase in fluorescence at 530 nm when the sample is excited at 485 nm. Fluorescence at 530 nm can be measured using a flow cytometer and is assumed to be proportional to the concentration of hydrogen peroxide in the cells (40, 41).

DCF, the oxidized fluorescent product of $DCFH_2$, is membrane permeable and can leak out of cells over time. Detecting slow hydrogen peroxide production over time can be difficult (42). To enhance retention of the fluorescent product, various analogues (improved versions of H2DCFDA) have been developed. For instance, carboxylated H2DCFDA analog, which has two negative charges at physiological pH, and its di(acetoxymethyl ester), which should more easily pass through membranes during cell loading. Upon oxidation and cleavage of the acetate and ester groups by intracellular esterases, both analogs form carboxydichlorofluorescein, with additional negative charges that should impede its leakage out of the cell. There is another analog called 5-(and-6)-chloromethyl-2', 7'-dichlorodihydrofluorescein diacetate, acetyl ester (CM-H2DCFDA)-a chloromethyl derivative of H2DCFDA. The CM-H2DCFDA passively diffuses into cells, where its acetate groups are cleaved by intracellular esterases and its thiol-reactive chloromethyl group reacts with intracellular glutathione and other thiols. Subsequent oxidation yields a fluorescent adduct that is trapped inside the cell, thus facilitating long-term studies.

The chemistry of the conversion is complex (Fig. 4.1) (43). Neither H_2O_2 nor $O_2^{\cdot-}$ can oxidize DCFH, but peroxyl, alkoxyl,

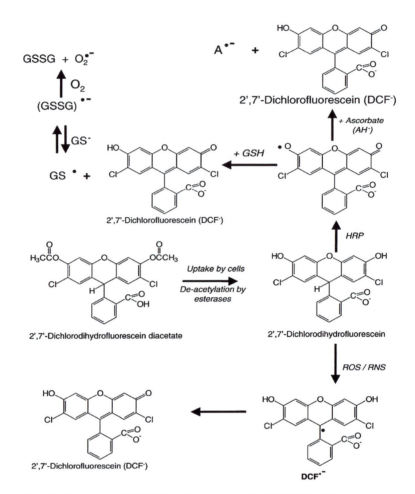

Fig. 4.1. (From Ref. 43). Conversion of DCFDA to a fluorescent product DCFDA is dichlorofluorescein diacetate, It is hydrolysed by cellular esterases to dichlorofluorescin (2′,7′-dichlorodihydrofluorescein), whose oxidation by several RS yields florescent DCF (dichlorofluorescein, more correctly called 2′7′-dichlorofluorescein) via an intermediate radical, DCF−. Peroxidases can also convert it into a phenoxyl radical that can interact with antioxidants such as ascorbate (AH−), reducing the phenoxyl radical and oxidizing ascorbate, or with GSH. GS resulting from the latter reaction can lead to O_2^- generation. The phenoxyl radical can also be recycled by NADH (not shown), producing NAD radical, which reacts rapidly with O_2 to produce O_2^-.

NO_2^-, carbonate (CO_3^{--}) and OH radicals can as well as peroxynitrite (44–46). DCFDA can only detect cellular peroxides efficiently if they are decomposed to radicals, for example, by transition metal ions. For instance, in bovine aortic endothelial cells, the generation of a signal from DCFDA upon addition of H_2O_2 required the uptake of extracellular iron from the medium (47). Horseradish peroxidase, myeloperoxidase and other heme proteins can also oxidize DCFH in the presence of H_2O_2 (indeed, DCFDA was first used in biology as a detector for H_2O_2 by adding horseradish peroxidase (48). Hence cellular peroxidase level and heme protein content are other variables to consider when interpreting studies with this probe (46, 48).

It follows that DCF fluorescence is an assay of generalized oxidative stress rather than of any particular RS, and is *not* a direct assay of H_2O_2, NO·, lipid peroxides, singlet O_2 or $O_2^{·-}$. One-electron oxidation of DCFH by various radicals and heme proteins is likely to produce intermediate radicals (Fig. 4.1), including phenoxyl radicals that can interact with such cellular antioxidants as GSH and ascorbate and with NADH to create more free radicals (49–51). Lawrence et al. (52) pointed out that cytochrome *c* is a powerful catalyst of DCFH oxidation, and so use of DCFDA to probe oxidative stress during apoptosis should be approached with caution, as a rise in cytosolic cytochrome *c* levels could result in a bigger "signal" without any change in cellular peroxide levels. Chromium (V), pyocyanin, mitoxantrone and ametantrone can directly oxidize DCFH and cause an artifactual signal, and the possibility of such direct oxidations must always be checked before using DCFDA to measure oxidative stress in cells exposed to various toxins (53, 54). Variation in cellular esterase content could also conceivably affect the use of DCFDA as a probe, but this issue has not been explored in the literature.

Some cell types have low esterase activity. This can have important implications for detecting ROS because DCFH-DA needs to be hydrolysed to $DCFH_2$ by cellular esterases. Careful consideration must be given to the levels of esterase activity in cells before using this probe. In cells with low esterase activity or where the esterases are sequestered in inaccessible parts of the cell, using the deacetylated probe $DCFH_2$ or some other assay is advised. For example, luminol- and lucigenin-dependent CL possess many of the advantages of fluorescent probes such as DCFH-DA. They are easy to use, sensitive, and inexpensive and can measure ROS generation in intact cells over time. They possess advantages over DCFH-DA in that they do not need to be cleaved by esterases and can be used in systems with low esterase activity. Also, cytochrome *c* is not able to catalyse chemiluminescent reactions and cannot interfere with ROS measurements during apoptosis.

The simplest technique to read samples is the fluorescence microplate reader, where data are presented as increases or decreases in relative. Bottom-reading machines have the advantage that the cells can be measured in situ without the need for trypsinization or cell scraping, processes that themselves generate cellular oxidative stress and result in artifactual changes in fluorescence. Plate readers measure total fluorescence, that is, they do not distinguish between intracellular and extracellular fluorescence from chemical reactions in the culture medium. We have already alluded to this problem in the case of DCFDA.

Flow cytometry offers the advantage of being able to measure the intracellular fluorescence of cells in the culture media. Quantitative data on the numbers of cells emitting fluorescence can be obtained rather than just relative fluorescence units. However, cells are required to be in suspension and require either scraping or trypsinization, which induce oxidative stress. Control experiments to optimize assay conditions must always be conducted to limit this

Confocal microscopy is a powerful tool; cells can be loaded with fluorescent dyes and viewed in real time in situ in culture chambers at 37°C. The intracellular location of RS can be visualised, and the role of mitochondrial, endoplasmic reticulum or lysosomal events in oxidative stress may be visualised using counter stains.

Some simple principles can be used as guidelines in understanding oxidative stress/oxidative damage in cell culture. Hydrogen peroxide generally crosses cell membranes readily (55). Thus catalase added outside cells can exert both intracellular and extracellular effects on H_2O_2 level, the former by "draining" H_2O_2 out of the cell by removing extracellular H_2O_2 and thus establishing a concentration gradient. In contrast, $O_2^{\cdot-}$ does not generally cross cell membranes readily (56). Thus if externally added superoxide dismutase is protective against an event in cell culture, be wary of what this means; it could be indicative of extracellular $O_2^{\cdot-}$-generating reactions. Similarly, neither the iron-chelating agent deferoxamine (which suppresses most, but not all, iron-dependent free radical reactions) nor the thiol antioxidant GSH enter cells easily, so again be wary if they have protective effects in short-term experiments: this is suggestive of extracellular effects (56). As an example, Clement et al. (57) showed that GSH protects against the cytotoxicity of dopamine simply because it reacts with dopamine oxidation products generated in the cell culture medium.

2. Materials

2.1. Equipment

1. Table centrifuge.
2. Flow cytometer.

2.2. Reagents and Supplies

1. 5-(and-6)-chloromethyl-2′, 7′-dichlorodihydrofluorescein diacetate, acetyl ester CM-H2DCFDA. Catalog number C6827 (Invitrogen/Molecular Probes). Do not dissolve products until immediately before use. Extremely air and light sensitive.
2. High quality anhydrous dimethylsulphoxide (DMSO), dimethylform-amide (DMF), or 100% ethanol.

3. Loading buffer such as simple physiological buffer PBS or HBSS.

4. Hydrogen Peroxide or tert-butyl hydroperoxide (TBHP) to a final concentration 50 µM (increase or decrease based on the sensitivity and response of the cells).

5. Six well Ultra Low Cluster Plate, Ultra Low Attachment, (Costar, #3471).

3. Methods

It is important to understand the limitations of using this probe. Although DCFH-DA is used to measure the concentration of hydrogen peroxide in cells, superoxide and NO generation are also capable of oxidizing DCFH$_2$ (49). The presence of SOD in the cytosol (Cu,Zn-SOD), mitochondria (Mn-SOD) and extracellular space (Ec-SOD) all convert superoxide into hydrogen peroxide, resulting in the accumulation of DCF in the cells. Inhibitors of SOD can be used to eliminate this source of hydrogen peroxide production. DCFH$_2$ may also be oxidized independently of hydrogen peroxide. Nitric oxide reacts with superoxide producing peroxynitrite. DCFH$_2$ can be directly oxidized to DCF by peroxynitrite. Inhibitors of NOS should be used to prevent NO mediated DCFH$_2$ oxidation (50). Some cell types have low esterase activity. This can have important implications for detecting ROS because DCFH-DA needs to be hydrolysed to DCFH$_2$ by cellular esterases. Careful consideration must be given to the levels of esterase activity in cells before using this probe.

3.1. General Guidelines

1. Remove cells from growth media and wash out from the serum of culture medium (*see* Note 1), leave cells for 10 min in PBS. If cells have been isolated ex vivo transfer these cells directly in prewarmed loading buffer. To minimized extracellular hydrolysis of the dye medium should not contain culture serum, primary and secondary amines, phenol red or other colometric dyes before and throughout the assay or dymethylform-amide (DFM).

2. The CM-H2DCFDA should be reconstituted only in anhydrous dimethylsulphoxide (DMSO), or dimethylform-amide (DMF), or 100% ethanol (*see* Note 2). Working solutions should be freshly prepared (*see* Note 3). Keep the solution in the dark and tightly sealed until ready to use and discard excess diluted probe at the end of the work session. These probes oxidize more readily in solution, and the presence of moisture will facilitate the decomposition of the dye. To prepare 10 mM

concentrated stocks of CM-H2DCFDA (MW: 577.8) add 8.6 µl DMSO into the original vial with 50 µg dye.

3. Resuspend cells in pre-warmed loading buffer containing the probe to provide a final working concentration of 1–10 µM dye. Remember, that CM-H2DCFDA concentration should be kept as low as possible to reduce potential artifacts from overloading, including incomplete hydrolysis, compartmentalization, and the toxic effects of hydrolysis by-products. The optimal working concentration for your application should be empirically determined. Incubate cells in the 6 well Ultra Low Cluster Plate at the optimal temperature for the cells. Generally, a loading time of 30–60 min is sufficient.

4. Wash out the loading buffer and plate the cells back to medium and incubate at the optimal temperature to allow a short recovery time for cellular esterases to hydrolyze the AM or acetate groups and then let the dye to be responsive to oxidation in the particular experiment's condition.

5. It is recommended to use positive control. To create positive controls, oxidative activity may be stimulated with hydrogen peroxide or tert-butyl hydroperoxide (TBHP) to a final concentration of 50 µM or with PMA.

6. Examine the intensity of fluorescence by flow cytometry. The redox state of the sample can be monitored by detecting the increase in fluorescence. Accumulation of DCF in cells may be measured by an increase in fluorescence at 530 nm when the sample is excited at 485 nm

7. First, examine the fluorescence of negative control which is an untreated loaded with dye cells maintained in a buffer. In healthy cells, oxygen radicals are eliminated by cellular enzymes and/or natural antioxidant and because of that reason healthy cells should exhibit a low level of fluorescence.

3.2. Results

3.2.1. Detection of ROS

To measure ROS generation by myeloid cells we used two dyes: DHE and DCFDA. DHE is selectively oxidized by superoxide anion, while the fluorescence of DCFDA indicates oxidation by hydrogen peroxide, peroxynitrite or hydroxyl radical. Superoxide anions can also contribute to DCFDA oxidation albeit at a lesser degree.

Freshly isolated splenocytes from tumor-bearing or from tumor-free immunized mice were loaded with these dyes and then labeled with anti-Gr-1-APC and CD11b-PE antibodies. The fluorescence of those dyes was evaluated within the population of gated double positive Gr-1$^+$CD11b$^+$ myeloid cells (Fig. 4.2). No difference in superoxide production (DHE oxidation) was found between two groups of cells, whereas the level of DFCDA oxidation by ImC from tumor-bearing mice was significantly (threefold)

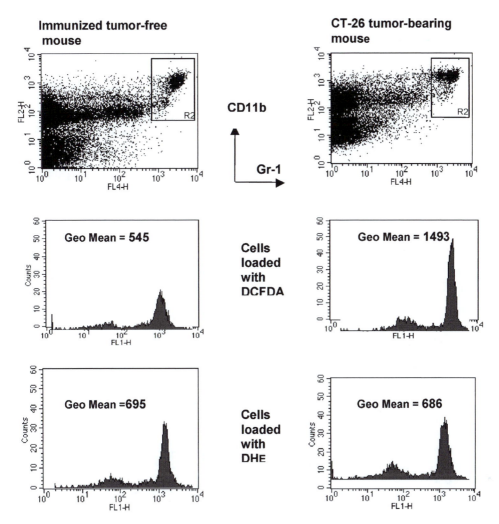

Fig. 4.2. Gr-1+CD11b+ myeloid cells from tumor-bearing mice demonstrate increased level of ROS. Splenocytes from tumor-bearing mice were incubated in serum-free medium at 37°C in the presence of DCFDA (2 μM, 30 min) or DHE (2 μM, 60 min), washed with cold PBS and then labeled with Gr-1-APC and CD11b-PE antibodies. After incubation on ice for 20 min cells were washed and analyzed by 3-color flow cytometry, using FACS Calibur (Becton Dickinson, Anaheim, CA). Two experiments with similar results were performed. The intensity of fluorescence (Geo Mean) in gated population of cells for each histogram is shown in Fig. 4.3.

higher than their counterparts from immunized tumor-free mice (Fig. 4.2). We compared the levels of DCFDA mediated fluorescence in Gr-1+CD11b+ ImC and Gr-1−CD11b+ macrophages in same spleens. ImC generated threefold to fourfold more ROS than Gr-1−CD11b+ macrophages.

When loaded with DCFDA Gr-1+, splenocytes isolated from tumor-bearing mice demonstrated more than two-fold higher proportion of the cells with bright fluorescence and threefold higher intensity of total fluorescence than Gr-1+ cells isolated from immunized mice even without additional stimulation with

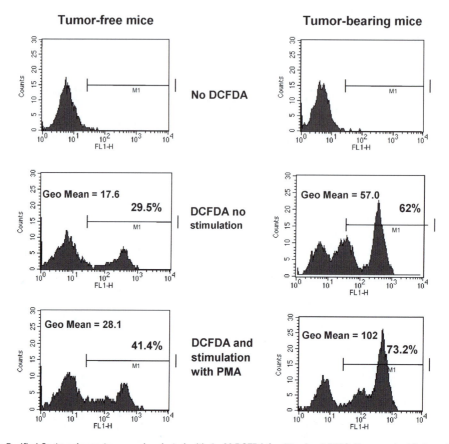

Fig. 4.3. Purified Gr-1+ splenocytes were incubated with 2 μM DCFDA for 15 min at 37°C, then washed twice with PBS. In the case of stimulation with PMA, cells were incubated at 37°C for 5 min with PMA (30 ng/ml), and then washed again with PBS. Intensity of fluorescence was measured by flow cytometry. Typical results of one of three performed experiments are shown. Proportion of cells with bright fluorescence was calculated. Geometric mean (Geo Mean) was used to calculate the total intensity of fluorescence.

PMA (Fig. 4.3). Activation of cells with PMA increased observed differences. Thus, Gr-1+ ImC from tumor-bearing mice had substantially higher levels of ROS production than ImC from tumor-free mice.

3.2.2. Nature of ROS

ROS may include different types of molecules from singlet oxygen to hydroxyl peroxide. We asked what type of ROS is produced by ImC. Gr-1+ cells were isolated from tumor-bearing mice and different oxygen species were neutralized using specific inhibitors or scavengers. The effect of these compounds was evaluated by flow cytometry using DCFDA. Catalase reduced ROS level in ImCs more than fourfold, indicating that H_2O_2 contributed greatly into overall level of ROS in these cells (Fig. 4.4). Uric acid had similar effects, suggesting that peroxynitrite could be a substantial part of ROS pool. However, the most noticeable differences were found in the effect of arginase inhibitor

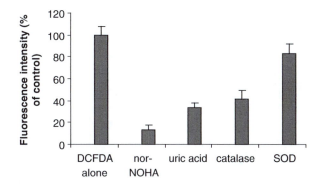

Fig. 4.4. Gr-1⁺ myeloid cells were isolated from control and tumor-bearing mice as described in Methods. The oxidative sensitive dye DCFDA was used for the measurement of ROS production by these cells. To block ROS production, Gr-1⁺ cells were incubated for 10 min at 37°C with different inhibitors followed by 20 min incubation at 37°C with DCFDA. The following reagents purchased from Calbiochem were used: Superoxide dismutase (SOD) – 200 units/ml, Catalase – 1,000 unit/ml, Peroxynitrite scavenger, Uric acid – 0.5 mM, Arginase inhibitor – Nor-NOHA (N-Hydroxy-nor-l-arginine, diacetate salt) – 2 µM. After incubation cells were washed in cold PBS and analyzing by flow cytometry. The level of fluorescence in ImC incubated in medium alone was used as a background.

Nor-NOHA. It decreased ROS levels in ImCs more than tenfold (Fig. 4.4). This strongly suggests that arginine metabolites play a critical role in generation of ROS in tumor-bearing mice derived ImCs. SOD did not significantly affect the levels of ROS (Fig. 4.4). This suggested a rather minor contribution of superoxide into the total ROS pool, which was consistent with lack of changes in DHE oxidation.

4. Notes

1. Serum-free media must be used since serum will contain endogenous esterase activity and de-esterified dichlorofluorescein (DCF) which is less permeable and will generate inconsistent data.

2. DCFDA enters cells and accumulates mostly in the cytosol. To avoid any cytotoxicity, cells should be loaded with DCFDA at low concentrations. With a variety of cell types, we have found loading at 1–10 µM for 45 min to 1 h is adequate. Higher levels of DCFDA or high light intensities can also result in an artifactual photochemical oxidation to fluorescent products that can be mistaken for ROS generation.

3. Because of oxidation, working solutions of CM-H2DCFDA should be freshly prepared right before performing the experiments.

References

1. Halliwell B, Gutteridge JM (1999) Free radicals in biology and medicine. Oxford University Press, Oxford

2. Sohal RS, Mockett RJ, Orr WC (2002) Mechanisms of aging: an appraisal of the oxidative stress hypothesis. Free Radic Biol Med 33:575–586

3. Halliwell B (2001) Role of free radicals in the neurodegenerative diseases: therapeutic implications for antioxidant treatment. Drugs Aging 18:685–716

4. Butterfield DA (2002) Amyloid beta-peptide (1–42)-induced oxidative stress and neurotoxicity: implications for neurodegeneration in Alzheimer's disease brain. Rev Free Radic Res 36:1307–1313

5. Liu Q, Raina AK, Smith MA, Sayre LM, Perry G (2003) Hydroxynonenal, toxic carbonyls, and Alzheimer disease. Mol Aspects Med 24:305–313

6. Hagen TM, Huang S, Curnutte J, Flower P, Martinez V, Wehr CM, Ames BN, Chisari F (1994) Extensive oxidative DNA damage in hepatocytes of transgenic mice with chronic active hepatitis destined to develop hepatocellular carcinoma. Proc Natl Acad Sci U S A 91:12808–12812

7. Chowienczyk PJ, Brett SE, Gopaul NK, Meeking D, Marchetti M, Russel-Jones DL, Anggard EE, Ritter JM (2000) Oral treatment with an antioxidant (raxofelast) reduces oxidative stress and improves endothelial function in men with type II diabetes. Diabetologia 43:974–977

8. Parthasarathy S, Santanam N, Ramachandran S, Meilhac O (2000) Potential role of oxidized lipids and lipoproteins in antioxidant defense. Free Radic Res 33:197–215

9. Beyer RE (1992) An analysis of the role of coenzyme Q in free radical generation and as an antioxidant. Biochem Cell Biol 70:390

10. Raha S, McEachern GE, Myint AT, Robinson BH (2000) Superoxides from mitochondrial complex III: the role of manganese superoxide dismutase. Free Radic Biol Med 29:170

11. Finkel T (2000) Redox-dependent signal transduction. FEBS Lett 476:52

12. Tammariello SP, Quinn MT, Estus S (2000) NADPH oxidase contributes directly to oxidative stress and apoptosis in nerve growth factor-deprived sympathetic neurons. J Neurosci 20:RC53

13. Suzuki Y, Ono Y, Hirabayashi Y (1998) Rapid and specific reactive oxygen species generation via NADPH oxidase activation during Fas-mediated apoptosis. FEBS Lett 425:209

14. Hampton MB, Orrenius S (1997) Dual regulation of caspase activity by hydrogen peroxide: implications for apoptosis. FEBS Lett 414:552

15. Creagh EM, Cotter TG (1999) Selective protection by hsp 70 against cytotoxic drug-, but not Fas-induced T-cell apoptosis. Immunology 97:36

16. Michiels C, Raes M, Toussaint O, Remacle J (1994) Importance of Se-glutathione peroxidase, catalase, and Cu/Zn-SOD for cell survival against oxidative stress. Free Radic Biol Med 17:235

17. Reed DJ (1990) Glutathione: toxicological implications. Annu Rev Pharmacol Toxicol 30:603

18. Lledias F, Rangel P, Hansberg W (1998) Oxidation of catalase by singlet oxygen. J Biol Chem 273:10630

19. Bai J, Rodriguez AM, Melendez JA, Cederbaum AI (1999) Overexpression of catalase in cytosolic or mitochondrial compartment protects HepG2 cells against oxidative injury. J Biol Chem 274:26217

20. Murphy TH, De Long MJ, Coyle JT (1991) Enhanced NAD(P)H:quinone reductase activity prevents glutamate toxicity produced by oxidative stress. J Neurochem 56:990

21. Mustacich D, Powis G (2000) Thioredoxin reductase. Biochem J 346:1–8

22. Kasahara Y, Iwai K, Yachie A, Ohta K, Konno A, Seki H, Miyawaki T, Taniguchi N (1997) Involvement of reactive oxygen intermediates in spontaneous and CD95 (Fas/APO-1)-mediated apoptosis of neutrophils. Blood 89:1748

23. Lundqvist-Gustafsson H, Bengtsson T (1999) Activation of the granule pool of the NADPH oxidase accelerates apoptosis in human neutrophils. J Leukoc Biol 65:196

24. Oishi K, Machida K (1997) Inhibition of neutrophil apoptosis by antioxidants in culture medium. Scand J Immunol 45:21

25. Woo CH, Eom YW, Yoo MH, You HJ, Han HJ, Song WK, Yoo YJ, Chun JS, Kim JH (2000) Tumor necrosis factor-alpha generates reactive oxygen species via a cytosolic phospholipase A2-linked cascade. J Biol Chem 275:32357–32362

26. Elchuri S, Oberley TD, Qi W, Eisenstein RS, Jackson Roberts L, Van Remmen H, Epstein CJ, Huang TT (2005) CuZnSOD deficiency leads to persistent and widespread oxidative damage and hepatocarcinogenesis later in life. Oncogene 24:367–380

27. Li Y, Huang TT, Carlson EJ, Melov S, Ursell PC, Olson JL, Noble LJ, Yoshimura MP, Berger C, Chan PH et al (1995) Dilated cardiomyopathy and neonatal lethality in mutant mice lacking manganese superoxide dismutase. Nat Genet 11:376–381

28. Evans MD, Dizdaroglu M, Cooke MS (2004) Oxidative DNA damage and disease: induction, repair and significance. Mutat Res 567:1–61

29. Kawanishi S, Hiraku Y, Pinlaor S, Ma N (2006) Oxidative and nitrative DNA damage in animals and patients with inflammatory diseases in relation to inflammation-related carcinogenesis. Biol Chem 387:365–372

30. Mori K, Shibanuma M, Nose K (2004) Invasive potential induced under long-term oxidative stress in mammary epithelial cells. Cancer Res 64:7464–7472

31. Radisky DC, Levy DD, Littlepage LE, Liu H, Nelson CM, Fata JE, Leake D, Godden EL, Albertson DG, Nieto MA et al (2005) Rac1b and reactive oxygen species mediate MMP-3-induced EMT and genomic instability. Nature 436:123–127

32. Kusmartsev S, Nefedova Y, Yoder D, Gabrilovich DI (2004) Antigen-specific inhibition of CD8+ T cell response by immature myeloid cells in cancer is mediated by reactive oxygen species. J Immunol 172:989–999

33. Corzo CA, Nagaraj S, Kusmartsev S, Gabrilovich D (2007) Role of reactive oxygen species in immune suppression in cancer. J Immunol 178:49.12

34. Kusmartsev S, Eruslanov E, Kübler H, Tseng T, Sakai Y, Su Z, Kaliberov S, Heiser A, Rosser C, Dahm P, Siemann D, Vieweg J (2008) Oxidative stress regulates expression of VEGFR1 in myeloid cells: Link to tumor-induced immune suppression in renal cell carcinoma. J Immunol 181:346–353

35. Dahlgren C, Karlsson A (1999) Respiratory burst in human neutrophils. J Immunol Methods 232:3

36. Mellqvist UH, Hansson M, Brune M, Dahlgren C, Hermodsson S, Hellstrand K (2000) Natural killer cell dysfunction and apoptosis induced by chronic myelogenous leukemia cells: role of reactive oxygen species and regulation by histamine. Blood 96:1961

37. Gyllenhammar H (1987) Lucigenin chemiluminescence in the assessment of neutrophil superoxide production. J Immunol Methods 97:209

38. Nourooz-Zadeh J (1999) Ferrous ion oxidation in presence of xylenol orange for detection of lipid hydroperoxides in plasma. Methods Enzymol 300:58

39. Ottonello L, Frumento G, Arduino N, Dapino P, Tortolina G, Dallegri F (2001) Immune complex stimulation of neutrophil apoptosis: investigating the involvement of oxidative and nonoxidative pathways. Free Radic Biol Med 30:161–169

40. Bass DA, Parce JW, Dechatelet LR, Szejda P, Seeds MC, Thomas M (1983) Flow cytometric studies of oxidative product formation by neutrophils: a graded response to membrane stimulation. J Immunol 130:1910

41. Royall JA, Ischiropoulos H (1993) Evaluation of 2,7-dichlorofluorescin and dihydrorhodamine 123 as fluorescent probes for intracellular H_2O_2 in cultured endothelial cells. Arch Biochem Biophys 302:348–355

42. Ubezio P, Civoli F (1994) Flow cytometric detection of hydrogen peroxide production induced by doxorubicin in cancer cells. Free Radic Biol Med 16:509

43. Halliwell B, Whitemann M (2004) Measuring reactive species and oxidative damage in vivo and in cell culture:how should you do it and do the results mean? Br J Pharmacol 142:231–255

44. Ischiropoulos H, Gow A, Thom SR, Kooy NM, Royall JA, Crow JP (1999) Detection of reactive nitrogen species using 2, 7-dichloro-dihydrofluorescein and dihydrorhodamine 123. Methods Enzymol 301:367–373

45. Bilski P, Belanger AG, Chignell CF (2002) Photosensitized oxidation of 2', 7'-dichlorofluorescin: singlet oxygen does not contribute to the formation of fluorescent oxidation product 2', 7'-dichlorofluorescein. Free Radic Biol Med 33:938–946

46. Ohashi T, Mizutani A, Mukarima A, Kojo S, Ishii T, Taketani S (2002) Rapid oxidation of dichlorodihydrofluorescin with heme and hemoproteins: formation of the fluorescein is independent of the generation of reactive oxygen species. FEBS Lett 511:21–27

47. Tampo Y, Kotamraju S, Chitambar RC, Kalivendi SV, Kezler A, Joesph J, Kalyanaraman B (2003) Oxidative stress-induced iron signaling is responsible for peroxide-dependent oxidation of dichlorodihydrofluorescein in endotheliel cells. Role of transferrin receptor-dependent iron uptake in apoptosis. Circ Res 92:56–63

48. Lebel CP, Ischiropoulos H, Bondy SC (1992) Evaluation of the probe 2', 7'-dichlorofluorescin as an indicator of reactive oxygen species formation and oxidative stress. Chem Res Toxicol 5:227–231

49. Rao KM, Padmanabhan J, Kilby DL, Cohen HJ, Currie MS, Weinberg JB (1992) Flow cytometric analysis of nitric oxide production in human neutrophils using dichlorofluorescein

diacetate in the presence of a calmodulin inhibitor. J Leukoc Biol 51:496

50. Kooy NW, Lewis SJ, Royall JA, Ye YZ, Kelly DR, Beckman JS (1997) Extensive tyrosine nitration in human myocardial inflammation: evidence for the presence of peroxynitrite. Crit Care Med 25:812–819

51. Rota C, Fann YC, Mason RP (1999) Phenoxyl free radical formation during the oxidation of the fluorescent dye 2', 7'-dichlorofluorescein by horseradish peroxidase. Possible consequences for oxidative stress measurements. J Biol Chem 274:28161–28168

52. Lawrence A, Jones CM, Wardman P, Burkitt MJ (2003) Evidence for the role of a peroxidase compound I-type intermediate in the oxidation of glutathione, NADH, ascorbate, and dichlorofluorescin by cytochrome c/H_2O_2. Implications for oxidative stress during apoptosis. J Biol Chem 278:29410–29419

53. Martin BD, Schoenhard JA, Sugden KD (1998) Hypervalent chromium mimics reactive oxygen species as measured by the oxidant-sensitive dyes 2', 7'-dichlorofluorescin and dihydrorhodamine. Chem Res Toxicol 11:1402–1410

54. O'Malley YQ, Reszka KJ, Britigan BE (2004) Direct oxidation of 2', 7'-dichlorodihydrofluorescein by pyocyanin and other redox-active compounds independent of reactive oxygen species production. Free Radic Biol Med 36:90–100

55. Henzler T, Steudle E (2000) Transport and metabolic degradation of hydrogen peroxide in *Chara corallina*: model calculations and measurements with the pressure probe suggest transport of $H(2)O(2)$ across water channels. J Exp Bot 51:2053–2066

56. Marla SS, Lee J, Groves JT (1997) Peroxynitrite rapidly permeates phospholipid membranes. Proc Natl Acad Sci U S A 94:14243–14248

57. Clement MV, Long LH, Ramalingam J, Halliwell B (2002) The cytotoxicity of dopamine may be an artefact of cell culture. J Neurochem 81:414–421

Chapter 5

Measurement of 8-Isoprostane in Exhaled Breath Condensate

Paolo Montuschi, Peter J. Barnes, and Giovanni Ciabattoni

Abstract

Oxidative stress is functionally involved in the pathophysiology of lung diseases including asthma and chronic obstructive pulmonary disease. 8-Isoprostane, which is derived from free radical-catalyzed peroxidation of arachidonic acid, is one of the most reliable biomarkers of oxidative stress. Exhaled breath condensate (EBC) is a completely noninvasive method for collecting airway secretions. We developed a specific and sensitive radioimmunoassay (RIA) that has been applied to the measurement of 8-isoprostane in EBC. This RIA for 8-isoprostane has been validated using high performance liquid chromatography. Measurement of 8-isoprostane in EBC is a useful noninvasive technique for exploring the role of oxidative stress in lung diseases. This technique might provide important insights into the understanding of the clinical pharmacology of antioxidants and might be useful for monitoring the effects of pharmacological therapy.

Key words: 8-isoprostane, Oxidative stress, Exhaled breath condensate, Radioimmunoassay, Reverse phase-high performance liquid chromatography

1. Introduction

Oxidative stress derives from increased production of reactive oxygen species and decreased antioxidant defences (1). Oxidative stress is functionally involved in the pathophysiology of lung diseases including asthma (2), chronic obstructive pulmonary disease (COPD) (3), and cystic fibrosis (4). The assessment of oxidant stress is relevant to the management of patients with respiratory diseases as it may indicate that pharmacological therapy is required in an early stage in the disease process. Moreover, monitoring of oxidative stress, a component of inflammation, might be useful in the follow-up of patients with respiratory disease and for

D. Armstrong (ed.), *Advanced Protocols in Oxidative Stress II*, Methods in Molecular Biology, vol. 594
DOI 10.1007/978-1-60761-411-1_5, © Humana Press, a part of Springer Science+Business Media, LLC 2010

guiding pharmacological therapy. Measurement of 8-isoprostane, a prostaglandin (PG)–like compound that is produced in vivo independently of cyclooxygenase (COX) enzymes, primarily by free radical-induced peroxidation of arachidonic acid (5), has emerged as one of the most reliable approaches to assess oxidative stress status in vivo (6, 7). By measuring 8-isoprostane in biological fluids, it is possible to explore the role of oxidative stress in the pathogenesis of human disease (6, 7). 8-Isoprostane concentrations are elevated in exhaled breath condensate (EBC) in patients with airway inflammatory diseases including asthma (8–10), COPD (11), and cystic fibrosis (12,13). EBC is a completely noninvasive method for collecting airway secretions and studying the composition of airway lining fluid (14–17).

In this article, we describe the analytical aspects of a radioimmunoassay (RIA) for 8-isoprostane that was developed in our laboratory; we summarize previous work that aimed to validate the RIA for 8-isoprostane in EBC using reverse phase-high performance liquid chromatography (RP-HPLC) (18); we provide examples to the applications of this RIA to the measurement of 8-isoprostane in EBC in patients with airway inflammatory diseases.

2. Materials

2.1. Equipment

1. Ecoscreen, Jaeger, Hoechberg, Germany.
2. Waters HPLC 484 UV detector, Waters Associates, Milford, Massachussets, USA.
3. Waters HPLC 600E pump controller, Waters Associates, Milford, Massachussets, USA.
4. C18, 125×4.6 mm, 5 µm, LiChrospher column, Merck, Darmstadt, Germany.
5. Packard Tri-Carb 1900 CA liquid scintillation counter, Packard, Meriden, CT, USA.
6. Speedvac evaporator linked with Savant RT490 refrigerated condensation trap, Thermo Fischer Scientific Inc., Waltham, MA, USA.

2.2. Reagents

1. 8-Isoprostane standard, Cayman Chemicals, Ann Arbor, MI, USA.
2. [^{3}H]8-Isoprostane (specific activity 22 Ci/mmol), custom synthesis, New England Nuclear Du Pont, Boston, Massachussets, USA, (original study); custom synthesis, Perkin Elmer, Shelton, CT, USA.
3. HPLC grade solvents, Merck, Darmstadt, Germany.

4. Human and bovine serum albumin, Sigma Chemical Co. St. Louis, MO, USA.

5. Instagel® scintillation liquid, Packard, Meriden, CT, USA.

6. Sep-Pak C18 cartridges, Waters Associates, Milford, Massachussets, USA.

7. Antiserum against 8-isoprostane, which was used in the RIA (Rab 1), was developed in our laboratory.

8. Anti-8-isoprostane serum (antiserum L9), Cayman Chemicals, Ann Arbor, MI, USA.

9. α-Amylase kit, Roche Diagnostics, Basel, Switzerland.

10. Thimerosal, Sigma Aldrich, Milan, Italy.

3. Methods

3.1. Collection of Exhaled Breath Condensate

EBC is collected by using a condensing chamber (19). Exhaled air enters and leaves the chamber through 1-way valves at the inlet and outlet, thus keeping the chamber closed. Subjects are instructed to breathe tidally through a mouthpiece connected to the condenser for 15 min while wearing a nose clip. 1–2.5 ml of EBC per subject is generally collected and stored at –70°C before 8-isoprostane measurements.

3.2. Radioimmunoassay for 8-Isoprostane

Antiserum against 8-isoprostane that was used in the RIA (Rab 1) was developed in our laboratory (*see* Notes 1–4) and was obtained by conjugating 8-isoprostane to human serum albumin using the carbodiimide method and a previously established protocol (2). Duplicates of EBC samples are analyzed. 250 µl of unextracted EBC sample is added to 1,250 µl of assay buffer (phosphate 0.025 M, pH = 7.5) containing approximately 2,500 dpm of [^{3}H]8-isoprostane, which is mixed with appropriately diluted antiserum and incubated for 24 to 30 h at 4°C (18) (*see* Note 5). Approximately, 40–45% binding of the labeled hapten is obtained when 8-isoprostane antiserum is used at a final dilution of 1:200,000. 8-Isoprostane standard displaces the binding of the homologous tracer in a linear fashion over the range from 2 to 250 pg/ml (21). Separation of antibody-bound from free [^{3}H]8-isoprostane is achieved by rapidly adding 0.1 ml of a 5% bovine serum albumin solution and 0.1 ml of a charcoal suspension (70 mg/ml) (*see* Note 6) and subsequent centrifugation at 4°C for 10 min at 5,000 rpm (3,000×*g*) (*see* Note 7) (21). Supernatant solutions containing antibody-bound 8-isoprostane are decanted directly into 10 ml scintillation liquid (*see* Note 8). Radioactivity is counted in a liquid scintillation counter (*see* Note 9). Data are processed using a computer that was programmed to correct for nonspecific binding (*see* Note 10).

3.3. Validation of Radioimmunoassay for 8-Isoprostane in Exhaled Breath Condensate

In a previous study, we performed a qualitative validation of RIA for 8-isoprostane in EBC through studies of immunological and chromatographic behavior of the measured 8-isoprostane-like immunoreactivity (18). Validation of RIA for 8-isoprostane in EBC was sought by (1) RP-HPLC purification of a pool of EBC samples with subsequent RIA of the eluted fractions for 8-isoprostane-like immunoreactivity; (2) simultaneous measurement of unextracted EBC samples with two 8-isoprostane antisera with different cross-reactivities.

3.3.1. Experimental Procedure

The experimental procedure included the following steps (1) collection of EBC samples that were pooled together (2) extraction of the EBC sample pool; (3) calculation of recovery by extracting 130,000 cpm of [^{3}H]8-isoprostane added to a volume of distilled water equal to that of the EBC sample; (4) RP-HPLC purification of 8-isoprostane-like immunoreactivity and its respective radiolabeled compounds; and (5) RIA of 8-isoprostane-like immunoreactivity in the RP-HPLC fractions eluted (18). Characteristics of the 8-isoprostane antiserum (Rab 1) have been described previously (21, 22). Cross-reactivity of antiserum Rab 1 with 8-iso-PGE$_2$ is 7.7%, whereas with other prostanoids it is < 0.3% (21). Radioactivity was counted in the single RP-HPLC fractions eluted containing [^{3}H]8-isoprostane.

1. EBC samples were collected as described above. Samples were stored at –70°C before 8-isoprostane extraction, RP-HPLC purification, and RIA, which was performed within two weeks after the collection of EBC samples. α-Amylase concentrations in EBC samples were measured by an in vitro colorimetric method using maltotriose with 2-chloro-p-nitrophenol to exclude salivary contamination. No α-amylase concentrations were detected in any sample indicating no salivary contamination of EBC.

2. 8-Isoprostane was extracted from a pool of EBC samples (23 ml total) obtained from a group of 16 subjects with different lung diseases. One aliquot of 1 ml of sample pool was measured unextracted to assess the total endogenous 8-isoprostane content in the EBC sample pool and to calculate recovery after extraction and RP-HPLC purification. 8-Isoprostane was extracted on Sep-Pak C18 cartridges following a procedure described previously (21).

3. Recovery was evaluated by a radioactive tracer and by calculating the recovery of endogenous 8-isoprostane. 130,000 cpm of [^{3}H]8-isoprostane was added to a volume of distilled water equal to the volume of EBC pool and subjected to extraction and RP-HPLC purification as the EBC sample pool. The single fractions were eluted by RP-HPLC and assayed for radioactivity. The recovery for endogenous 8-isoprostane was calculated dividing the total 8-isoprostane immunoreactivity measured after extraction and HPLC purification by the total

amount of 8-isoprostane measured before extraction and RP-HPLC purification. This was calculated multiplying the mean 8-isoprostane-like immunoreactivity in an unextracted aliquot by the total volume of the EBC sample pool.

4. RP-HPLC purification and RIA for 8-isoprostane in EBC included (1) RP-HPLC ultraviolet profiling of standard 8-iso-prostane; (2) RP-HPLC purification of the solvent system (blank) to rule out any carry-over after injection of the standard into RP-HPLC; (3) RP-HPLC purification of 8-isoprostane-like immunoreactivity in the extracted EBC sample pool; (4) a second blank; (5) RP-HPLC purification of the sample containing [^{3}H]8-isoprostane; (6) RIA for 8-isoprostane-like immunoreactivity in the HPLC fractions eluted; and (7) counting for radioactivity in the HPLC fractions eluted containing [^{3}H] 8-isoprostane (18).

5. 8-Isoprostane standard (0.5 µg) was injected into RP-HPLC to identify the corresponding peak without collecting the fractions eluted. The peak of 8-isoprostane standard eluted at 13 min (Fig. 5.1a). Peak width for 8-isoprostane ranged from 2 to 3 min. The solvent system (blank) was injected into the RP-HPLC and purified. The fractions eluted were collected every min for 30 min at a flow rate of 1 ml/min for RIA of 8-isoprostane-like immunoreactivity to exclude the possible carry-over effect after standard UV profiling. The extractes EBC pool was recovered with 100 µl of methanol and subjected to RP-HPLC on C18, 125 × 4.6 mm, 5 µm, LiChrospher column, with the solvent system acetonitrile/water/acetic acid 27:73:0.18 by volume. 1 min samples were collected for 30 min (18).

A second blank was performed to rule out carry-over after the EBC sample pool purification.

The sample containing [^{3}H]8-isoprostane was subjected to RP-HPLC and the fractions eluted were collected to calculate recovery as described above. All RP-HPLC purifications were performed isocratically.

6. Each RP-HPLC fraction was vacuum-dried in a Speedvac evaporator linked with a Savant refrigerated condensation trap and recovered with 1 ml buffer. Aliquots of fraction eluted were then analyzed for 8-isoprostane-like immunoreactivity by RIA or for radioactivity. 8-Isoprostane-like immunoreactivity in EBC was measured by a RIA developed in our laboratory using a specific anti-8-isoprostane (Rab 1) serum (21–23). The detection limit of the RIA for 8-isoprostane considering the final dilution of the sample in the RIA incubation volume was 10 pg/ml. The IC_{50} was 39.8 pg/ml (18).

A second anti-8-isoprostane serum (L9) with different cross-reactivity was used for validation of 8-isoprostane-like immuno-reactivity measurements in EBC (18). Antiserum L9, which is

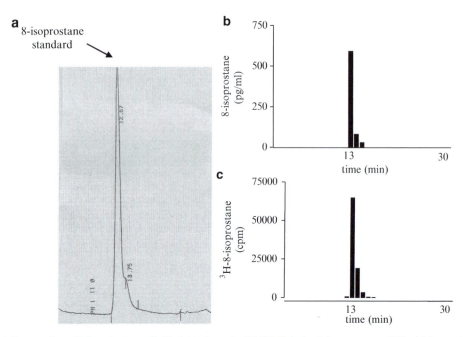

Fig. 5.1. Reverse phase-high performance liquid chromatography (RP-HPLC) and radioimmunoassay (RIA) of 8-isoprostane in exhaled breath condensate (EBC). (**a**) 8-Isoprostane peak identified by ultraviolet profile of authentic standard. Retention time for 8-isoprostane was 12.67 min. (**b**) 8-Isoprostane-like immunoreactivity in a pool of EBC samples. Fractions eluted were collected every min for 30 min. Flow rate was 1 ml/min. 8-isoprostane-like immunoreactivity in each aliquot was measured by RIA. An estimated total amount of 1,056 pg of 8-isoprostane-like immunoreactivity was extracted and purified by RP-HPLC (see text). (**c**) [^{3}H]8-Isoprostane recovery after extraction and RP-HPLC purification of a distilled water sample containing 130,000 cpm of [^{3}H]8-isoprostane. Fractions eluted were assayed for radioactivity. Recovery for [^{3}H]8-isoprostane was 66.7%. All RP-HPLC purifications were performed isocratically (modified with permission from ref. 18) See text for details.

commercially available and is used in an enzyme immunoassay kit for 8-isoprostane, was obtained by coupling 8-isoprostane to keyhole limpet hemocyanin (24) and following a previously described immunization protocol (25). In particular, 8-isoprostane-like immunoreactivity was measured in 12 EBC unextracted samples by RIAs using two 8-isoprostane antisera (Rab 1 and L9) that have different cross- reactivities (18).

Characteristics of the L9 antiserum and validation with negative ion chemical ionization-gas chromatography/ mass spectrometry have been described previously (21). Antiserum L9 has a cross-reactivity of 0.08% with 8-iso-PGE$_2$ and <0.03% with other prostanoids (21, 23).

RIA for 8-isoprostane was performed as described above. The intra-assay ($n=6$) and interassay ($n=8$) coefficients of variation for 8-isoprostane were ±2.0% and ±2.9% at 2 pg/ml, the lowest standard concentration and ±3.7% and ±10.8% at 250 pg/ml, the highest standard concentration (18).

3.3.2. Results and Interpretation of Results

1. RP-HPLC separation of a pool of EBC samples and RIA with 8-isoprostane antiserum Rab 1 showed a single peak of immu-

noreactivity that co-eluted with 8-isoprostane standard, indicating that the unknown 8-isoprostane-like immunoreactivity recognized by the antiserum has identical chromatographic behavior with 8-isoprostane standard (Fig. 5.1a, b). RP-HPLC separation of the sample containing [^{3}H]8-isoprostane also showed a single peak of radioactivity that was coincident with the 8-isoprostane peak, which was identified by ultraviolet profiling of authentic standard (18) (Fig. 5.1c). Recovery for [^{3}H]8-isoprostane was 66.7%. Mean concentration of 8-isoprostane-like immunoreactivity in an unextracted aliquot of the EBC sample pool was 48.0 pg/ml (18). An estimated amount of 1,056 pg of 8-isoprostane-like immunoreactivity was extracted (18). The recovery for endogenous 8-isoprostane-like immunoreactivity calculated after extraction and RP-HPLC purification of the EBC sample pool was 66.3%. This value was similar to that obtained using [^{3}H]8-isoprostane to calculate recovery.

2. Using two 8-isoprostane antisera (Rab 1 and L9) with different cross-reactivities to measure the same EBC samples, similar concentrations of 8-isoprostane-like immunoreactivity were measured (limits of agreement: 4.5 pg/ml and –4.1 pg/ml, standard error of limits of agreement:1.08 pg/ml, n=12) (18). 8-Isoprostane-like immunoreactivity was undetectable in all the RP-HPLC eluted fractions of blanks indicating no carry-over after standard and sample purification.

3. To summarize, using RP-HPLC, we demonstrated identical chromatographic behavior of 8-isoprostane-like immunoreactivity in EBC with 8-isoprostane standard. These findings indicate that the immunoreactive material detected in EBC is represented by authentic 8-isoprostane, although the possible existence of different compounds that have a similar chromatographic behaviour in certain solvent systems cannot be definitively excluded. The specificity of RIA for 8-isoprostane in EBC was further demonstrated by the simultaneous measurement of the same unextracted EBC samples with different 8-isoprostane antisera as every antiserum has a unique immunological profile of cross-reactivity. Cross-reactivity of 8-isoprostane antiserum (Rab 1) used in this RIA with endogenous 8-iso-PGE$_2$ (7.7%) is unlikely to have biological relevance as (1) similar concentrations of 8-isoprostane (limits of agreement: 4.5 pg/ml and –4.1 pg/ml) were obtained with 8-isoprostane antiserum L9 that has a cross-reactivity of 0.08% with 8-iso-PGE$_2$; (2) retention time of 8-iso-PGE$_2$ (31–33.5 min) in very similar HPLC conditions (26) is different from that of 8-isoprostane (retention time: 13 min) and no 8-isoprostane-like immunoreactivity was detected in the fractions eluted corresponding to the 8-iso-PGE$_2$ peak. The recovery calculated by [^{3}H]8-isoprostane was similar to the recovery of the endogenous compound (18).

Taken together, this evidence indicates the specificity of the RIA for 8-isoprostane in the EBC discussed above.

3.4. RIA Measurement
of 8-Isoprostane in
EBC in Patients with
Inflammatory Airway
Diseases

1. Using this RIA, we measured 8-isoprostane concentrations in EBC in 20 healthy nonatopic nonasthmatic children, 20 atopic nonasthmatic children, 30 steroid-naive atopic children with stable mild intermittent asthma, and 25 atopic children with stable mild-to-moderate persistent asthma who were treated with inhaled corticosteroids (9). 8-Isoprostane concentrations were detectable in all healthy children [15.5 (14.1–17.5) pg/15 min, median and interquartile range] and were increased in both steroid-naive [29.8 (26.0–34.3) pg/15 min, $p < 0.001$] and steroid-treated atopic asthmatic children [33.0 (28.5–35.8) pg/15 min, $p < 0.001$], but not in atopic nonasthmatic children [15.8 (13.9–20.1) pg/15 min, $p = 0.52$] (9) (Fig. 5.2). Asthmatic children had higher concentrations of 8-isoprostane in EBC (steroid-naïve children, $p < 0.001$; steroid-treated children, $p < 0.001$) than atopic nonasthmatic children (Fig. 5.2). There was no difference in 8-isoprostane concentrations between the two groups of asthmatic children ($p = 0.36$) (9), indicating that 8-isoprostane is relatively resistant to inhaled corticosteroid treatment (Fig. 5.2).

2. We studied the short-term effect of nonselective and selective COX-2 inhibition on the concentrations of eicosanoids, including 8-isoprostane, in EBC in patients with stable COPD (27). Two studies were performed. The first study was a double blind, crossover, randomized, placebo-controlled study with

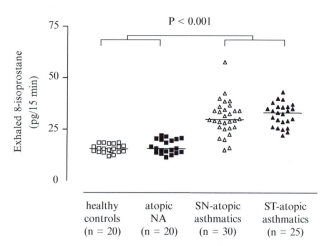

Fig. 5.2. 8-Isoprostane concentrations in exhaled breath condensate (EBC) in healthy children (*open squares*), atopic nonasthmatic children (*filled squares*), steroid-naïve (SN) atopic asthmatic children (*open triangles*), and steroid-treated (ST) children with atopic asthma (*filled triangles*). 8-Isoprostane values are expressed as picograms produced during 15 min of breathing. Median values are shown with horizontal bars (modified with permission from ref. 9).

ibuprofen, a nonselective COX inhibitor. Fourteen patients with stable severe COPD were randomized to receive either oral ibuprofen (400 mg every 6 h) or matched placebo for 2 days (27). The second study was open labeled and uncontrolled. After a two week run-in period, seventeen patients with stable moderate COPD were given oral rofecoxib (25 mg once daily for 5 days), a selective COX-2 inhibitor (27). 8-Isoprostane concentrations in EBC were neither affected by ibuprofen [pretreatment: 47.9 (40.5–51.9) pg/ml; posttreatment: 42.2 (34.9–60.0) pg/ml, $p=0.64$] (Fig. 5.3a) nor rofecoxib treatment (pretreatment: 39.5 (33.0–46.5) pg/ml; posttreatment: [39.5 (34.8–43.5) pg/ml, $p=0.93$]) (Fig. 5.3b), whereas PGE_2

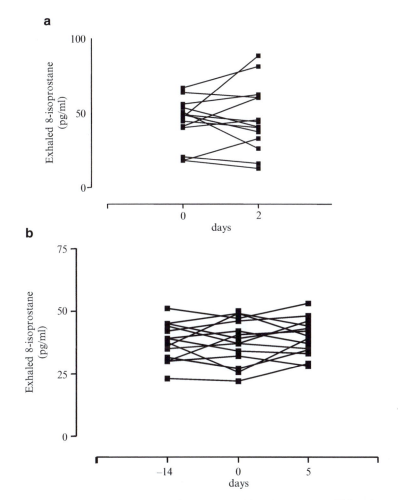

Fig. 5.3. (a) 8-Isoprostane concentrations in exhaled breath condensate (EBC) before (day 0) and after treatment (day 2) with oral ibuprofen (400 mg four times a day) for 2 days in patients with COPD ($n = 14$). (b) 8-Isoprostane concentrations in EBC in patients with COPD ($n = 16$) at baseline (day −14), before (day 0) and after treatment with oral rofecoxib 25 mg once a day for 5 days (day 5). Values are expressed as medians (modified with permission from ref. 27).

concentrations in EBC were reduced after treatment with ibuprofen [pretreatment: 93.5 (84.0–105.5) pg/ml; posttreatment: 22.0 (15.0–25.5) pg/ml, $p < 0.0001$]. The lack of effect of ibuprofen and rofecoxib on 8-isoprostane concentrations in EBC indicates that these compounds are primarily formed independently of the COX pathway (27).

Measurement of 8-isoprostane in EBC provides a tool to explore the role of oxidative stress in the pathogenesis of inflammatory airway diseases including asthma and COPD and to assess the effects of pharmacological therapy (16, 17, 27, 28).

4. Conclusions

The RIA for 8-isoprostane in EBC described in this article is a sensitive and specific tool for exploring the role of oxidative stress in the pathophysiology of lung diseases and for noninvasive assessment of lung oxidative stress. Measurement of 8-isoprostane in EBC might be useful for a better understanding of the clinical pharmacology of antioxidants and for monitoring the effects of anti-inflammatory therapy. Quantitative comparisons between RIA and mass spectrometry measurements are required.

5. Notes

1. Reconstitute lyophilized aliquots of 8-isoprostane antiserum (100 μl) with 100 μl of distilled water.

2. Prepare 10 ml of standard diluent consisting of 0.9% sodium chloride solution containing 0.1% bovine serum albumin and thimerosal diluted 1:5,000.

3. Dilute the reconstituted aliquot of 8-isoprostane antiserum with 10 ml of standard diluent to obtain the antiserum stock solution.

4. 8-Isoprostane antiserum has been made by Professor Giovanni Ciabattoni and is not commercially available.

5. To perform the RIA for 8-isoprostane, add 250 μl of EBC sample to each test tube. Add 1,250 μl of RIA buffer (phosphate 0.025 M, pH = 7.5) containing approximately 2,500 dpm of [^{3}H]8-isoprostane and 8-isoprostane antiserum to reach a final dilution of 1:200,000 in the RIA incubation volume (1.5 ml) and to obtain approximately 40–45% binding of the labeled hapten. Incubate for 24–30 h at 4°C.

6. Separate the antibody-bound from free [^{3}H]8-isoprostane by rapidly adding 0.1 ml of a 5% bovine serum albumin solution and 0.1 ml of a charcoal suspension (70 mg/ml).

7. Centrifuge at 4°C for 10 min at 5,000 rpm ($3,000 \times g$).

8. Decant supernatant solutions containing antibody-bound 8-isoprostane directly into a vial containing 10 ml of Instagel® scintillation liquid.

9. Count radioactivity in a liquid scintillation counter.

10. 8-Isoprostane concentrations are calculated using a software on the basis of the 8-isoprostane standard curve.

Acknowledgments

This work was funded by Catholic University of the Sacred Heart, "Fondi di Ateneo" 2007–2009.

References

1. Halliwell B, Grootveld M (1987) The measurement of free radical reactions in humans. FEBS Lett 213:9–14

2. Montuschi P, Barnes PJ (2006) Isoprostanes and asthma. Drug Discov Today: Ther Strateg 3:287–292

3. Rahman I, Adcock IM (2006) Oxidative stress and redox regulation of lung inflammation in COPD. Eur Respir J 28:219–242

4. Cantin AM, White TB, Cross CE, Forman HJ, Sokol RJ, Borowitz D (2007) Antioxidants in cystic fibrosis. Conclusions from the CF antioxidant workshop, Bethesda, Maryland, November 11–12. Free Radic Biol Med 42:15–31

5. Morrow JD, Hill KE, Burk RF, Nammour TM, Badr KF, Roberts LJ, II. (1990) A series of prostaglandin F_2-like compounds are produced in vivo in humans by a non-cyclooxygenase, free radical-catalyzed mechanism. Proc Natl Acad Sci USA 87:9383–9387

6. Montuschi P, Barnes PJ, Roberts LJ, II (2004) Isoprostanes: markers and mediators of oxidative stress. FASEB J 18:1791–1800

7. Montuschi P, Barnes PJ, Roberts LJ, II (2007) Insights into oxidative stress: the isoprostanes. Curr Med Chem 14:703–717

8. Montuschi P, Corradi M, Ciabattoni G, Nightingale J, Kharitonov SA, Barnes PJ (1999) Increased 8-isoprostane, a marker of oxidative stress, in exhaled condensate of asthma patients. Am J Respir Crit Care Med 160:216–220

9. Mondino C, Ciabattoni G, Koch P, Pistelli R, Trové A, Barnes PJ, Montuschi P (2004) Effects of inhaled corticosteroids on exhaled leukotrienes and prostanoids in asthmatic children. J Allergy Clin Immunol 114: 761–767

10. Shahid SK, Kharitonov SA, Wilson NM, Bush A, Barnes PJ (2005) Exhaled 8-isoprostane in childhood asthma. Respir Res 6:79

11. Montuschi P, Collins JV, Ciabattoni G, Lazzeri N, Corradi M, Kharitonov SA, Barnes PJ (2000) Exhaled 8-isoprostane as an in vivo biomarker of lung oxidative stress in patients with COPD and healthy smokers. Am J Respir Crit Care Med 162:1175–1177

12. Montuschi P, Kharitonov SA, Ciabattoni G, Corradi M, van Rensen L, Geddes DM, Hodson ME, Barnes PJ (2000) Exhaled 8-isoprostane as a new non invasive biomarker of oxidative stress in cystic fibrosis. Thorax 55:205–209

13. Lucidi P, Ciabattoni G, Bella S, Barnes PJ, Montuschi P (2008) Exhaled 8-isoprostane and prostaglandin E_2 in patients with stable and unstable cystic fibrosis. Free Radic Biol Med 45:913–919

14. Montuschi P (2002) Indirect monitoring of lung inflammation. Nat Rev Drug Discov 1:238–242

15. Montuschi P, Barnes PJ (2002) Analysis of exhaled breath condensate for monitoring airway inflammation. Trends Pharmacol Sci 23:232–237

16. Montuschi P (ed) (2004) New perspectives in monitoring lung inflammation: analysis of exhaled breath condensate. CRC, Boca Raton, FL

17. Montuschi P (2007) Analysis of exhaled breath condensate in respiratory medicine: Methodological aspects and potential clinical applications. Ther Adv Respir Dis 1:5–23

18. Montuschi P, Ragazzoni E, Valente S, Corbo G, Mondino C, Ciappi G, Ciabattoni G (2003) Validation of 8-isoprostane and prostaglandin E_2 measurements in exhaled breath condensate. Inflamm Res 52:502–506

19. Montuschi P, Barnes PJ (2002) Exhaled leukotrienes and prostaglandins in asthma. J Allergy Clin Immunol 109:615–620

20. Ciabattoni G (1987) Production of antisera by conventional techniques. In: Patrono C, Peskar BA (eds) Radioimmunoassay in basic and clinical pharmacology, Handbook of experimental pharmacology. Springer, Berlin, pp 23–68

21. Wang Z, Ciabattoni G, Créminon C, Lawson J, Fitzgerald GA, Patrono C, Maclouf J (1995) Immunological characterization of urinary 8-epi-prostaglandin $F_{2\alpha}$ excretion in man. J Pharmacol Exp Ther 275:94–100

22. Ciabattoni G, Pugliese F, Spaldi M, Cinotti GA, Patrono C (1979) Radioimmunoassay measurement of prostaglandins E_2 and $F_{2\alpha}$ in human urine. J Endocrinol Invest 2:173–182

23. Montuschi P, Currò D, Ragazzoni E, Preziosi P, Ciabattoni G (1999) Anaphylaxis increases 8-epi-prostaglandin $F_{2\alpha}$ release from guinea-pig lung in vitro. Eur J Pharmacol 365:59–64

24. Pradelles P, Grassi J, Maclouf J (1990) Enzyme immunoassays of eicosanoids using acetylcholinesterase. Methods Enzymol 187:24–34

25. Vaitukaitis J, Robbins JB, Nieschlag E, Ross GT (1971) A method for producing specific antisera with small doses of immunogen. J Clin Endocrinol Metab 33:988–991

26. Morrow JD, Scruggs J, Chen Y, Zackert WE, Roberts LJII (1998) Evidence that the E2-isoprostane, 15–E2t-isoprostane (8-isoprostaglandin E2) is formed in vivo. J Lipid Res 39:1589–1593

27. Montuschi P, Macagno F, Parente P, Valente S, Lauriola L, Ciappi G, Kharitonov SA, Barnes PJ, Ciabattoni G (2005) Effects of cyclo-oxygenase inhibition on exhaled eicosanoids in patients with COPD. Thorax 60:827–833

28. Montuschi P (2004) Isoprostanes, prostanoids, and leukotrienes in exhaled breath condensate. In: Montuschi P (ed) New perspectives in monitoring lung inflammation: analysis of exhaled breath condensate. CRC Press, Boca Raton, FL, pp 53–66

Chapter 6

Electron Paramagnetic Resonance Oximetry and Redoximetry

Guanglong He

Abstract

Reactive oxygen/nitrogen species (ROS/RNS) have been increasingly recognized as important mediators and play a number of critical roles in cell injury, metabolism, disease pathology, diagnosis, and clinical treatment. Electron paramagnetic resonance (EPR) spectroscopy enables the spectral information at certain spatial position, and, from the observed line-width and signal intensity, the localized tissue oxygenation, and tissue redox status can be determined. We applied in vivo EPR oximetry and redoximetry technique and implemented its physiological/pathophysiological applications, along with the use of biocompatible lithium pthalocyanine (liPc) and nitroxide redox sensitive probes, on in vivo tissue oxygenation and redox profile of the ischemic and reperfused heart in living animals. We have observed that the hypoxia during myocardial ischemia limited mitochondrial respiration and caused a shift of tissue redox status to a more reduced state. ROS/RNS generated at the beginning of reperfusion not only caused a shift of redox status to a more oxidized state which may contribute to the postischemic myocardial injury, but also a marked suppression of in vivo tissue O_2 consumption in the postischemic heart through modulation of mitochondrial respiration based on alterations in enzyme activity and mRNA expression of NADH dehydrogenase (NADH-DH) and cytochrome c oxidase (CcO). In addition, ischemic preconditioning was found to be able to markedly attenuate postischemic myocardial hyperoxygenation with less ROS/RNS generation and preservation of mitochondrial O_2 metabolism, due to conserved NADH-DH and CcO activities. These studies have demonstrated that EPR oximetry and redoximetry techniques have advanced to a stage that enables in-depth insight in the process of ischemia reperfusion injury.

Key words: Free radicals, EPR spectroscopy, Oximetry, Redox status, Redoximetry, Ischemia/Reperfusion, Ischemic Preconditioning

1. Introduction

Free radical plays an important role in mediating the process of physiology/pathophysiology (1). It is also a potential killer of biological molecules, cells, and tissues. Under normal physiological conditions, the producing and scavenging of free radicals is under

D. Armstrong (ed.), *Advanced Protocols in Oxidative Stress II,* Methods in Molecular Biology, vol. 594
DOI 10.1007/978-1-60761-411-1_6, © Humana Press, a part of Springer Science+Business Media, LLC 2010

a tight dynamic balance. Any shift in the balance either due to increasing production or increasing scavenging of free radicals can cause severe problems like causing diseases and aging. So, monitoring of free radicals level is particularly important, and EPR spectroscopy technique has found its unique application in detection and quantification of free radical-related processes.

Since the intrinsically produced radicals are low in their concentration and short in their life, the exteriorly administered radicals were used as the spin probe to perform EPR measurements. Among the exteriorly administered radicals, lithium phthalocyanine (LiPC) is among the most important because of its narrow line-width and high sensitivity to oxygen presence and this property makes this group of radicals suitable for oxygen oximetry technique. Another group of exterior radicals being extensively used is nitroxide radicals due to their biocompatibility and bioreducibility. By measuring the reduction to hydroxylamine, the biometabolic behavior of the biological sample could be determined.

EPR oximetry measures oxygen partial pressure (PO_2) by measuring the line-width of the EPR spectrum of a given radical at a given spatial position. Since molecular oxygen is paramagnetic, it produces line-width broadening caused by spin–spin interaction of oxygen with the probe, and the magnitude of this broadening is a function of the partial pressure of oxygen. So, with the use of suitable oximetry probes, this technique can measure tissue oxygenation or oxygen tension.

EPR redoximetry measures the redox status by monitoring the reduction of an exteriorly administered radical such as nitroxide on the decay of its signal intensity. Since nitroxide is readily reduced by the tissue equivalent antioxidants, the reducing agents, so the decay of the nitroxide signal provides a noninvasive way to measure tissue redox status.

It has been demonstrated that NO generated from endothelial NO synthase (eNOS) inhibits mitochondrial respiration by peroxynitrite $(ONOO^-)^-$ mediated nitration of mitochondrial complex I and IV (2–5). The inhibited mitochondrial respiration increases tissue PO_2 after reperfusion (2). This increased tissue PO_2 may potentiate the increase of ROS through electron leakage on mitochondrial respiratory chain (6–8). On the other hand, mitochondrial respiration controls tissue redox balance by controlling the utilization of reduced substrates such as NAD(P)H. Inhibition of electron flux of mitochondrial subcomponents will switch cellular redox status to a more reduced state with excessive accumulation of NAD(P)H (9, 10). This reduced state may also potentiate formation of ROS/RNS upon reoxygenation (11). Therefore, alterations in tissue redox status, oxygenation, and formation of ROS/RNS occur in the ischemic and reperfused myocardium and are of central importance in the pathogenesis of postischemic injury.

Therefore, there is a need of in vivo techniques to measure tissue redox status, oxygenation, and formation of ROS. We used localized EPR spectroscopy to measure in vivo tissue redox status and tissue PO_2 by introducing bioreducible spin probe nitroxide (12–14) and oxygen sensitive probe LiPc (15). EPR spectroscopy has advantages over other techniques. The implanted/injected probes can move freely with the beating heart, and myocardial redox status and PO_2 can be monitored repetitively after the chest is closed. As an additional evidence for the variation of tissue redox status and oxygenation, we used localized fluorometry to measure in vivo tissue NAD(P)H level and formation of ROS (16). Autofluorescence spectroscopy is an accepted method for measuring NAD(P)H in isolated hearts and myocardial tissue (17, 18). The majority of measured NAD(P)H is believed to originate from the tricarboxylic acid cycle (TCA) (19). This technique has been used to monitor relative NAD(P)H concentration and thus, mitochondrial redox status. The dye hydroethidine (HE) is a noncharged fluorescent probe specifically sensitive to superoxide (O_2^-), ONOO$^-$, and hydroxyl radical ($\cdot$OH), but not to hydrogen peroxide (H_2O_2) (20, 21). After reacting with O_2^-, HE forms oxyethidine and probably react with other oxidants to form ethidine (ET) (22).

However, in vivo assessment of myocardial tissue oxygenation and redox status has been challenging. The standard oxygen electrode techniques suffer from motion artifacts and nonrepeatability if the electrode is inserted into the myocardium of a beating heart, though this technique can be applied to measure tissue oxygen consumption by monitoring coronary oxygenation in the effluent/affluent in large animals (23, 24). The development of EPR spectroscopy techniques using oxygen sensitive probes such as LiPc and redox sensitive probes such as nitroxides, has provided a fast and accurate method for monitoring tissue PO_2 and redox status in various organs and tissues in vivo (25, 26). Doppler blood flow measurement, in conjunction with in vivo EPR measurement of tissue oxygenation has enabled in vivo assessment of myocardial oxygen consumption and mitochondrial function.

As an example, we also studied the impact of ischemic preconditioning (IPC) on the in vivo tissue oxygenation and redox status on a mouse heart regional ischemia and reperfusion model. There are two phases of protection afforded by IPC, e.g., the early phase and the late phase (27). Both share some common signaling pathways as well as some distinct mechanisms. In untreated or IPC mice, EPR oximetry and redoximetry were applied to monitor cardiac tissue PO_2 and along with laser Doppler flow measurements, tissue redox status, tissue perfusion, and oxygen consumption were determined. We showed that in the early phase, IPC attenuates postischemic myocardial hyeroxygenation, preserves tissue oxygen consumption, and mitochondrial function through protection of the mitochondrial respiratory chain and this is associated with infarct size reduction in the postischemic heart.

One of the challenges in measuring tissue redox status, is to find an appropriate index to indicate the nonequilibrated/time-dependent characteristics. NAD(P)H was reported to be a redox indicator during myocardial ischemia/reperfusion (28, 29). However, our data demonstrated that NAD(P)H level was higher in the postischemic state than in the preischemic state and this would imply that the redox status was still on the reducing side after reperfusion, if NAD(P)H was considered as the redox indicator. Therefore, NAD(P)H can only be considered as a redox indicator during the nonequilibrated ischemic period. Formation of ROS was also considered as an indicator of redox status. Again, our data indicated that even during ischemia there was a burst of ROS which would imply that the tissue redox status was on the oxidative side during ischemia. Therefore, ROS can only be considered as a redox indicator during the nonequilibrated reperfusion period. These data may reflect the nonequilibration of tissue redox status and oxygenation during ischemia and reperfusion. However, if we take both NAD(P)H and ROS into consideration, we come to the conclusion that during ischemia even though there was a burst of ROS, the NAD(P)H level dominated the tissue into the reducing side. Similarly, during reperfusion, even though the NAD(P)H level was slightly higher than that of normal tissue, the larger burst of ROS dominated the tissue into oxidative side. Actually, tissue GSH/GSSG level and reduction rate of nitroxide gave a better indication of the equilibrated tissue redox status.

There were several challenges of using EPR spin probes to measure tissue redox status. One was to determine whether the applied spin probes were reduced or oxidized. Another was to determine the ratio of the reduction vs. tissue washout. From the reoxidation experiments, it was clearly shown that the applied spin probes were reduced and the tissue washout was at most 10% of the total applied amount, which was within the range of the standard error (10% of 0.042/min was 0.004/min that was the SE value). The reduction rates of nitroxide were averaged over 30 min due to the requirement of a period of time for the calculation. Therefore, the time resolution was limited in the current experiments. Future study is warranted to design such a redox probe with fast decay rate and high cellular retention to allow redox measurement in a narrow time window. One of the challenges of localized spectroscopy was its limitation to a single point site; therefore, results were dependent on tissue heterogeneity. By carefully implanting or injecting the EPR and fluorescence probes, one can obtain the localized information from the core of the area at risk. Future development is warranted to image the whole risk area to allow three dimensional mapping of oxygen and fluorescence intensity.

Therefore, EPR spectroscopy and fluorometry provided direct evidence of the alterations of tissue redox status, oxygenation, and formation of ROS in the ischemic and reperfused myocardium. The high reduction rate of nitroxide in the ischemic myocardium was consistent with the low tissue PO_2, low level of ROS, high NAD(P)H content, and high level of GSH/GSSG. The low reduction rate of nitroxide in the postischemic myocardium was consistent with the high tissue PO_2, high level of ROS, low NAD(P)H content, and low level of GSH/GSSG. The limited mitochondrial respiration due to hypoxia during ischemia caused an increase of tissue NAD(P)H content and therefore shifted the tissue to a more reduced state. The formation of ROS after reperfusion contributed to the oxidative tissue redox status. The altered redox status and formation of ROS/RNS may eventually contribute to the postischemic myocardial injury.

EPR oximetry with the recently developed LiPc probe provides a sensitive technique for repetitive measurement of PO_2 in capillaries, tissues and cells. Our studies are among the first to measure in vivo myocardial PO_2 using this technique. To detect myocardial PO_2 directly, EPR oximetry has advantages over microelectrode oximetry. The implanted probe can move freely with the beating heart, and the myocardial PO_2 can be monitored repetitively after the chest is closed. The baseline of PO_2 in the myocardium was 8.6–10.1 mmHg before coronary ligation (30), which is consistent with prior studies (31, 32). The nonzero myocardial PO_2 during coronary ligation suggests the existence of collateral circulation, as also shown by the nonzero myocardial blood flow.

Coronary endothelial dysfunction was reported in postischemic myocardium, but increased in vivo NO production occurs during the early period of reperfusion. eNOS-derived NO with its derivative, peroxynitrite, suppress oxygen consumption through CcO and NADH-DH regulation of mitochondrial respiration, resulting in marked myocardial hyperoxygenation following reperfusion. This myocardial hyperoxygenation may be a critical factor influencing postischemic remodeling.

We have also investigated the effects of IPC on myocardial oxygenation, perfusion and mitochondrial function in the postischemic heart. After ischemia and reperfusion in IPC hearts, regional blood flow was improved compared to that in non-IPC hearts. Tissue PO_2 overshot preischemic baseline values in non-IPC hearts but this overshoot was not seen with IPC. This overshoot with higher levels of myocardial oxygenation in postischemic myocardium suggested that oxygen consumption is decreased by the process of ischemia/reperfusion injury. These observations are consistent with prior reports that ROS/RNS, formed upon postischemic reperfusion inhibit mitochondrial oxygen consumption and electron transport (2, 30). In our own studies, we have observed for the first time in the regional ischemia and reperfusion mouse heart model, that IPC attenuates in vivo myocardial hyperoxygenation status in the postischemic myocardium. The attenuation of the hyperoxygenation after regional ischemia and reperfusion may have profound effect on the postischemic cardiac remodeling since oxygenation is a key mediator of the development and maturation of cardiac myocytes (33). Furthermore, the higher blood flow and lower PO_2 with IPC indicates that IPC attenuates postischemic myocardial hyperoxygenation, preserves higher levels of O_2 consumption in postischemic myocardium. We also observed that IPC prevented the loss of the activities of the critical mitochondrial electron transport enzymes NADH-DH and CcO. This study suggests that IPC attenuates in vivo postischemic myocardial hyperoxygenation by improving myocardial oxygen consumption and decreasing ROS/RNS generation after regional ischemia reperfusion, which in turn preserves mitochondrial function and electron transport enzyme activity. This preservation of mitochondrial function and oxygen consumption contributes to myocardial protection. Further studies of the detailed mechanisms by which IPC induces suppression of pathological ROS/RNS formation and how this process relates to other pathways of protection are warranted.

These studies have demonstrated that EPR oximetry and redoximetry techniques have been demonstrated to be powerful techniques enabling the measurement of in vivo myocardial tissue oxygenation and redox status. These techniques should be particularly valuable in the dissection of the role of ROS/RNS in the process of ischemia reperfusion injury and can potentially provide valuable therapeutic interventions to cure the coronary heart disease.

2. Materials

2.1. Equipment

1. Mouse respirator (Harvard Apparatus).
2. Laser Doppler perfusion monitor (Moor Instruments).
3. Optic suction probe (P10d, Moor Instruments).
4. Custom-made L-band EPR spectrometer.
5. 1.4F Millar tip transducer catheter (model SPR-261) connected to a PowerLab system.
6. Ratiometric fluorometer (Radnoti Glass; Monrovia, CA).
7. HPLC separation ESA solvent delivery system (Chelmsford, MA).
8. Chemiluminescence (Denville Scientific Inc.).

9. AlphaImagerTM high performance gel documentation & image analysis system (model 3300, Alpha Innotech Co. San Leandro, CA).

2.2. Reagents

1. *N*-nitro-L-arginine methyl ester (Sigma).

2. 2,2,5,5-tetramethyl-3-carboxylpyrrolidine-*N*-oxyl (PCA) (Sigma Chemical Co.).

3. Evans blue (Sigma).

4. 2,3,5-triphenyltetrazolium chloride (TTC, Sigma).

5. *N,N*-dimethylacetamide (Acros Organics).

6. Peroxynitrite (Cayman Chemical Co.).

7. $KHCO_3$.

8. KH_2PO_4.

9. Tetrabutylammonium hydrogensulfate.

10. Ethylenediaminetetraacetic acid.

11. HPLC-grade acetonitrile.

12. Sucrose.

13. EGTA.

14. Reduced cytochrome c ($60\,\mu M$, Sigma).

15. Tris–HCl buffer (20 mM, pH 8.0).

16. NADH ($150\,\mu M$, Sigma).

17. Coenzyme Q_1 ($100\,\mu M$, Sigma).

2.3. Supplies

1. Tris–glycine polyacrylamide gradient gels.

2. BCA assay (PIERCE Biotechnology).

3. Ketamine, xylazine and atropine.

4. 7-0 silk suture.

5. PBS.

6. 0.2-μm membrane filter.

7. Hepes buffer.

8. Protease-inhibitor cocktail (1:40, Sigma).

9. 3-μm symmetry C18 column (3.9×150 mm inner diameter, Waters Corporation, Milford, MA).

3. Methods

3.1. Animals

Male wild-type C57BL/6 and eNOS$^{-/-}$ mice were purchased from Jackson Laboratory (Bar Harbor, Maine). Wild-type mice were randomly selected to be administered 1 mg/ml of

N-nitro-L-arginine methyl ester (Sigma) in drinking water for 3 days before experiment (L-NAME-treated group). All procedures were performed with the approval of the Institutional Animal Care and Use Committee at The Ohio State University, Columbus, OH, and conformed to the *Guide for the Care and Use of Laboratory Animals* (NIH publication No.86-23, revised 1985).

3.2. In Vivo Ischemia Reperfusion Mouse Model

The in vivo ischemia reperfusion mouse model was performed with a technique similar to that described previously (2, 34). Mice were anesthetized with ketamine (55 mg/kg) plus xylazine (15 mg/kg). Atropine (0.05 mg SC) was administered to reduce airway secretion. Animals were intubated and ventilated with room air (tidal volume 250 μl, 120 breath/min) with a mouse respirator (Harvard Apparatus). Rectal temperatures were maintained at 37°C by a thermo heating pad. After thoracotomy, the left anterior descending (LAD) coronary artery was ligated with a 7-0 silk suture. After 30 min of ischemia, the occlusion was released, and reperfusion was confirmed visually.

To further confirm the LAD occlusion, myocardial tissue blood flow was monitored. After thoracotomy, an optic suction probe (P10d, Moor Instruments) connected to a laser Doppler perfusion monitor (Moor Instruments) was placed on the area at risk and blood flow was monitored before, during, and after coronary occlusion (34).

IPC was introduced by three cycles of 5 min ischemia, followed by 5 min reperfusion. The final reperfusion period of the IPC protocol was extended to 15 min in order to ensure the stabilization of tissue PO_2. The IPC+I/R mice were then subjected to 30 min LAD occlusion followed by either 60 min of reperfusion (for EPR oximetry and redoximetry measurements) or 24 h of reperfusion (for measurements of infarct size) similar to I/R mice.

3.3. Measurement of In Vivo Tissue Redox Status with Localized EPR

In vivo EPR measurements of tissue redox status were performed using 2,2,5,5-tetramethyl-3-carboxylpyrrolidine-*N*-oxyl (PCA) (Sigma Chemical Co.) as the spin probe with a three-line EPR spectrum. The nitroxide solutions were prepared in PBS and kept frozen until use. About 5 μl of 10 mM PCA solution was intramuscularly injected as a bolus into the area at risk after thoracotomy (*see* Note 1). Then EPR spectra were acquired before, during, and after ischemia with a surface loop resonator placed on top of the heart. The lower field peak-height was monitored with time to determine the rate of reduction.

3.4. In Vivo Localized EPR Oximetry

LiPc was used as the probe for EPR oximetry (35, 36). The O_2 response of LiPc showed good linearity from 0 to 150 mmHg with a sensitivity of 7.25 mG/mmHg. After thoracotomy and exposure of the heart, ~10 μg of LiPc crystals loaded in a 27-gauge needle was implanted into the midmyocardium in the area at risk (*see* Note 2). After 30 min equilibration, the mouse was placed into the custom-made L-band EPR spectrometer with its heart under the resonator loop (37, 38). EPR spectra of LiPc crystals were acquired with following parameters: frequency 1.1 GHz, microwave power 16 mW, modulation field 0.0045 mT, and scan width 0.2 mT. The position of the implanted crystals was confirmed by histology.

3.5. Measurement of Myocardial Hemodynamics and Infarct Size

After anesthesia, the right common carotid artery was cannulated with a 1.4F Millar tip transducer catheter (model SPR-261) connected to a PowerLab system for monitoring of mean arterial blood pressure (MABP) and heart rate (HR). Rate pressure product (RPP) was calculated as: $RPP = MABP \times HR$ (mm Hg/min). In all groups, similar basal heart rates were observed with values around 300 bpm, typical for anesthetized mice.

In order to measure the infarct size, mouse heart was subjected to 30 min LAD occlusion and 24 h reperfusion. Then mice were reventilated and LAD was reoccluded. About 1.5 ml of 4.0% Evans blue (Sigma) was injected from the inferior vena cave to delineate the nonischemic myocardium from that of the ischemic myocardium. Then mice were euthanized and hearts were excised and cut into four transverse slices. The slices were stained with 1.5% 2,3,5-triphenyltetrazolium chloride (TTC, Sigma) to determine the infarct area (IF). Then the slices were photographed under a microscope (Nikon) to determine area of LV, area at risk (AAR), and IF by computerized planimetry. Infarct size was expressed as percentage of the AAR in LV.

3. Measurement of Myocardial Tissue Blood Flow

After thoracotomy, the fourth rib was removed. An optic suction probe (P10d, Moor Instruments) connected to a Laser Doppler Perfusion Monitor (Moor Instruments) was placed on the area at risk. Before and during coronary occlusion and reperfusion, regional myocardial blood flow was monitored continuously and presented as a percentage of the baseline before ischemia.

3.7. In Vivo Fluorescence Measurement of Tissue NAD(P)H Content

In vivo autofluorescence of the sum of tissue NADH and NADPH, NAD(P)H, in the area at risk were measured within a black box that excluded the ambient light. A 6-mm diameter fiber-optic bundle that contained both excitation and emission fibers was carefully positioned adjacent to and directed toward the area at risk. To avoid noise from other organs, the whole body of the mouse was covered with a black cloth with one small opening (3×3 mm) at the area at risk. The proximal end of the fiber-optic cable was connected to a ratiometric fluorometer (Radnoti Glass; Monrovia, CA). Fluorescence was excited using a 150-W xenon arc lamp filtered through one of four alternating excitation filters. A single excitation filter with a specific designated ultraviolet range was used to excite NAD(P)H at 330 nm with a band width of 80 nm.

The emission range for NAD(P)H was 470 ± 5 nm. The signals were averaged over 5 s, recorded and graphed using a modified program (IOTECH with software from Strawberry Tree).

3.8. HPLC Analysis of NAD(P)H Levels

In order to quantify NAD(P)H levels measured from autofluorescence, HPLC analysis of tissue NAD(P)H levels were followed on tissues collected from the area at risk before ischemia, at the end of 25 min ischemia, and at the end of 20 min reperfusion. Then heart samples were ground to fine powder under liquid nitrogen and extracted and homogenized in ice cold perchloric acid (0.4 mol/L). The denatured protein was pelleted and reserved for protein analysis. The acid extract was neutralized with equal volumes of 0.4 mol/L $KHCO_3$. Each extract was subjected to nucleotide analysis using gradient ion-pair reversed-phase liquid chromatography. HPLC separation was performed using an ESA (Chelmsford, MA) solvent delivery system with a 3-μm symmetry C18 column (3.9×150-mm inner diameter, Waters Corporation, Milford, MA). Separation was performed by reverse-phase chromatography using an isocratic mobile phase consisting of buffer A (35 mmol/L KH_2PO_4, 6 mmol/L tetrabutylammonium hydrogensulfate, pH 6.0, 125 mmol/L ethylenediaminetetraacetic acid) and buffer B (a mixture of buffer A and HPLC-grade acetonitrile in a ratio of 1:1, v/v), filtered through a 0.2-μm membrane filter and helium degassed. The flow rate was set at 1.0 ml/min and detection was performed at 260 nm using an ESA variable wavelength UV/V absorbance detector.

3.9. In Vivo Localized Fluorescence Measurement of ROS

Fluorescence dye HE stock solution was made in N,N-dimethylacetamide (Acros Organics). About 20 μl of 200 μM HE in PBS solution was injected intramuscularly as a bolus into the area at risk 5 min before ischemia. The fluorescence measurements were followed immediately after the injection. The ET excitation filter was set at 515 ± 20 nm, and the ET emission at 590 ± 25 nm.

3.10. HPLC Measurement of Tissue GSH/GSSG

At the end of 30 min ischemia and 60 min reperfusion, animals were euthanized and myocardial tissue in the area at risk were excised and weighed. Heart tissue was ground in liquid nitrogen and homogenized in 0.5 ml of 200 mM methanesulfonic acid containing 5 mM diethylenetriaminepentaacetic acid (pH < 2.0) using a dounce glass homogenizer. Then tissue homogenate was centrifuged for 30 min at 19,068 g at 4°C. The supernatant was diluted 1:1 with mobile phase and stored frozen at −80°C for GSH assay. The GSH assays were performed with HPLC (39). Samples were separated on a Polaris 5 μm, 0.4×20 cm C-18 column eluted with a mobile phase of 50 mM NaH_2PO_4, 0.05 mM octanesulfonic acid, and 2% acetonitrile adjusted to pH 2.7 with phosphoric acid and a flow rate of 1 ml/min. An ESA CoulArray detector was used with the guard cell set at +950 mV, electrode 1 at +400 mV, and electrode 2 at +880 mV. Full-scale output was set at $100\,\mu A$ and peak areas were analyzed using the CoulArray software (ESA, Chelmsford, MA). A standard curve was obtained using a $10\,\mu M$ to 1 mM solution of GSH, from which GSH concentrations were determined. The GSH values were expressed as nmols/g of tissue wet weight.

3.11. Activities of CcO and NADH-DH

Frozen myocardial tissue, obtained from risk region, was homogenized in ice-cold Hepes buffer (3 mM, pH 7.2) containing sucrose (0.25 M), EGTA (0.5 mM) and protease-inhibitor cocktail (1:40, Sigma). CcO activity was measured in the presence of phosphate buffer (50 mM, pH 7.4) and reduced cytochrome c ($60\,\mu M$, Sigma) (40). NADH-DH activity was measured in the presence of Tris–HCl buffer (20 mM, pH 8.0), NADH ($150\,\mu M$, Sigma), and coenzyme Q_1 ($100\,\mu M$, Sigma) (41). The extinction coefficients, ε 550 nm $= 18.5$ mM^{-1}cm^{-1} for cytochrome c and ε 340 nm $= 6.22$ mM^{-1}cm^{-1} for NADH were utilized for activity calculation. Protein concentration of the tissue homogenate was measured by BCA assay (PIERCE Biotechnology).

3.12. Western Blot Analysis of Protein Expression of NADH-DH and CcO

Proteins of the homogenate were subjected to electrophoresis on 4–20% Tris–glycine polyacrylamide gradient gels. Anti-OxPhos Complex I subunit 39 kDa and Anti-OxPhos Complex IV subunit VIb (Invitrogen, Carlsbad, CA) and anti-GAPDH was used to confirm equal loading (Santa Cruz Biotechnology, Santa Cruz, CA). The secondary antibodies were conjugated with horseradish peroxidase and the protein was detected by use of enhanced chemiluminescence (Denville Scientific Inc.). The blots were imaged and quantified in an AlphaImagerTM high performance gel documentation & image analysis system (model 3300, Alpha Innotech Co. San Leandro, CA).

3.13. Immunohisto-chemistry for Nitrotyrosine

Mouse heart injected with peroxynitrite (1 mM, 0.05 ml, Cayman Chemical Co.) into the left ventricle was taken as the positive control. Immunohistochemistry was performed as described previously by our laboratory (4). Briefly, the formalin-fixed paraffin sections were incubated with rabbit polyclonal antinitrotyrosine antibody (Upstate 1:400), then with the biotinylated secondary, and again with the tertiary, ExtrAvidin alkaline phosphatase (1:800). Color was developed using fast red.

4. Results

1. In vivo myocardial tissue reduction capability
   After injection of PCA into the area at risk, EPR spectrum before, during and after ischemia was collected every 60 s

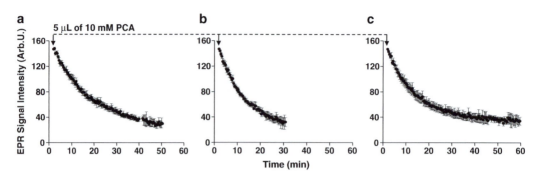

Fig. 6.1. In vivo measurement of tissue reduction rate of PCA. After thoracotomy and exposure of the heart, 5 μl of 10 mM PCA PBS solution was injected into the area at risk and EPR spectroscopy was followed. The decay of EPR signal intensity was shown as the open circles while the single exponential decay fitting curves were shown as the solid lines: (**a**) before; (**b**) during; (**c**) after 30 min LAD occlusion. $N = 7$.

for up to 1 h. The peak-height of the lowest field EPR line was monitored with time. The time-dependent profile of the EPR signal was shown in Fig. 6.1. A faster decay of the EPR signal was observed in the ischemic tissue and a slower decay was observed in the postischemic tissue relative to that of the preischemic tissue.

In order to confirm that the decay of EPR signal was due to the reduction of the applied probe, 5 μl of 10 mM potassium ferricyanide ($K_3Fe(CN)_6$) was injected to the same spot 10 min after the injection of PCA. $K_3Fe(CN)_6$ was a known standard oxidant to oxidize the reduced form of PCA, hydroxylamine, back to its nitroxide form. It was observed that after injection of $K_3Fe(CN)_6$, nearly 90% of the reduced PCA signal was restored. This confirmed that the decay of PCA signal was mainly due to tissue reduction of this probe.

Then the decay of EPR signal intensity was fitted with a single exponential curve. The reduction rate constants of PCA in the preischemic, ischemic, and postischemic myocardium in the area at risk were 0.042 ± 0.004, 0.084 ± 0.015, and 0.028 ± 0.004 min^{-1} as shown in Fig. 6.2. A fast reduction rate was observed during ischemia relative to that in the preischemic and postischemic time. The reduction rate was the lowest in the postischemic myocardium.

2. Variation of tissue NAD(P)H level

From the autofluorescence measurement, it was observed that tissue NAD(P)H level was increased from 38.3 ± 4.5 nmol/mg protein before ischemia to 90.0 ± 9.6 nmol/mg protein at the end of 25 min ischemia, and 50.9 ± 6.5 nmol/mg protein after 20 min reperfusion (Fig. 6.3). This data, together with the measurement of redox status in the previous section, strongly suggested that during ischemia, tissue redox status shifted to a reducing state which could be due to high tissue NAD(P)H content and low tissue PO$_2$.

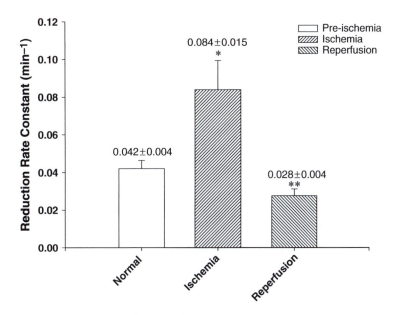

Fig. 6.2. Reduction rate constants of PCA. The decay of EPR signal in Fig. 6.1 was fitted with a single exponential decay curve to obtain the reduction rate constants of PCA (**a**) before, (**b**) during, and (**c**) after LAD occlusion. $N=7$. *ischemia vs. normal, $P<0.01$, **reperfusion vs. normal, $P<0.01$.

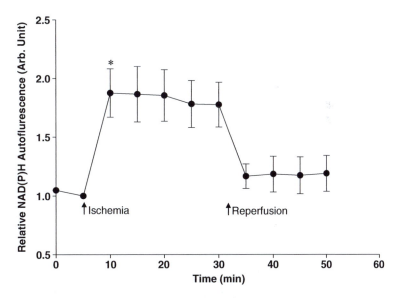

Fig. 6.3. Fluorescence and HPLC measurements of tissue NAD(P)H content in the area at risk. Tissue NAD(P)H level was increased from 38.3 ± 4.5 nmol/mg protein before ischemia to 90.0 ± 9.6 nmol/mg protein at the end of ischemia, and 50.9 ± 6.5 nmol/mg protein after reperfusion. $N=4$. *ischemia vs. pre-ischemia, $P<0.05$.

3. Alterations of the formation of ROS in the area at risk

As shown in Fig. 6.4, fluorescence measurement showed that there was a small peak (peaked at 7.5 min with a peak value of 1.07 ± 0.02 of the baseline) observed during the first 5 min

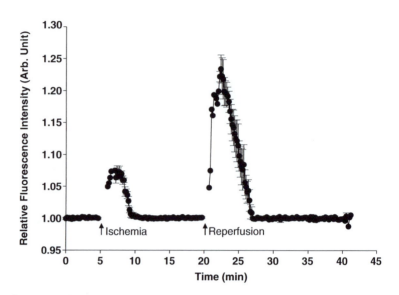

Fig. 6.4. In vivo measurements of the burst formation of ROS with fluorometry. After thoracotomy and exposure of the heart, 20 μl of 200 μM HE was injected intramuscularly in the area at risk 5 min before ischemia. Then fluorescence was measured before ischemia, during 15 min LAD occlusion, and after reperfusion. The measurements were repeated on four animals and the averaged results were shown in the figure. There appeared a burst formation of fluorescence signal within the first 5 min of ischemia. Upon reperfusion, there was another larger burst of fluorescence signal.

of ischemia and it was decreased to baseline level after the burst. Upon reperfusion, a larger peak was observed for a period of 7 min with a peak value of 1.24 ± 0.20 of the baseline.

4. Myocardial tissue PO_2

   After 5 min of the implantation of LiPc into the area at risk, EPR oximetry was performed. As shown in Fig. 6.5, in wild-type, L-NAME-treated, and eNOS$^{-/-}$ mice, myocardial tissue PO_2 dropped from baseline values of 8.6 ± 0.7, 10.1 ± 1.2, and 10.0 ± 1.2 mmHg to 1.4 ± 0.6, 2.3 ± 0.9, and 3.1 ± 1.4 mmHg respectively ($P < 0.001$ within each group) at 30 min of coronary ligation. Reperfusion resulted in a marked hyperoxygenation state during reperfusion ($P < 0.001$ vs baseline within each group). However, the hyperoxygenation level was significantly lower in L-NAME-treated (17.4 ± 1.6 mmHg) and eNOS$^{-/-}$ (20.4 ± 1.9 mmHg) mice than that in wild-type (46.5 ± 1.7 mmHg) mice ($P < 0.001$). A transient peak of myocardial PO_2 (46.5 ± 1.7 mmHg) was observed at 12.5 min of reperfusion followed by a constant high level in wild-type, but not in L-NAME-treated or eNOS$^{-/-}$ mice.

   To identify whether the myocardial hyperoxygenation during reperfusion was due to reactive hyperemia, myocardial tissue blood flow was measured in mice. Myocardial tissue blood flow was reduced to 14.2 ± 1.2 %, 12.8 ± 0.9 %, and 14.2 ± 1.8 %

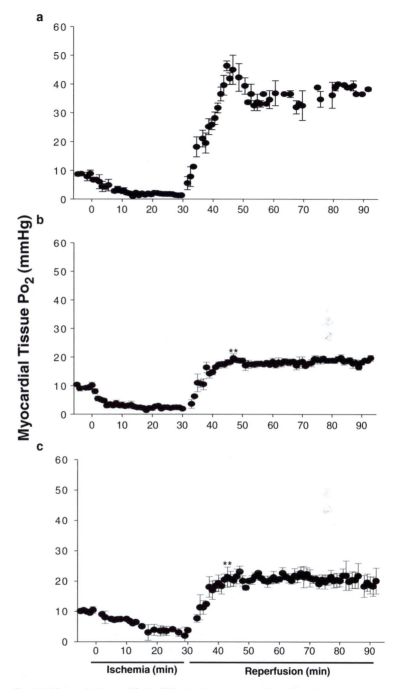

Fig. 6.5. Myocardial tissue PO$_2$ by EPR oximetry in mice subjected to 30 min of coronary ligation followed by 60 min of reperfusion. The O$_2$ probe was implanted in the myocardium at risk prior to the introduction of ischemia. (**a**) Marked hyperoxygenation was induced by reperfusion in postischemic myocardium in wild-type mice. (**b, c**) Reperfusion-induced hyperoxygenation was suppressed in L-NAME-treated (**b**) and eNOS$^{-/-}$ (**c**) mice. $N=7$. **$P<0.001$ vs. wild-type.

at the end of 30 min of coronary ligation compared to the values at preischemic baseline. The flow was restored to 92.8 ± 2.8 %, 81.8 ± 6.2 %, and 86.0 ± 3.8 % of the baseline levels during the first 3 min of reperfusion in WT, L-NAME-treated, and eNOS$^{-/-}$ mice, respectively. No difference in myocardial blood flow was observed among the three groups suggesting that the marked increase in PO$_2$ during the initial period of reperfusion was not due to hyperemia.

In a separate group of mice, IPC was implemented before the index ischemia and then EPR oximetry was performed. As shown in Fig. 6.6, 30 min after the implantation of LiPc crystals into the area at risk, the baseline values of tissue PO$_2$ in the preischemic state were measured as 16.3 ± 0.7 and 15.1 ± 0.8 mmHg in I/R and IPC+I/R mice, respectively. Decreases of tissue PO$_2$ were clearly observed during each 5 min ischemic period followed by increases with each reperfusion phase. In both groups, tissue PO$_2$ values dropped rapidly following the onset of ischemia to less than 2 mmHg within 10 min and then gradually decreased to values of 0.7 ± 0.2 and 1.2 ± 0.3 mmHg in I/R and IPC+I/R mice, respectively. Upon reperfusion, the values of PO$_2$ increased rapidly in both groups. In I/R mice, tissue PO$_2$ increased and overshot basal values, with 26.2 ± 0.8 mmHg seen at the end

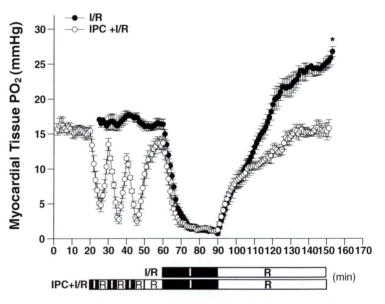

Fig. 6.6. In vivo measurement of myocardial tissue PO$_2$ with EPR oximetry in IPC hearts. Mice were subjected to a 30 min LAD occlusion followed 60 min reperfusion (I/R) or three cycles of 5 min LAD occlusion followed by 5 min reperfusion (the last reperfusion was prolonged to 15 min), and 30 min ischemia and 60 min reperfusion (IPC+I/R). Tissue PO$_2$ overshot the baseline values in I/R group but this was attenuated in the IPC+I/R group. *I/R vs. IPC+I/R at the end of 60 min reperfusion, $N=7$/group.

of 60 min reperfusion period. In IPC+I/R mice, the PO$_2$ increased back to preischemic values reaching 15.6 ± 1.5 mmHg at 60 min reperfusion. In contrast to the I/R mice no overshoot in myocardial oxygenation was seen.

5. Measurements of myocardial hemodynamics and infarct size
   Myocardial RPP was considered as an index of cardiac tissue oxygen demand. Myocardial RPP was decreased slightly from a value of $20.2 \pm 2.3 \times 10^3$ mmH/min before ischemia to $16.2 \pm 2.1 \times 10^3$ mmH/min after 60 min reperfusion. However, the difference was not significant since RPP represented the global function of the heart while the ischemic injury in our study was only in the area at risk. The measurement of hemodynamics indicated that the postischemic myocardial injury was only partially manifested by the global function of the heart such as RPP. Infarct size was measured as 31.0 ± 2.5 % of the LV 24 h after reperfusion.

6. Tissue GSH/GSSG levels in the postischemic myocardium
   GSH/GSSG level was increased by 48% in the ischemic myocardium and decreased by 29% in the postischemic myocardium compared to that of the preischemic tissue. This data confirmed that tissue redox status shifted to a reducing state in the ischemic myocardium and to an oxidized state in the postischemic myocardium.

7. Mitochondrial NADH-DH and CcO activity
   To investigate the mechanism(s) by which ROS/RNS regulate mitochondrial O$_2$ consumption, the activities of NADH-DH and CcO were measured. Baseline values of the activities of NADH-DH and CcO in the sham group were measured as 0.23 ± 0.01 and 0.44 ± 0.03 µmole/min/mg protein, respectively (Fig. 6.7). The activities of NADH-DH and CcO at the end of 60 min reperfusion were 0.17 ± 0.01 and

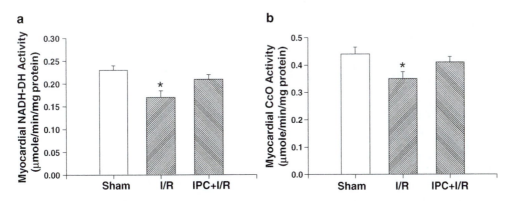

Fig. 6.7. Activities of mitochondrial enzymes NADH-DH and CcO in sham, I/R and IPC + I/R mice. The activities of NADH-DH and CcO were significantly lower in I/R mice than that in sham mice (*$P < 0.05$). However, activities in IPC + I/R mice were restored to the values comparable to that of sham group. $N = 7$/group.

$0.35 \pm 0.03\,\mu$mol/min/mg protein in I/R group, and 0.21 ± 0.01 and 0.41 ± 0.02 µmole/min/mg protein in IPC + I/R mice, respectively. Compared to sham group, the activities of NADH-DH and CcO were significantly lower in I/R group ($P < 0.05$) but not in IPC + I/R group.

8. mRNA expressions of CcO and NADH-DH

To explore the regulation mechanisms of myocardial O_2 consumption at the gene level, mRNA expressions of CcO and NADH-DH were detected by real-time PCR. Sham group was taken as the baseline control. At 60 min of reperfusion, significant up-regulation of CcO mRNA expression was detected in wild-type mice ($P < 0.01$), but not in L-NAME-treated or eNOS$^{-/-}$ mice ($P > 0.05$). The extent of CcO mRNA up-regulation was significantly lower in L-NAME-treated and eNOS$^{-/-}$ compared to that in wild-type mice ($P < 0.05$). No marked difference for NADH-DH mRNA expression was found among all the groups ($P > 0.05$).

9. Immunohistochemical staining of nitrotyrosine

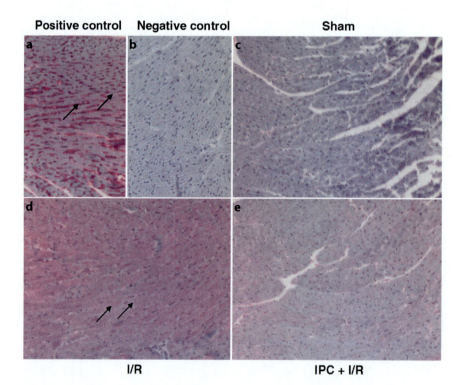

Fig. 6.8. Immunohistochemical staining of nitrotyrosine, a biomarker of peroxynitrite formation. Sections were fixed and prepared as described in the Subheading 6.3 section and photomicrographs are obtained at ×200 magnification. Positive control with direct injection of peroxynitrite (**a**), strong nitrotyrosine staining in the vascular beds was present (*red color, arrows*). Negative control (**b**), no nitrotyrosine staining (*red*) was present. Only weak staining in hearts of sham group (**c**) was seen, while strong red staining was present in the I/R heart (**d**) (*arrows*). In the IPC + I/R heart only weak staining was seen (**e**).

Immunohistochemical staining of nitrotyrosine, a biomarker of peroxynitrite formation, was performed to demonstrate the burst formation of ROS/RNS at the beginning of reperfusion (4, 42). Nitrotyrosine staining showed very prominent red coloration in the vascular beds in the positive control group that was infused with peroxynitrite (see arrows in Fig. 6.8). There was no observable nitrotyrosine staining in the negative control group. Only very weak red staining for nitrotyrosine was seen in the sham group. In contrast, strong nitrotyrosine staining was observed with a homogeneous distribution throughout the myocytes in the I/R group (see arrows in the figure), however, staining was much weaker in the IPC + I/R group.

5. Notes

1. When applying redoximetry using nitroxides as the redox sensitive probes, control experiments have to be performed first to confirm the reduction reaction of the injected nitroxides. In our studies, using $Fe_2(CN)_6$, we confirmed that PCA is reduced to hydroxylamine when injected into the myocardium of a beating mouse heart.

2. When applying oximetry using LiPc as the oxygen sensitive probe, an equilibration time has to be given for the crystals to be equilibrated after the initial implantation. In our studies, EPR oximetry was performed at 5 min and at 30 min after the implantation of the probes.

References

1. Zweier JL, Kuppusamy P, Lutty GA (1988) Measurement of endothelial cell free radical generation: evidence for a central mechanism of free radical injury in postischemic tissues. Proc Natl Acad Sci U S A 85(11):4046–4050

2. Zhao X, He G, Chen YR et al (2005) Endothelium-derived nitric oxide regulates postischemic myocardial oxygenation and oxygen consumption by modulation of mitochondrial electron transport. Circulation 111(22): 2966–2972

3. Lizasoain I, Moro MA, Knowles RG et al (1996) Nitric oxide and peroxynitrite exert distinct effects on mitochondrial respiration which are differentially blocked by glutathione or glucose. Biochem J 314(Pt 3):877–880

4. Wang P, Zweier JL (1996) Measurement of nitric oxide and peroxynitrite generation in the postischemic heart. Evidence for peroxynitrite-mediated reperfusion injury. J Biol Chem 271(46):29223–29230

5. Wolin MS, Xie YW, Hintze TH (1999) Nitric oxide as a regulator of tissue oxygen consumption. Curr Opin Nephrol Hypertens 8(1):97–103

6. Ohnishi ST, Ohnishi T, Muranaka S et al (2005) A possible site of superoxide generation in the complex I segment of rat heart mitochondria. J Bioenerg Biomembr 37(1):1–15

7. Ferdinandy P, Schulz R (2003) Nitric oxide, superoxide, and peroxynitrite in myocardial ischaemia-reperfusion injury and preconditioning. Br J Pharmacol 138(4):532–543

8. Moncada S, Erusalimsky JD (2002) Does nitric oxide modulate mitochondrial energy generation and apoptosis? Nat Rev Mol Cell Biol 3(3):214–220

9. Beltran B, Mathur A, Duchen MR et al (2000) The effect of nitric oxide on cell respiration: A key to understanding its role in cell survival or death. Proc Natl Acad Sci U S A 97(26):14602–14607

10. Cleeter MW, Cooper JM, Darley-Usmar VM et al (1994) Reversible inhibition of cytochrome c oxidase, the terminal enzyme of the mitochondrial respiratory chain, by nitric oxide. Implications for neurodegenerative diseases. FEBS Lett 345(1):50–54

11. Klawitter PF, Murray HN, Clanton TL et al (2002) Reactive oxygen species generated during myocardial ischemia enable energetic recovery during reperfusion. Am J Physiol Heart Circ Physiol 283(4):H1656–H1661

12. Zweier JL, Chzhan M, Ewert U et al (1994) Development of a highly sensitive probe for measuring oxygen in biological tissues. J Magn Reson B 105(1):52–57

13. Swartz HM, Dunn JF (2003) Measurements of oxygen in tissues: overview and perspectives on methods. Adv Exp Med Biol 530:1–12

14. Halpern HJ, Yu C, Peric M et al (1994) Oxymetry deep in tissues with low-frequency electron paramagnetic resonance. Proc Natl Acad Sci U S A 91(26):13047–13051

15. Liu KJ, Gast P, Moussavi M et al (1993) Lithium phthalocyanine: a probe for electron paramagnetic resonance oximetry in viable biological systems. Proc Natl Acad Sci U S A 90(12):5438–5442

16. Stoner JD, Angelos MG, Clanton TL (2004) Myocardial contractile function during postischemic low-flow reperfusion: critical thresholds of NADH and O_2 delivery. Am J Physiol Heart Circ Physiol 286(1):H375–H380

17. Brandes R, Bers DM (1996) Increased work in cardiac trabeculae causes decreased mitochondrial NADH fluorescence followed by slow recovery. Biophys J 71(2):1024–1035

18. Riess ML, Camara AK, Chen Q et al (2002) Altered NADH and improved function by anesthetic and ischemic preconditioning in guinea pig intact hearts. Am J Physiol Heart Circ Physiol 283(1):H53–H60

19. Eng J, Lynch RM, Balaban RS (1989) Nicotinamide adenine dinucleotide fluorescence spectroscopy and imaging of isolated cardiac myocytes. Biophys J 55(4):621–630

20. Al-Mehdi AB, Shuman H, Fisher AB (1997) Intracellular generation of reactive oxygen species during nonhypoxic lung ischemia. Am J Physiol 272(2 Pt 1):L294–L300

21. Budd SL, Castilho RF, Nicholls DG (1997) Mitochondrial membrane potential and hydroethidine-monitored superoxide generation in cultured cerebellar granule cells. FEBS Lett 415(1):21–24

22. Zhao H, Kalivendi S, Zhang H et al (2003) Superoxide reacts with hydroethidine but forms a fluorescent product that is distinctly different from ethidium: potential implications in intracellular fluorescence detection of superoxide. Free Radic Biol Med 34(11): 1359–1368

23. Tanoue Y, Herijgers P, Meuris B et al (2002) Ischemic preconditioning reduces unloaded myocardial oxygen consumption in an in-vivo sheep model. Cardiovasc Res 55(3): 633–641

24. An J, Camara AK, Rhodes SS et al (2005) Warm ischemic preconditioning improves mitochondrial redox balance during and after mild hypothermic ischemia in guinea pig isolated hearts. Am J Physiol Heart Circ Physiol 288(6):H2620–H2627

25. Zhu X, Zuo L, Cardounel AJ et al (2007) Characterization of in vivo tissue redox status, oxygenation, and formation of reactive oxygen species in postischemic myocardium. Antioxid Redox Signal 9(4):447–455

26. Swartz HM, Bacic G, Friedman B et al (1994) Measurements of pO2 in vivo, including human subjects, by electron paramagnetic resonance. Adv Exp Med Biol 361:119–128

27. Bolli R (1996) The early and late phases of preconditioning against myocardial stunning and the essential role of oxyradicals in the late phase: an overview. Basic Res Cardiol 91(1):57–63

28. Angelos MG, Kutala VK, Torres CA et al (2006) Hypoxic reperfusion of the ischemic heart and oxygen radical generation. Am J Physiol Heart Circ Physiol 290(1):H341–H347

29. Zuo L, Clanton TL (2005) Reactive oxygen species formation in the transition to hypoxia in skeletal muscle. Am J Physiol Cell Physiol 289(1):C207–C216

30. Shen W, Xu X, Ochoa M et al (1994) Role of nitric oxide in the regulation of oxygen consumption in conscious dogs. Circ Res 75(6):1086–1095

31. Al-Obaidi MK, Etherington PJ, Barron DJ et al (2000) Myocardial tissue oxygen supply and utilization during coronary artery bypass surgery: Evidence of microvascular no-reflow. Clin Sci (Lond) 98(3):321–328

32. Trochu JN, Bouhour JB, Kaley G et al (2000) Role of endothelium-derived nitric oxide in the regulation of cardiac oxygen metabolism: implications in health and disease. Circ Res 87(12):1108–1117

33. Roy S, Khanna S, Bickerstaff AA et al (2003) Oxygen sensing by primary cardiac fibroblasts: a key role of p21(Waf1/Cip1/Sdi1). Circ Res 92(3):264–271

34. Zhu X, Liu B, Zhou S et al (2007) Ischemic preconditioning prevents in vivo hyperoxygenation in postischemic myocardium with preservation of mitochondrial oxygen consumption. Am J Physiol Heart Circ Physiol 293(3):H1442–H1450

35. Swartz HM, Boyer S, Brown D et al (1992) The use of EPR for the measurement of the concentration of oxygen in vivo in tissues under physiologically pertinent conditions and concentrations. Adv Exp Med Biol 317:221–228

36. Ilangovan G, Zweier JL, Kuppusamy P (2004) Mechanism of oxygen-induced EPR line broadening in lithium phthalocyanine microcrystals. J Magn Reson 170(1):42–48

37. Hirata H, He G, Deng Y et al (2008) A loop resonator for slice-selective in vivo EPR imaging in rats. J Magn Reson. 190(1):124–134

38. He G, Evalappan SP, Hirata H et al (2002) Mapping of the B1 field distribution of a surface coil resonator using EPR imaging. Magn Reson Med 48(6):1057–1062

39. Kuppusamy P, Li H, Ilangovan G et al (2002) Noninvasive imaging of tumor redox status and its modification by tissue glutathione levels. Cancer Res 62(1):307–312

40. Chen YR, Deterding LJ, Tomer KB et al (2000) Nature of the inhibition of horseradish peroxidase and mitochondrial cytochrome c oxidase by cyanyl radical. Biochemistry 39(15):4415–4422

41. Gong X, Xie T, Yu L et al (2003) The ubiquinone-binding site in NADH:ubiquinone oxidoreductase from *Escherichia coli*. J Biol Chem 278(28):25731–25737

42. Teng RJ, Ye YZ, Parks DA et al (2002) Urate produced during hypoxia protects heart proteins from peroxynitrite-mediated protein nitration. Free Radic Biol Med 33(9): 1243–1249

Chapter 7

Measurement of Mitochondrial Membrane Potential and Proton Leak

Gaetano Serviddio and Juan Sastre

Abstract

The major component of mitochondrial electrochemical potential gradient of protons is the mitochondrial membrane potential ($\Delta\Psi$), and hence it is a suitable parameter for assessment of mitochondrial function. Dissipation of the mitochondrial membrane potential causes uncoupling of the electron transport through the respiratory chain and the phosphorylation reaction for ATP synthesis (proton leak). Proton leak functions as a regulator of mitochondrial reactive oxygen species (ROS) production and its modulation by uncoupling proteins, which may be involved in pathophysiology. In this report, we describe the assays for mitochondrial membrane potential and proton leak, which require a TPP+ electrode and a Clark electrode. The determination of mitochondrial peroxide production with homovanillic acid is also included.

Key words: Mitochondrial membrane potential, Mitochondrial proton leak, TPP+ electrode, Uncoupling proteins, Mitochondrial peroxide production

1. Introduction

Determination of the electrochemical potential gradient of protons across the mitochondrial inner membrane provides clear-cut experimental evidence to support the chemiosmotic theory of energy transduction (1). The major component of the electrochemical gradient is the mitochondrial membrane potential ($\Delta\Upsilon$), and therefore it is a suitable parameter for assessment of mitochondrial function. The distribution of lipid-penetrable ions has been used for the determination of $\Delta\Upsilon$ (2). These ions include fluorescent probes (rhodamine 123, tetramethylrhodamine methyl ester, JC-1 (3)), radiolabeled probes ([14C]tetraphenylphosphonium, [3H]methyltriphenylphosphonium (4)), and unlabeled probes (tetraphenylphosphonium (TTP+), and dibenzyldimethyl

D. Armstrong (ed.), *Advanced Protocols in Oxidative Stress II*, Methods in Molecular Biology, vol. 594
DOI 10.1007/978-1-60761-411-1_7, © Humana Press, a part of Springer Science+Business Media, LLC 2010

ammonium (DDA$^+$) (5)). These compounds permeate the mitochondrial membranes and distribute in accordance with the
Nernst equation. The concentration of the unlabeled probe (ion
activity) in the medium can be determined by an ion-selective
electrode. This allows determining the changes of the probe accumulation and thus the membrane potential, continuously and in
the absolute scale of millivolts. DDA$^+$ has been used as a membrane potential indicator (6), However, it has been replaced by
TTP$^+$ because the latter permeates membranes about 15 times
faster than DDA$^+$ (5).

Dissipation of the mitochondrial membrane potential causes
uncoupling of the electron transport through the respiratory
chain and the phosphorylation reaction for the synthesis of ATP
(proton leak). Multiple parameters could degrade the proton
gradient across the membrane, such as classical uncouplers (7),
fatty acids (8) and uncoupling proteins (9). The measurement of
proton leak is important for the evaluation of mitochondrial oxygen consumption and energy homeostasis. Moreover, in recent
years it has been clearly demonstrated that proton leak functions
as a regulator of mitochondrial reactive oxygen species (ROS)
production (10, 11). Indeed, an increase in proton leak mediated by uncoupling proteins leads to decreased mitochondrial
peroxide production (12, 13). However, this protective mechanism may be a double edged sword in liver diseases because up-
regulation of uncoupling protein 2 may compromise the capacity
to respond to additional acute energy demands, such as ischemia-
reperfusion (13).

Assessment of the overall kinetics of proton leak reactions
requires simultaneous determinations of oxygen consumption
and mitochondrial membrane potential (14). The following sections describe the procedure for isolation of mitochondria from
rat liver, together with the assays for mitochondrial membrane
potential, proton leak, and peroxide production.

2. Materials

2.1. Equipments

1. Clark-type oxygen electrode (Hansatech Instruments,
   Narborough Road, Pentney, King's Lynn, Norfolk, England
   PE32 1JL) equipped with Oxygraph Plus Version 1.01
   (2005).

2. TTP$^+$ electrode (WPI Europe Liegnitzer Strasse 15 D-10999
   Berlin, Germany).

3. Refrigerated Centrifuge Haeraeus Biofuge Stratos with #3335
   type rotor max load $8 \times 800 \times g$ max 20,500 rpm; #3336 type
   rotor max load $16 \times 30 \times g$ max 20,500 rpm; using Oak Ridge
   Centrifuge Tube FEP #3114-0050 and #3114-0010 (Nalge
   Nunc International).

4. Spectrophotometer Beckman Coulter DU 640B (Beckman Coulter S.p.A. Via Roma, 108 – Palazzo F1 Centro Cassina Plaza 20060 – Cassina De' Pecchi – Milano, Italy).

5. pH-meter Basic 20 (Crison (Crison Strumenti, SpA Via Villa Negro Ovest, 22 I-41012 Carpi Modena, Italy)).

6. Balance Adventurer Ohaus and Balance Sartorius (Chemie) (Chemie S.A.S. Via S. Pertini, 23/25 70010, Valenzano, Bari, Italy).

7. Magnetic agitator module.

8. Ice machine for ice production.

9. Thermostated water bath.

10. 30–50 ml Potter (with loose and tight-fitting Teflon pestles) in homogenizers TYP RW20 DZM.n IKA Labortechnik (IKA® Werke GmbH & Co. KG Janke & Kunkel-Str. 10 D-79219 Staufen Germany/Deutschland) (*see* Note 1).

2.2. Reagents and Supplies

1. Sucrose ≥ 99.5%: Sigma (St. Louis, MO), Code S-9378, MW = 342.30.

2. Albumin from bovine serum: Sigma (St. Louis, MO), Code A3294-100G.

3. Ethylene diamine tetraacetic acid disodium salt dihydrate (EDTA): Sigma (St. Louis, MO), Code E-5134, MW = 372.2.

4. Tris: Roche (F. Hoffmann-La Roche Ltd Group Headquarters Grenzacherstrasse 124 CH 4070 Basel Switzerland), Code 46149521, MW = 121.14.

5. Potassium hydroxide: Sigma (St. Louis, MO), Code P-5958, MW = 56.11.

6. Hydrochloric acid 37%: Riedel – de Haen (Fluka), Code 30721.

7. Magnesium chloride hexahydrate: Sigma (St. Louis, MO), Code M-2670, MW = 203.30.

8. K_2HPO_4: Fluka (St. Louis, MO), Code 60355, MW = 174.18.

9. KH_2PO_4: Sigma (St. Louis, MO), Code 30407, MW = 136.09.

10. Succinic acid: Sigma (St. Louis, MO), Code S-7501, MW = 118.09.

11. Pyruvic acid: Fluka (WGK, Germany), Code 15940, MW = 88.06.

12. Adenosine 5′-diphosphate sodium salt: Sigma (St. Louis, MO), Code A-2754, MW = 427.20.

13. L-Glutamic acid: BDH ITALIA s.r.l., Code 371024 T, MW = 147.02.

14. L-(–)-Malic acid: Sigma (St. Louis, MO), Code M-7397 MW = 134.09.

15. Tetraphenylphosphonium chloride: Merck (Merck Serono S.p.A. Via Casilina, 125-00176 Roma), Code 8.08244.0010, MW = 374.85.

16. Guanosine 5′-diphosphate disodium salt: Sigma-Aldrich (St. Louis, MO), Code 51060, 100 mg, MW = 487.16.

17. Malonic acid: Sigma (St. Louis, MO), Code M-1750, MW = 104.06.

18. Carbonyl cyanide phenylhydrazone (CCCP): Sigma (St. Louis, MO), Code C-2920, MW = 204.62.

19. Valinomycin: Sigma (St. Louis, MO), Code V0627, MW = 374.85.

20. Antimycin A from Streptomyces sp.: Sigma (St. Louis, MO), Code A-8674, MW = 541.6.

21. Oligomycin from Streptomyces diastatochromogenes: Sigma (St. Louis, MO), Code O-4876.

22. Rotenone: Sigma (St. Louis, MO), Code R8875, MW = 394.42.

23. Potassium cyanide (KCN): Fluka (St. Louis, MO), Code 60178, MW = 65.12.

24. Ethanol 96: Carlo Erba (Carlo Erba Reagenti SpA Strada Rivoltana km 6/7 I-20090 Rodano (MI), Code no. 308649, MW = 46.070).

25. Microfuge tubes 1.5–2 ml (Eppendorf).

26. Natural, Graduated Pipette Tips 2-10 μl ; 20–200 μl ; 200–1,000 μl.

27. 27 Pipettes 2–20 μl; 10–100 μl; 20–200 μl; 200–1,000 μl; Eppendorf.

28. Pipettes 0.2–2 μl; 2–10 μl Gilson.

29. Glass Beaker: 10 ml, 50 ml, 100 ml, 500 ml, 1 l.

30. Pasteur glass.

31. Sterile gauze.

32. Cylinders graduated glass.

33. Magnetic stirrer.

34. Magnetic stirring bars.

35. Variable automatic pipettes 10, 20, 100, 200, 1,000 μl Eppendorf.

36. Hamilton syringes 10, 25, 50, 100 μl mod. 701N.

2.2.1. Mitochondrial Isolation buffer

Isolation buffer: 0.25 M sucrose, 0.5% bovine serum albumin, 5 mM K–EDTA pH 7.4, 10 mM Tris–HCl pH 7.4.

2.2.2. Mitochondrial Respiration Buffer

Respiration buffer: 0.2 M sucrose, 3 mM $MgCl_2$, 1 mM K–EDTA pH 7.4, 10 mM potassium phosphate buffer (KPi) pH 7.0

2.2.3. Substrates and
Inhibitors (See Note 2)

1. 100 mM K–EDTA pH 7.4
2. 1 M Tris–HCl pH 7.4
3. 100 mM KPi 100/1 mM K–EDTA pH 7.0
4. 100 mM $MgCl_2$
5. 1 M K-Succinate pH = 7.4 dissolved in H_2O
6. 1 M K-Glutamate pH = 7.4 dissolved in H_2O
7. 1 M K-Malate pH = 7.4 dissolved in H_2O
8. 50 mM ADP dissolved in H_2O (concentration determined by spectrophotometer, $\varepsilon_{260nm} = 15.4$).
9. 1 M Potassium cyanide pH = 7.0 dissolved in H_2O
10. 1 mM TPP^+ dissolved in H_2O
11. 2.5 mM Rotenone dissolved in ethanol
12. 2 µg/µl Oligomycin dissolved in ethanol
13. 100 µM Antimycin dissolved in ethanol
14. 0.5 M Malonate dissolved in H_2O

2.2.4. Mitochondrial
Peroxide Production

1. Hepes-phosphate buffer: 0.1 mM EGTA, 5 mM KH_2PO_4, 3 mM $MgCl_2$, 145 mM KCl, 30 mM HEPES in distilled purified water: pH should be adjusted to 7.4 with 20% KOH.
2. Glycine–EDTA buffer: 2 M Glycine, 50 mM EDTA, 2.2 M NaOH, in distilled purified water.
3. Horseradish peroxidase (grade I): Diluted with hepes-phosphate buffer at 6 U/ml.
4. Homovanillic acid (4-hydroxy-3-methylphenylacetic acid): 1 mM homovanilic acid in hepes-phosphate buffer as stock solution. Final concentration in the assay medium will be 0.1 mM.
5. Succinate: 1 M succinic acid in hepes-phosphate buffer as stock solution; pH should be adjusted to 7.4 with 20% KOH. Final concentration in the assay medium will be 10 mM.
6. Malate: 0.25 M malic acid in hepes-phosphate buffer as stock solution; pH should be adjusted first to 6 with 1 M KOH and finally to 7.4 with 0.1 M KOH. Final concentration in the assay medium will be 2.5 mM.
7. Pyruvate: 0.5 M pyruvic acid in hepes-phosphate buffer as stock solution; pH should be adjusted first to 6 with 1 M KOH and finally to 7.4 with 0.1 M KOH. Final concentration in the assay medium will be 5 mM.
8. Hydrogen peroxide: stock commercial solution of H_2O_2 (33.3% w/v).

3. Methods

3.1. Procedure
for Isolation
of Mitochondria

The present protocol for preparation of mitochondria from rat liver is a modification of the method described by Kun and co-workers (15), based on the typical differential centrifugation procedure (*see* Fig. 7.1).

1. After removal, the liver is weighed out and immediately washed in 50 ml ice-cold mitochondrial isolation buffer (MIB). The fresh tissue is finely chopped with scissors in small fragments and washed three times to remove blood and connective tissue. The fragments are then transferred to a Potter homogenizer (Teflon pestles) in MIB (1:10 w/v) (*see* Notes 1, 3, and 4). The homogenate is centrifuged at $700 \times g$ for 10 min at +4°C to remove nuclei, intact cells and other non-sub cellular tissues and the supernatant recovered.

2. The supernatant is then filtered through four sterile gauzes, centrifuged at $6,700 \times g$ for 10 min at +4°C and the resulting pellet is carefully recovered and re-suspended in the same volume of MIB, followed by a centrifugation at $21,800 \times g$ for 10 min at +4°C.

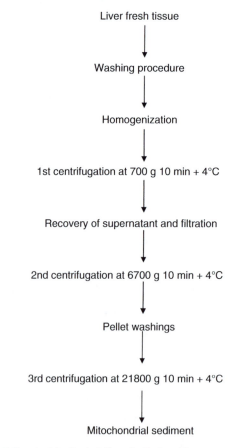

Liver fresh tissue

↓

Washing procedure

↓

Homogenization

↓

1st centrifugation at 700 g 10 min + 4°C

↓

Recovery of supernatant and filtration

↓

2nd centrifugation at 6700 g 10 min + 4°C

↓

Pellet washings

↓

3rd centrifugation at 21800 g 10 min + 4°C

↓

Mitochondrial sediment

Fig. 7.1. General flowchart for the isolation of mitochondria.

3. The resulting pellet, representing the mitochondrial sediment, is gently re-suspended in cold MIB, adding 2 μl of MIB per mg of mitochondrial pellet in order to have a final concentration of 40–50 mg mitochondrial protein per ml (*see* Note 5).

3.2. Mitochondrial Membrane Potential

3.2.1. Principle

The present protocol for determination of mitochondrial membrane potential (ΔΨ) from rat liver is a modification of the method described by Brand (4). In this method, membrane potential is determined by the distribution of a lipid-penetrable ion tetraphenylphosphonium (TPP⁺) within the mitochondrial matrix. The concentration of TPP⁺ is determined by a TPP⁺ sensitive electrode.

The following procedure describes the measurement of mitochondrial membrane potential using succinate as substrate in state 4 of respiration (without ADP) in the presence of rotenone, a selective inhibitor of complex I. Determination of membrane potential in mitochondria during state 4 of respiration may be also conducted using glutamate-malate as complex I-linked substrates or in state 3 with the addition of 100 μM ADP.

3.2.2. Procedure (Fig. 2)

1. Isolation of mitochondria as indicated above (Subheading 7.3.1).

2. Add 1 ml of mitochondrial respiration buffer (MRB) in a glass reaction vessel equipped with a TPP⁺ electrode, previously connected to a 37°C thermostated water bath (*see* Note 6).

3. Add 1 μl of rotenone (final concentration 2.5 μM, step 1 of Fig. 7.2) (*see* Notes 2 and 7).

4. Add 1 mg of mitochondrial protein (if the initial concentration of mitochondrial suspension is 50 mg/ml, add 20 μl; step 2).

5. Prepare a standard curve of TPP: add three times 1 μl of TPP⁺ (final concentration 3 μM, step 3).

6. Add 2 μl of succinate (2 mM step 4) (*see* Notes 2 and 7).

7. Add 1 μl of KCN (final concentration 1 mM) to block the respiratory chain and to end the analysis (**step 5** of Fig. 7.2) (*see* Note 8).

8. Calculation of the MMP: we use a modified Nernst equation, assuming a liver mitochondrial volume of 1.25 μl/mg of protein (16), as follows:

$$\Delta\psi(\text{mV}) = -61.5\log\left(\frac{[TPP]_i}{[TPP]_e}\right),$$

where $[TPP]_i$ is the concentration of TPP⁺ in the mitochondrial matrix and $[TPP]_e$ is the concentration of TPP⁺ outside the mitochondria.

(a) $[TPP]_i$. Use the standard curve as reported in Fig. 7.2. Measure the difference of the distance (cm) between baseline (line 1 of the figure) and lines 2, 3 and 4, and interpolate

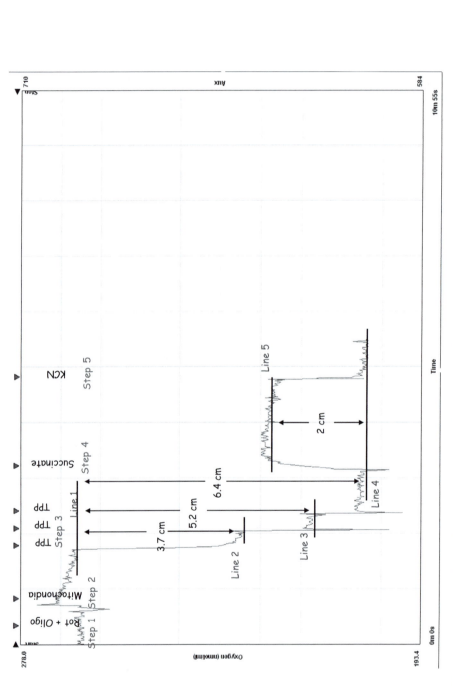

Fig. 7.2. Measurement of membrane potential in mitochondria from rat liver. The assay is performed using succinate as substrate, pre-incubating mitochondria with rotenone as complex I inhibitor. The assay allows measurement of $\Delta\Psi$ in state 4 of mitochondrial respiration (without ADP).

them in a polynomial curve as follows: line 2 – line 1 = 1 μM TPP+, line 3 – line 1 = 2 μM TPP+, line 4 – line 1 = 3 μM TPP+. By the addition of substrates to the reaction, TPP+ enters mitochondria and by interpolating the distance between line 5 and line 4 with the standard curve we obtain $[TPP]_i$. The result obtained should be expressed per mitochondrial volume (mg/ml of mitochondrial proteins × 1.25).

(b) $[TPP]_c$. 3 μM-$[TPP]_i$ expressed as μM.

3.3. Determination of Mitochondrial Proton Leak

3.3.1. Principle

The measurement of membrane potential dependence on proton leak activity in isolated mitochondria is based on the protocol described by Porter and Brand (17). Assessment of the overall kinetics of proton leak reactions requires simultaneous determinations of oxygen consumption and mitochondrial membrane potential. Oxygen consumption is determined by the Clark electrode; membrane potential is determined by the TPP+ electrode as reported above. The kinetics of proton conductance are measured in non-phosphorylating mitochondria to avoid interference of any change in the rate of respiration driving phosphorylation.

3.3.2. Procedure (Fig. 3)

1. Isolation of mitochondria as above indicated (Subheading 7.3.1).

2. Add 1 ml of mitochondrial respiration buffer (MRB) in the oxygraphic cell equipped with a Clark's and TPP+ electrodes, connected to a 37°C thermostated water bath (*see* Notes 9 and 10).

3. Add 1 μL of rotenone (final concentration 2.5 μM) and 1 μL oligomycin (step 1 of Fig. 7.3) (*see* Notes 2 and 7).

4. Add 1 mg of mitochondrial protein (if the initial concentration of mitochondrial suspension is 50 mg/ml, add 20 μl; step 2 of Fig. 7.3).

5. Prepare a standard curve of TPP+ (see above; step 3 of Fig. 7.3).

6. Oxygen consumption and TPP modification are simultaneously recorded after the addition of 10 μl succinate (step 4 of Fig. 7.3) (*see* Notes 2 and 7).

7. Titration of malonate: add 0.6 μl malonate during continuing recording as reported in step 5 of Fig. 7.3 until a complete inhibition is obtained, approximately at 3.0 mM (step 5 of Fig. 7.3) (each addition increases the malonate concentration by 0.3 mM) (*see* Notes 2 and 7).

8. Add 1 μl of KCN (final concentration 1 mM) to completely block the respiratory chain and to end the analysis (step 6 of Fig. 7.3) (*see* Note 8).

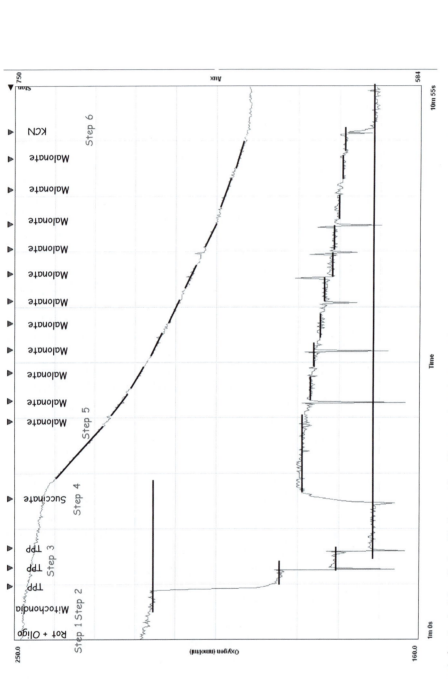

Fig. 7.3. Measurement of proton leak in mitochondria from rat liver. The assay is performed using succinate as substrate, pre-incubating mitochondria with rotenone as complex I inhibitor and olygomicin as complex V inhibitor. The upper trace represents the oxygen consumption, while the lower trace corresponds to the TPP+ uptake.

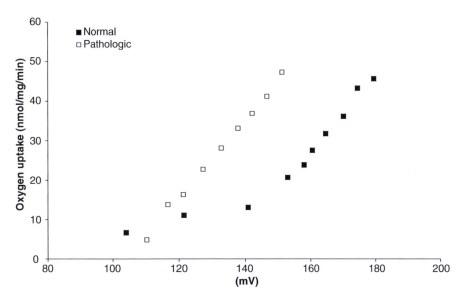

Fig. 7.4. Proton leak as a function of the mitochondrial membrane potential under physiological conditions (*filled squares*) and pathological conditions (*open squares*). Mitochondria were incubated with succinate and rotenone in presence of oligomycin.

9. Measurement of proton leak: variation of membrane potential is measured for each addition of malonate and expressed per mitochondrial volume as mentioned in Subheading 7.3.2. The oxygen uptake is also measured at the same times as slope of the curve (the available software for oxygraphic apparatus provides automatically this rate as nmolesO$_2$/ml/min). This value is normalized for mitochondrial protein concentration. Membrane potential and oxygen uptake for each point are interpolated in a graph as reported in Fig. 7.4.

3.4. Determination of Mitochondrial Peroxide Production

3.4.1. Principle

The method described here to determine the rate of mitochondrial peroxide generation is a modification of the method described by Barja (18) using the conditions described by Ruch et al. (19) to stop the reaction. In this method, the generation of hydrogen peroxide is measured following the reaction between H$_2$O$_2$ and homovanillic acid catalyzed by peroxidase, which forms a H$_2$O$_2$–homovanillic acid dimer that exhibits fluorescence at 420 nm when stimulated at 312 nm. In the original method described by Barja, the formation of the dimer was followed kinetically, however, in the present method the reaction is stopped at different times and at a basic pH to reach maximal fluorescence of the dimer.

Determination of peroxide production in mitochondria also requires incubation of isolated mitochondria with succinate as complex I-linked substrate or with pyruvate and malate as complex II-linked substrates. For each type of substrates, measurements should be performed in duplicate at 5, 10, and 15 min after the beginning of the incubation.

3.4.2. Procedure

1. Isolation of mitochondria as indicated previously (*see* Subheading 7.3.1).

2. After the last centrifugation in the previous step, remove the supernatant and add 20 μl of phosphate buffer per mg of mitochondrial pellet in order to have a final concentration around 500 mg of protein per ml.

3. Prepare six tubes for each substrate (succinate or pyruvate/ malate), two for each incubation time, this being 5, 10 and 15 min.
   (a) For succinate as substrate, add the following (*see* Note 11):
      - 3 μl of peroxidase
      - 20 μl of homovanillic acid
      - 20 μl of succinate
      - 1,457 μl of phosphate buffer (bubbled with air for at least 5 min using a Pasteur pipette just before adding it to the tube).
   (b) For pyruvate/malate as substrates, add the following:
      - 3 μl of peroxidase
      - 20 ml of homovanillic acid
      - 20 ml of pyruvate
      - 20 ml of malate
      - 1,457 μl of phosphate buffer (bubbled with air for at least 5 min using a Pasteur pipette just before adding it to the tube).

4. Put the mitochondrial suspension and the tubes containing the reagents in the bath at 37°C for 5 min with shaking.

5. Add 500 μl of mitochondrial suspension to each tube and incubate at 37°C shaking for 5, 10, or 15 min. Stop the incubation at these times by adding 1 ml of ice-cold glycine–EDTA buffer to each tube and put the tubes quickly on ice.

6. Keep an aliquot of the mitochondrial suspension to determine its protein content.

7. Spin the tubes at 13,000×*g* for 15 min. Keep the supernatants for fluorescence measurement.

8. Prepare a standard curve of hydrogen peroxide using the following final concentrations: 0, 1.0, 1.5, and 2.0 nmol/ml. The working solution of hydrogen peroxide should be prepared from the commercial stock solution of hydrogen peroxide (33.3% w/v) as follows: first a $1/10^3$ dilution in hepes-phosphate buffer is performed and the precise H_2O_2 concentration is measured using its absorbance at 230 nm $[\varepsilon(H_2O_2) = 72.4 \text{ mM}^{-1}\text{cm}^{-1}]$; subsequently a $1/10^5$ dilution is performed as the working solution to be used for the standard tubes. Caution: hydrogen peroxide solutions should be

protected from light and used immediately after dilution from the stock commercial solution. The precise concentration of hydrogen peroxide in the standard tubes can be calculated using the following formula:

$$\text{nmoles } H_2O_2/mL = \frac{A_{230} \times \mu L H_2O_2 \times 10^4}{\varepsilon_{H_2O_2} \times 1,000\,\mu L}$$

9. Measure the fluorescence of supernatants at 312 nm of excitation and 420 nm of emission using a slit width of 5.0.

10. Calculate using the standard curve as reference.

4. Results

Regarding the determination of mitochondrial membrane potential, Fig. 7.2 shows an example of this measurement in mitochondria isolated from rat liver. In this case, the calculation of $[TPP]_i$ would be as follows: line 2 – line 1 = 3.7 cm; line 3 – line 1 = 5.2 cm; line 4 – line 1 = 6.4 cm; standard curve: $y = 0.073x^2 + 0.0027x + 0.001$; line 5 – line 4 = 2 cm; from the curve we obtain 0.2964 that should be referred to the mitochondrial volume. Assuming that we have 1 mg of proteins in the vessel, $[TPP]_i$ is 370.5.

$[TPP]_e$ is calculated as: $3\,\mu M\text{-}[TPP]_i$ expressed as μM, and in this case it would be: $3 - 0.2964 = 2.7036$.

Concerning the determination of proton leak in mitochondria, Fig. 7.3 shows this measurement in rat liver mitochondria incubated with succinate as substrate, in presence of rotenone as complex I inhibitor and olygomicin as complex V inhibitor.

When mitochondria are incubated with succinate and rotenone in presence of oligomycin, Complex I and ATPase are inhibited and the proton gradient produced by succinate oxidation cannot be dissipated; as a consequence, any loss of membrane potential is due to the proton leak across the mitochondrial membrane. Succinate oxidation is progressively suppressed by titration with malonate. Under these conditions, the dissipation of membrane potential is entirely due to the proton leak and measures the proton leak as a function of the mitochondrial membrane potential. In physiological conditions (*see* Fig. 7.4), mitochondria exhibit a bi-phasic relationship between the rate of respiration and the extent of the membrane potential. At relatively low respiratory rate, a linear increase of the membrane potential is measured up to around 150 mV. Further increase of the respiratory activity results in a lower rate of enhancement of the membrane potential that remains within 180 mV. In pathological conditions (*see* Fig. 7.4), mitochondria show an increased slope, consistently with uncoupling.

5. Notes

1. Potter–Elvehjem homogenizers consist of Teflon pestles that are driven into a glass vessel manually or mechanically at very low rotation rate (<250 rpm).

2. All substrates and inhibitors are stored at –20°C, except for TPP⁺ that is stored at room temperature.

3. In the first phases of liver mitochondria isolation, time is of great importance for maintaining mitochondria tightly coupled.

4. All the isolation procedures should be performed "on ice."

5. Mitochondria should not be vortexed before analyses.

6. It is advisable to frequently change the Teflon membrane in the glass reaction vessel.

7. During experiments, substrates and inhibitors should be kept on ice and vortexed before use.

8. Wash the glass reaction chamber with distilled water before each assay; after using rotenone, it is advisable to utilize ethanol.

9. The speed of the stirrer of oxygraphic apparatus should be <60 rpm.

10. Avoid electromagnetic source near the oxygraphic apparatus during experiments.

11. In order to determine the rate of mitochondrial peroxide at complex III, 2 μM rotenone (final concentration) should be present -together with complex-II linked substrates- to avoid the black flux of electrons towards complex I. Since rotenone is dissolved in pure ethanol, a control should be performed by adding the same amount of ethanol to the mitochondrial suspension without any substrate.

Acknowledgments

The authors thank Dr. N. Capitanio for critical suggestions and Drs. F. Bellanti and R. Tamborra for technical assistance.

The authors also acknowledge the financial support obtained by Grants from Ministerio de Educación y Ciencia (SAF2006-06963 and Consolider CSD-2007-00020) to J. S.

References

1. Mitchell P, Moyle J (1969) Estimation of membrane potential and pH difference across the cristae membrane of rat liver mitochondria. Eur J Biochem 7:471–484

2. Hoek JB, Nicholls DG, Williamson JR (1980) Determination of the mitochondrial proton-motive force in isolated hepatocytes. J Biol Chem 255:1458–1464

3. Plasek J, Vojtiskova A, Houstek J (2005) Flow-cytometric monitoring of mitochondrial depolarisation: from fluorescence intensities to millivolts. J Photochem Photobiol B 78:99–108

4. Brand MD (1995) The measurement of mitochondrial proton motive force. In: Brown GC, Cooper EC (eds) Bioenergetics: A practical approach. IRL, Oxford, pp 39–62

5. Kamo N, Muratsugu M, Hongoh R, Kobatake Y (1979) Membrane potential of mitochondria measured with an electrode sensitive to tetraphenyl phosphonium and relationship between proton electrochemical potential and phosphorylation potential in steady state. J Membr Biol 49:105–121

6. Shinbo T, Kamo N, Kurihara K, Kobatake Y (1978) A PVC-based electrode sensitive to DDA+as a device for monitoring the membrane potential in biological systems. Arch Biochem Biophys 187:414–422

7. Terada H (1990) Uncouplers of oxidative phosphorylation. Environ Health Perspect 87:213–218

8. Kohnke D, Ludwig B, Kadenbach B (1993) A threshold membrane potential accounts for controversial effects of fatty acids on mitochondrial oxidative phosphorylation. FEBS Lett 336:90–94

9. Boss O, Muzzin P, Giacobino JP (1998) The uncoupling proteins, a review. Eur J Endocrinol 139:1–9

10. Negre-Salvayre A, Hirtz C, Carrera G, Cazenave R, Troly M, Salvayre R, Penicaud L, Casteilla L (1997) A role for uncoupling protein-2 as a regulator of mitochondrial hydrogen peroxide generation. FASEB J 11:809–815

11. Echtay KS, Esteves TC, Pakay JL, Jekabsons MB, Lambert AJ, Portero-Otin M, Pamplona R, Vidal-Puig AJ, Wang S, Roebuck SJ, Brand MD (2003) A signalling role for 4-hydroxy-2-nonenal in regulation of mitochondrial uncoupling. EMBO J 22:4103–4110

12. Nicholls DG, Locke RM (1984) Thermogenic mechanisms in brown fat. Physiol Rev 64(1):1–64

13. Serviddio G, Bellanti F, Tamborra R, Rollo T, Capitanio N, Romano AD, Sastre J, Vendemiale G, Altomare E (2008) UCP2 induces mitochondrial proton leak and increases susceptibility of NASH liver to ischemia/reperfusion injury. Gut 57(7):957–65

14. Brand MD (1990) The proton leak across the mitochondrial inner membrane. Biochim Biophys Acta 1018:128–133

15. Kun E, Kirsten E, Piper WN (1979) Stabilization of mitochondrial functions with digitonin. Methods Enzymol 55:115–118

16. Brown GC, Brand MD (1988) Proton/electron stoichiometry of mitochondrial complex I estimated from the equilibrium thermodynamic force ratio. Biochem J 252(2): 473–479

17. Porter RK, Brand MD (1993) Body mass dependence of H+leak in mitochondria and its relevance to metabolic rate. Nature 362:628–630

18. Barja G (1998) Measurement of mitochondrial oxygen radical production. In: Yu BP (ed) Methods in aging research. CRC, Boca Raton, FL, pp 533–548

19. Ruch W, Cooper PH, Baggiolini M (1983) Assay of H_2O_2 production by macrophages and neutrophils with homovanillic acid and horse-radish peroxidase. J Immunol Methods 63(3):347–57

Chapter 8

Determination of Erythrocyte Fragility as a Marker of Pesticide-Induced Membrane Oxidative Damage

Bechan Sharma, Devendra K. Rai, Prashant Kumar Rai, S.I. Rizvi, and Geeta Watal

Abstract

Erythrocytes are readily available cells and a good model system to study the health status of individuals with pathologic complications. It can also serve as a meaningful target to study toxicant/xenobiotic-induced damages. We have prepared different concentrations of a carbamate pesticide (carbofuran) and carried out experiments to determine its toxicity on erythrocytes in terms of mean erythrocyte fragility (MEF). We observed a significant alteration in the osmotic fragility upon treatment with carbofuran. In our earlier studies we have observed a good correlation between OF and OS in diabetic subjects. Study reveals OF as a potential biomarker of oxidative membrane damage in pathologic conditions as well as toxicant/xenobiotic/pesticide-induced oxidative membrane damage to erythrocytes.

Key words: Xenobiotics, Erythrocyte, Osmotic fragility, Diabetes

1. Introduction

Toxicity of many xenobiotics is associated with the production of free radicals, which are not only toxic themselves but also are responsible for various pathophysiological complications. For example, Alzheimer's and Parkinson's diseases are caused by reactive oxygen species (ROS)-mediated neuronal damage (1, 2). Extended occupational exposure to many environmental chemicals including heavy metals such as Pb, Cd, and pesticides may also cause oxidative stress, as a mechanism underlying the adverse effects in biological systems (3). Pesticide induced oxidative stress has been a focus of toxicological researches for last one decade as a possible mechanism of toxicity. Different classes of pesticides induce production of ROS and oxidative damage (4).

D. Armstrong (ed.), *Advanced Protocols in Oxidative Stress II*, Methods in Molecular Biology, vol. 594
DOI 10.1007/978-1-60761-411-1_8, © Humana Press, a part of Springer Science+Business Media, LLC 2010

Available reports suggest that the enzymes activities associated with antioxidant defense are altered by insecticides/pesticides both in vitro and in vivo (5, 6). Induction of membrane lipid peroxidation has also been one of the contributing factors in inducing pesticide oxidative stress (7). In blood, normal erythrocyte function depends largely upon the integrity of erythrocyte membrane. The normal erythrocyte membrane is remarkable for its durability but the membrane's tolerance gets compromised under certain pathologic conditions. The study of erythrocyte fragility (EF) pattern gives an idea about membrane's integrity; therefore, erythrocyte fragility is frequently used as a measure of erythrocyte tensile strength. EF and OS are also known to be altered in diabetes mellitus (8, 9). The alteration has been associated with changes in membrane fluidity and increased oxidative stress. It depends on the movement of water into the cell and is related to cellular deformability. The factors, which determine EF, are complicated and include oxidative stress (10), cation content (11), cellular age (12), individual age, and changes in cell shape (13). The importance of EF estimation lies in the fact that this hematological parameter gives the information about the total status of red cell metabolism and membrane stability (14). In a recent study in patients on peritoneal dialysis and hemodialysis, it has been shown that there is a good correlation between erythrocyte hemolysis under hypotonic condition (osmotic fragility) and oxidative stress (10). As, EF can be easily monitored and it gives an overall estimate of effects of oxidative injury to the membrane due to preponderance of oxidizable fatty acids and protein motifs in the erythrocyte membrane, the current protocol provides a simple method in determining xenobiotic induced oxidative damage to erythrocyte membrane. The salient features of the protocol are illustrated in Notes.

2. Materials

2.1. Reagents

1. Stock solution of NaCl (5% in Phosphate buffer, pH 7.0).
   - 5 g NaCl, 1.365 g NaH_2PO_4, 0.243 g Na_2HPO_4, is dissolved in 100 ml of distilled water.
2. Heparin containing vials.
3. Human blood (from healthy volunteers).

2.2. Equipments

1. ELICO UV-VIS Double beam spectrophotometer.
2. Sigma bench top cooling centrifuge.

3. Methods

3.1. Sample Preparation

5 ml of venous blood is collected in vials containing heparin (10 U/ml) and stirred gently.

3.2. Serial Dilution of NaCl

Stock solution of NaCl is diluted as mentioned in the following table:

S. No.	Volume of 5% NaCl stock (ml)	Volume of distilled water (ml)	Final concentration of NaCl (%)
1.	0.0	50	0
2.	0.5	49.5	0.1
3.	1.0	49.0	0.2
4.	1.5	48.5	0.3
5.	2.0	48.0	0.4
6.	2.5	47.5	0.5
7.	3.0	47.0	0.6
8.	3.5	46.5	0.7
9.	4.0	46.0	0.8
10.	4.5	45.5	0.9

3.3. Procedure

1. Dissolve the two concentrations of carbofuran, 50 and 100 g/l carbofuran in 1% (v/v) DMSO. 500 µl of each is taken and incubated with 5.0 ml blood at room temperature (26 ± 2°C) for 30 min. The control blood sample contains equal volume of DMSO (0.1%, v/v) in the place of the pesticide.

2. The 50 µl aliquot of blood from each incubated sample is taken and mixed with 5 ml NaCl solution of each dilution. For the patients with pathologic complications 50 µl of blood is directly mixed with 5 ml NaCl at different dilutions.

3. The blood is mixed in NaCl by gently inverting the tubes (each 3 times).

4. The tubes are incubated for 30 min at room temperature by keeping the tubes in upright position.

5. The contents of tube are remixed after incubation by gentle inversion.

6. Centrifuged the contents of tubes at $5,000 \times 9$ for 5 min at room temperature.

7. Supernatant is collected and the optical density is recorded at 540 nm wavelength against distilled water used as a blank.

8. The experiments are run in the replicates of three or more in order to obtain statistically reliable data.

4. Results

The concentrations of NaCl corresponding to 50% hemolysis of erythrocytes for each treatment group are calculated by extrapolating the results of Fig. 8.1 in order to measure the effect of pesticide treatment on the erythrocyte stability. The mean NaCl concentrations corresponding to 50% hemolysis of erythrocytes (expressed as mean erythrocyte fragility, MEF) in the three groups such as control, CF 1, and CF2 were found to be 0.66, 0.675, and 0.68, respectively. The results suggest that carbofuran induced erythrocyte hemolysis at both the doses.

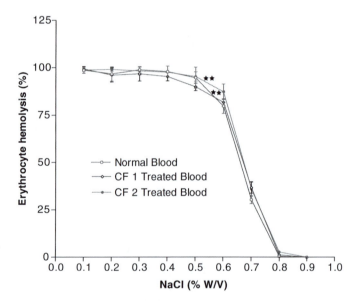

Fig. 8.1. The erythrocyte hemolysis is determined as described in Subheading 8.3. The highest O.D. is taken as 100% hemolysis and rest are calculated with respect to it. Mean erythrocyte fragility (MEF) is 50% hemolysis. The concentrations of NaCl corresponding to MEF for each treatment group are calculated by extrapolating the results from the individual erythrocyte fragility curves in order to measure the effect of pesticide treatment on the erythrocyte stability. The data are shown in the table (*inset*). CF1 and CF2 represent carbofuran concentrations (50 and 100 g/l) in 1% (v/v) DMSO, respectively.** Indicates the value being significantly different from control group at $p < 0.01$.

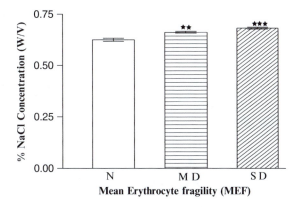

Fig. 8.2. *N* represents the MEF from erythrocyte of Normal individual. MD and SD, however are the MEF from the blood of Mild diabetic and severe diabetic individual respectively. Signs ** and *** depicts that the value is significantly different from normal at $p < 0.01$ and, $p < 0.001$ respectively.

Figure 8.2 shows the alteration in erythrocyte OF during diabetes mellitus. Significantly, we observe a correlation between MEF and severity of diabetes as evidenced from fasting blood sugar (FBS).

4.1. Precautions

1. Blood should be collected under cold conditions.

2. The processing should be carried out within 2 h of collection.

3. Blood should be withdrawn using with a broad gauge needle.

4. Mixing of blood should be gentle in order to avoid the accidental hemolysis of erythrocytes.

5. After mixing blood with buffer, the tubes should not be shaken.

6. Timings of incubations must be monitored strictly, otherwise, it would lead to poor and inconclusive result.

7. Temperature of incubation should be maintained consistently.

8. The experiment should preferably be carried out in centrifuge tube to eliminate the requirement of transferring the contents from test tube to centrifuge tube for centrifugation.

5. Notes

1. The merit of protocol lies in the fact that it can be easily performed in any clinical laboratory and it does not involve very specific and expensive reagents or sophisticated equipments.

2. The experiment gives a first hand estimate of status of membrane.

3. The changes are clearly distinguishable and so the effects of treatment can be clearly made out.

4. The results may be used as an early marker of xenobiotic toxicity to human and animal systems.

Acknowledgments

DKR and PKR are thankful to CSIR and ICMR, respectively, for the award of SRF to them.

References

1. Butterfield DA (2002) Amyloid beta-peptide (1–42)-induced oxidative stress and neurotoxicity: implications for neurodegeneration in Alzheimer's disease brain. Rev Free Rad Res 36:1307–1313

2. Butterfield DA, Lauderback CM (2002) Lipid peroxidation and protein oxidation in Alzheimer's disease brain: potential causes and consequences involving amyloid beta-peptide-associated free radical oxidative stress. Free Rad Biol Med 32:1050–1060

3. Akhgari M, Abdollahi M, Kebryaeezadeh A (2003) Biochemical evidence for free radical-induced lipid peroxidation as a mechanism for subchronic toxicity of malathion in blood and liver of rats. Hum Exp Toxicol 22: 205–211

4. Bagchi D, Hassoun EA, Bagchi M, Stohs SJ (1993) Protective effects of antioxidants against eldrin–induced hepatic lipid peroxidation, membrane microviscosity DNA damage and excretion of urinary lipid metabolites. Free Rad Biol Med 15:217–222

5. Gultekin F (2000) The effect of organophosphate insecticide chlorpyrifos-ethyl on lipid peroxidation and antioxidant enzymes (*in vitro*). Arch Toxicol 74:533–538

6. Gultekin F, Delibas N, Yasar S, Kilinc I (2001) In vivo changes in antioxidant systems and protective role of melatonin and a combination of vitamin C and vitamin E on oxidative damage in erythrocytes induced by chlorpyrifos-ethyl in rats. Arch Toxicol 75:88–96

7. Rai DK, Sharma B (2007) Carbofuran induced oxidative stress in mammalian brain. Mol Biotechnol 37:66–71

8. Suhail M, Rizvi SI (1987) Red cell membrane $(Na^+ + K^+)$-ATPase in diabetes mellitus. Biochem Biophys Res Comm 146:179–186

9. Uzum A, Topark O, Gumustak MK, Ciftci S, Sen S (2006) Effect of vitamin E therapy on oxidative stress and erythrocyte osmotic fragility in patients on peritoneal dialysis and hemodialysis. J Nephrol 19:739–745

10. West IC (2001) Radicals and oxidative stress in diabetes. Dia Med 47:171–180

11. Mentzer WC, Clark MR (1983) In: Novontony A (ed) Biomembranes pathological membranes, vol 2. Plenum, New York, p 79

12. Rifkind JM, Araki K, Hadley EC (1983) The relationship between the osmotic fragility of human erythrocytes and cell age. Arch Biochem Biophys 15:582–589

13. Chan TK, Lacelle PL, Weed RI (1975) Slow phase hemolysis in hypotonic electrolyte solutions. J Cell Physiol 85:47–57

14. US Govt. Printing Office (1963) Department of army clinical hematology. US Govt. Printing Office, Washington, DC, p 427

Chapter 9

Using *N,N,N',N'*-tetramethyl-*p*-phenylenediamine (TMPD) to Assay Cyclooxygenase Activity In Vitro

Nenad Petrovic and Michael Murray

Abstract

Prostaglandin endoperoxide synthase (PGH synthase), also known as cyclooxygenase (COX), was identified over 30 years ago and is the key enzyme in the pathway by which arachidonic acid is converted to the range of biologically active lipid mediators known as the prostanoids that participate in numerous physiological processes. The need for the development of new and improved COX inhibitors as potential therapeutics also drives the need for rapid, reliable, and inexpensive assays of COX activity. Colorimetric assays are often the preferred methods of enzyme analysis since they may be readily adapted to simple microplate formats that require relatively inexpensive and widely available instrumentation. The use of *N,N,N',N'*-tetramethyl-*p*-phenylenediamine (TMPD) in high throughput microplate assays of COX activity could become the approach of choice in the screening of potential therapeutics that inhibit COX activity in vivo. Considering that TMPD is also a potential substrate for most, if not all, heme peroxidases, it is anticipated that this agent could find increasing application in the future.

Key words: Enzyme assays, Cyclooxygenases, TMPD, *N,N,N',N'*-tetramethyl-*p*-phenylenediamine

1. Introduction

1.1. Role of COX in Human Health and Disease

COX catalyzes the conversion of arachidonic acid to prostaglandin (PG) G_2 and PGH_2. PGH_2 is subsequently converted to a variety of eicosanoids that include PGE_2, PGD_2, PGF_{2a}, prostacyclin (PGI_2), and thromboxane A_2 (TXA_2). The profile of prostanoids produced within particular cells is dependent on the enzymes that are expressed in a cell-specific manner. For example, endothelial cells primarily produce PGI_2, whereas platelets mainly produce TXA_2, because these cells express high levels of prostacyclin synthase and thromboxane synthase, respectively (1).

Two COX isoforms have been identified and are named COX-1 and COX-2 based on the order in which they were discovered.

D. Armstrong (ed.), *Advanced Protocols in Oxidative Stress II*, Methods in Molecular Biology, vol. 594
DOI 10.1007/978-1-60761-411-1_9, © Humana Press, a part of Springer Science+Business Media, LLC 2010

Both COX isoenzymes are integral membrane glycoproteins that are present as homodimers in the membranes of the nucleus and endoplasmic reticulum. COX-1 and COX-2 have subunit molecular weights of 70 and 72 kDa. Both COX isozymes have similar tertiary structures and perform essentially the same catalytic reaction (2). Under normal conditions most tissues express COX-1, with very low to undetectable levels of COX-2, but the COX-2 isoform is inducible by stress- and growth-related stimuli. The prostanoids produced by COX-2 have generally been shown to have pro-inflammatory roles. In contrast, most of the homeostatic functions of prostanoids are mediated by the constitutively expressed COX-1. COX-1 and COX-2 also differ in their mRNA stability and in the preferential utilization of arachidonic acid substrate pools in tissues (3–5).

The COX isoenzymes and their eicosanoid products play important functional roles in many physiological processes. In humans, prostaglandins are involved in diverse functions ranging from the inflammatory response and blood clotting to ovulation and initiation of labor. In spite of the fact that drugs such as aspirin have been used widely for over a century, and are now known to inhibit COX activity, the role of the COX in normal and disease physiology has only become clearer relatively recently. In 1971 Vane published his seminal observations indicating that the ability of nonsteroidal antiinflammatory drugs (NSAIDs) to suppress inflammation is probably due to their inhibitory actions on COX enzymes (6). Such inhibition suppresses production of proinflammatory PGs at the site of injury. Following this discovery, scientists and clinicians have used NSAIDs to dissect the critical role of the COX enzymes, and the eicosanoids generated by these pathways, in normal physiology and disease states.

Due to the documented importance of effective COX inhibition in treating numerous pathological conditions, the search for new and improved COX inhibitors with greater specificity is ongoing. Consequently, design of more convenient screening assays for COX activity is of great importance.

1.2. The Cox Reaction

Arachidonic acid, the most important physiological precursor of eicosanoid biosynthesis, is stored in membrane phospholipids in an esterified form and can be released intracellularly by the hydrolytic action of phospholipases A_2 (PLA_2) and/or phospholipase C (PLC). Within the cell arachidonic acid is converted by a range of enzymes, including COX, to biologically active eicosanoid derivatives. COX actually catalyzes two sequential enzymatic reactions the *bis*-oxygenation of arachidonic acid to prostaglandin G_2 (PGG_2; the cyclooxygenase reaction) and the reduction of PGG_2 to prostaglandin H_2 (PGH_2; the peroxidase reaction). Thus, COX catalyzes the addition of molecular oxygen at carbon 11 of arachidonic acid to form a peroxy intermediate, followed by

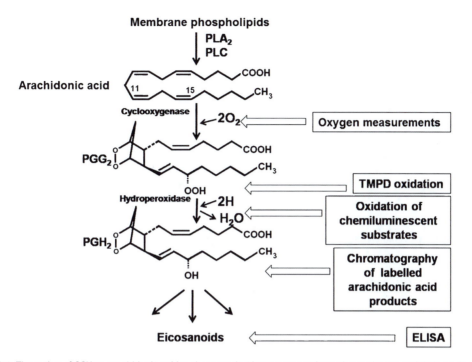

Fig. 9.1. The action of COX on arachidonic acid and assays that have measured reaction pathways and their products.

rearrangement to the cyclic endoperoxide and introduction of a second molecule of oxygen at carbon 15 to form prostaglandin G_2 (PGG_2). Conversion of the endoperoxide PGG_2 to the corresponding alcohol (PGH_2) employs glutathione as the reducing cofactor. PGH_2 may then be further metabolized to a range of eicosanoid derivatives (prostaglandins, prostacyclins and thromboxanes) by the appropriate synthase (Fig. 9.1).

1.3. Methods Used in the Detection of COX Activity

Various methods for the determination of COX activity have been developed and are based on the dual enzymatic activities exhibited by the enzyme: cyclooxygenation and peroxidation. Some of these assays measure reaction intermediates, whereas others detect the final products (eicosanoids) of pathways in which COX catalyzes the rate-limiting step (Fig. 9.1). A range of analytical methodologies have been applied to the measurement of COX activity. Thus, oxygen consumption during cyclooxygenation has been measured using an oxygraph equipped with an oxygen electrode. Various artificial electron donors have been applied in the measurement of the peroxidation reaction; these generate colored, fluorescent or chemiluminescent products. The peroxidation reaction requires a second substrate which is co-oxidized during hydroperoxide reduction. COX activity may also be estimated from the turnover of radioactive arachidonic acid or by measuring its conversion to labeled PGs (Fig. 9.1). Typically, in vitro assays of COX activity contain the most efficient substrate, arachidonic acid, an electron

Table 9.1
Methods used to measure COX activity

COX reaction	Detection method	References
Cyclooxygenation	Oxygen consumption	(7)
Peroxidation	5-Phenyl-4-pentenyl hydroperoxide (PPHP) reduction (colorimetry)	(8)
	N,N,N',N'-tetramethyl-p-phenylenedi-amine (TMPD) oxidation (colorimetry)	(9)
	10-Acetyl-3,7-dihydroxyphenoxazine oxidation (fluorometry)	(10)
	Luminol oxidation (luminescence)	(11)
Cyclooxygenation and peroxidation	Chromatography of radioactively labelled products	(12)
Prostaglandin synthesis	Enzyme-linked immunoassay or radioimmunoassay	(13–15)

donor molecule (usually a low concentration of phenol, about 1 mM, or more specialized electron donors) and hematin. COX contains Fe^{3+}-protoporphyrin IX as a cofactor which may dissociate from the protein during its purification, resulting in a mixture of apo- and holo-enzymes. Therefore, hematin is added to the reaction mixture (usually at a concentration of 1 µM) in order to reconstitute maximal enzyme activity. The principal methods for the assay of COX activity are listed in Table 9.1.

1.4. Direct Measurement of PG Production

The most widely used approach for the estimation of total COX activity in live cells and tissues is the quantification of prostanoids. Since the COX reaction is considered to be the rate-limiting factor in prostanoid synthesis (16–18), measurement of PGE_2 formation (the major prostanoid synthesized by the pathway) gives a relative indication of intracellular COX activity. Enzyme immunoassays (EIA) are based on the competition of binding of specific tracer prostanoids, including prostanoid–acetylcholinesterase conjugates, and those in test samples with antiprostanoid antibodies. Radioimmunoassays (RIA) measure the binding of antiprostanoid antibodies to radioactively labeled prostanoids (tracer) and prostanoids in test sample. The quantity of tracer molecules bound to the antibodies is inversely proportional to the concentration of competing (unlabeled) prostanoids in the sample. The specific interaction is then detected either by measuring enzyme activity in antibody complexes (EIAs), or radioactivity in RIAs (19, 20).

1.5. Ex Vivo Prostaglandin Measurements

Individual COX activity may also be determined in patients by use of the ex vivo whole blood assay (15). In these assays, PGE_2 or thromboxane B_2 levels in the whole blood (challenged with lipopolysaccharide in vitro) are measured by EIA. This approach can be used to assess the efficacy of selective COX-2 inhibitors in clinical trials.

These assays detect prostanoids in intact cells or tissues and are not intended to characterize the COX enzymes. More specific systems for the determination of COX activity in vitro have employed RIAs to measure PGE_2 formation in microsomal membrane protein preparations containing both COX and PGE_2 synthase (21). Several PGE_2 synthases have been identified, with the microsomal PGE_2 synthase-1 emerging as a key enzyme in the formation of PGE_2. This assay is relatively complex and is dependent on the quality of the microsomal preparations, because this influences the PGE_2 synthase-1 activity of the sample.

1.6. Detecting Radiolabeled COX-Derived Products of Arachidonic Acid

An early approach for the direct measurement of COX activity in vitro involved the biotransformation of $[1-^{14}C]$ labeled arachidonic acid. After the reaction, unmetabolized arachidonic acid was separated from the products by column chromatography on silica gel 60. The COX products were then eluted with organic solvents and the radioactivity in the samples was quantified (12). Adaptation of high-performance liquid chromatographic methods, often in conjunction with combinations of mobile phases and gradient elution, has enabled the resolution of complex mixtures of PGs and other eicosanoids.

1.7. Oxygen Consumption

Oxygen consumption during the first steps of the COX reaction may be monitored using an oxygraph equipped with an oxygen electrode. The reaction is initiated with arachidonate and the initial rate of oxygen consumption is charted on graph paper (7); the initial rate of oxygen consumption decreases during the reaction because of oxygen incorporation into arachidonic acid. This assay is continuous and has the advantage that it directly reflects COX activity, but is not suitable for high-throughput assays, requires complex instrumentation and uses relatively high amounts of enzyme.

Most of the currently used methods for the measurement of COX activity detect the enzyme's peroxidase activity in which PGG_2 is reduced to PGH_2.

1.8. Co-oxidation Reactions

An important property of COX is that a range of suitable substrates can replace PGG_2 in the peroxidation step. Thus, COX-mediated peroxidation activity may be determined conveniently by the inclusion of reducing substrates that generate products that are readily quantified. Reduction of 5-phenyl-4-pentenyl hydroperoxide (PPHP) generates the corresponding 5-phenyl-4-pentenyl alcohol (PPA; Fig. 9.2) (8) that may be separated by reverse-phase C_{18} HPLC and quantified using its absorbance at 252 nm.

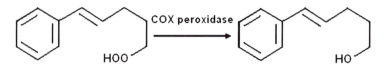

Fig. 9.2. Detection of COX peroxidase activity with PHPP.

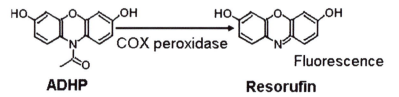

Fig. 9.3. Detection of COX peroxidase activity with ADHP.

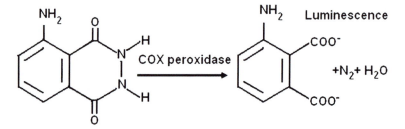

Fig. 9.4. Detection of COX peroxidase activity with Luminol.

Similarly, the reduction of PGG_2 by ADHP (10-acetyl-3,7-dihydroxyphenoxazine) produces the highly fluorescent compound resorufin. ADHP itself is colorless and non-fluorescent, but is converted during co-oxidation to the red colored, highly fluorescent product resorufin (Fig. 9.3). ADHP was initially used to detect horseradish peroxidase activity (9) and has been adapted by Cayman Chemical (Ann Arbor MI, USA) for screening COX activity in biological samples.

Co-oxidation of the chemiluminescent substrate luminol is also used to detect the peroxidase activity of COX; light emission is recorded with a luminometer (10) (Fig. 9.4).

The artificial electron donor N,N,N',N'-tetramethyl-p-phenylenediamine (TMPD) undergoes co-oxidation by PGG_2 to a blue product (oxidized TMPD) that has an absorbance maximum at 590 nm (Fig. 9.5). TMPD is a readily oxidizable compound that serves as a reducing cosubstrate for heme peroxidases (22). TMPD undergoes one-electron oxidation by heme peroxidase with two moles of TMPD oxidized per mole of hydroperoxide reduced by the peroxidase.

The method has been described previously (9, 23), and subsequently modified for small volumes added in a microplate

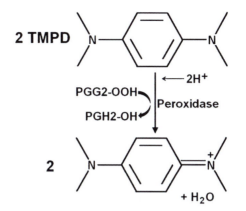

Fig. 9.5. Detection of COX peroxidase activity with TMPD.

setup (24). TMPD can be used with both crude (cell lysates/tissue homogenates) and purified enzyme preparations, although it should be noted that antioxidants in the sample may interfere with the assay.

Although this is an indirect method, the oxidation of TMPD has been shown to accurately reflect the rate of conversion of arachidonic acid to PGH_2. The TMPD assay has been used to evaluate potential new COX inhibitors in a microplate format (26). Peroxidase-mediated TMPD co-oxidation has been developed commercially by Cayman Chemical. Consequently, the effects of detergents on COX activity (27) and the characterization of COX-2 specific inhibitors DuP 697 and NS-398 (28), as well as the marketed COX-2 inhibitor celecoxib (29) have been undertaken using the microplate-adapted TMPD assay. The use of TMPD to detect peroxidase activity in polyacrylamide gels illustrates the further potential of this agent in the detection of a range of hemoproteins (25).

2. Materials

2.1. Equipment

1. Standard 96-well clear plastic plates. They could be obtained from a variety of suppliers.

2. Visual range spectrophotometer that could accommodate 96-well plates.

3. Automatic multi-plate mixer (shaker) (optional).

2.2. Reagents

1. N,N,N',N'-tetramethyl-p-phenylenediamine (TMPD) – Sigma Aldrich.

2. Recombinant COX-2 (positive control or the source of enzyme in substrate analysis) – Cayman Chemical.

3. PUFA – (arachidonic and eicosapentaenoic) – Cayman Chemicals.

4. Complete COX assay kit alternative (including TMPD) – Cayman Chemicals.

5. Hematin – Sigma Aldrich.

6. Dimethyl sulfoxide (DMSO) – Sigma Aldrich.

7. Tris(hydroxymethyl)aminomethane plus HCl – Sigma Aldrich.

3. Methods

3.1. Performance of the TMPD Assay

1. Stock solutions – TMPD stock solution (20 mM) was made in dimethyl sulfoxide (DMSO) whereas hematin stock solution (40 mM) was made in 1 M NaOH. PUFA were purchased from Cayman Chemical as premade solutions in ethanol (*see* Note 1). The COX-2 standard was also purchased from Cayman (*see* Note 2).

2. Reaction mixture and conditions – All reagents were mixed and reactions initiated by the addition of a combination of arachidonic acid (final concentration 50 μM, *see* Note 3) and TMPD (final concentration 100 μM, *see* Note 4). By varying substrate concentrations from 0.56 to 36 μM, we have found that 10 μM is sufficient in this assay (Fig. 9.6). Although very short reaction times (up to 2 min) have been used in reported studies, we have found that 10 min gives reproducibly good results and is technically much easier to perform (Fig. 9.6). K_m values for the COX reaction were consistent (289.1±9.5 nM) over reaction times between 2 and 20 min (Fig. 9.7). Reaction mixture also incorporates 1 μM hematin,

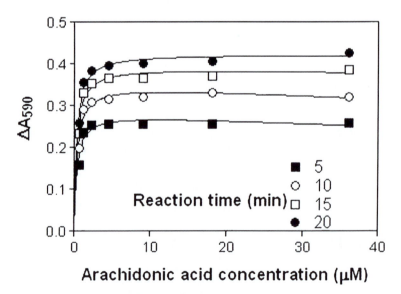

Fig. 9.6. Dependence of TMPD co-oxidation on arachidonic acid concentration and reaction time.

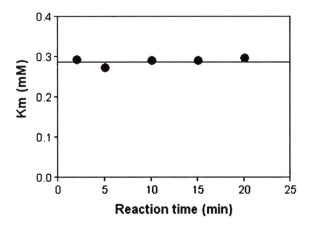

Fig. 9.7. Stability of K_m values for COX activity over a range of reaction times.

Tris/HCl buffer (100 mM, pH 8) in a final volume of 0.2 mL (*see* Notes 5 and 6).

3. Measuring of the oxidized TMPD absorbance – TMPD oxidation is monitored spectrophotometrically in a 96-well-plate format at 590 nm (*see* Notes 7 and 8).

4. Data interpretation – The absorbance (590 nm) in the absence of PUFA should be subtracted from the absorbance in the presence of PUFA and the net enzyme activity (expressed as the change in OD at 590 nm, ΔA_{590}) is then calculated. The reaction rate can be determined using the TMPD extinction coefficient (adjusted for the path length of the solution in the microplate) of $0.00826\ \mu M^{-1}$ (9). One unit is defined as the amount of enzyme that will catalyze the oxidation of 1.0 nmol of TMPD per minute at 25 C. COX activity in units is calculated as follows:

$$\text{COX activity} = \Delta A_{590} \div 10\ \text{min} \div 0.00826\mu\ M^{-1} \div 0.01^{\#}\ \text{mL} \times 1{,}000 \div 2^{\$} = \text{nmol/min/mL (U/mL)}$$

ΔA_{590} is the difference in absorbance between wells containing complete mixture and control wells without PUFA

[\$]Two molecules of TMPD are required for the reduction of PGG_2 to PGH_2

[#]0.01 mL is the enzyme sample volume

4. Results

4.1. Testing COX Substrates with TMPD Assay

In addition to the testing of potential COX inhibitors, the TMPD assay may also be used to evaluate alternate COX substrates. Although arachidonic acid is the most widely studied COX substrate, other types of polyunsaturated fatty acids (PUFA), such

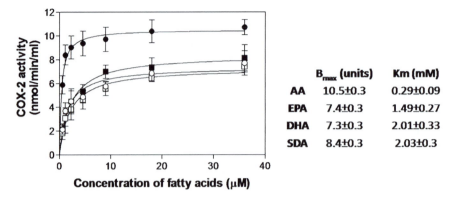

	B_{max} (units)	Km (mM)
AA	10.5±0.3	0.29±0.09
EPA	7.4±0.3	1.49±0.27
DHA	7.3±0.3	2.01±0.33
SDA	8.4±0.3	2.03±0.3

Fig. 9.8. In vitro activity of bovine COX-2 determined using AA, SDA, DHA and EPA as substrates. Enzyme kinetic parameters (V_{max} and K_m) were calculated by non-linear regression: *filled circle* AA; *open circle* EPA; *filled square* SDA; *open square* DHA ((30), permission granted by Blood journal).

as ω-3 PUFA, are also converted by COX into the corresponding prostanoids. Our findings suggest that ω-6 and ω-3 PUFA compete for enzymes involved in PUFA biotransformation including COX ((30), *see* Note 9). It is widely believed that PUFA bioconversion enzymes have a greater affinity for ω-3 PUFA so that their biotransformation is favored when the dietary ω-3 PUFA intake is high. Such an effect could lead to the "competitive inhibition" of ω-6 PUFA metabolism by ω-3 PUFA. In accord with this possibility, increased intake of the ω-3 PUFA eicosapentaenoic acid (EPA) in man has been shown to suppress the synthesis of arachidonic acid-derived eicosanoids and to increase the formation of the analogous EPA-derived mediators (31). Our data suggest that, at least in the case of COX-2, the enzyme apparently has a lower, rather than higher, affinity for ω-3 PUFA (Fig. 9.8). These findings are in accord with the mechanism proposed by Malkowski et al. (32) where a low rate of EPA oxygenation by COX-1 was observed and attributed to the presence of the additional double bond in the molecule, such that the ω-3 PUFA adopted a "strained" conformation in the COX-1 active site. Using the TMPD assay we have shown that all three ω-3 PUFA that were examined – EPA, docosahexaenoic acid (DHA) and stearidonic acid (SDA) – are less efficient substrates than the optimal substrate arachidonic acid (AA) (Fig. 9.8).

5. Notes

1. PUFA are light, oxidants and temperature-sensitive. Make small aliquots in organic solvents (usually ethanol or dimethyl sulfoxide), store them at –20 C and use them before their

expiration date. Diluted aliquots should be made fresh and used within a day. TMPD and hematin stock solutions should be stored similarly.

2. If the source of COX is tissue or cell extract preparation should be cleared by centrifugation (15 min at 10,000 × g) to minimize turbidity of the sample.

3. If possible, perform the experiments with PUFA at subdued light.

4. TMPD undergoes spontaneous oxidation. Include controls without PUFA and subtract their OD from the values obtained in sample wells.

5. Add all of the ingredients except the PUFA into the wells first.

6. Initiate reaction by adding PUFA last.

7. Since COX assay is completed relatively quickly, use of multi-channel-pipettor will assure that all the reactions started at the approximately same time.

8. Use of automatic multiplate mixer (shaker) is advisable to assure proper mixing of the sample at the beginning of the reaction.

9. Summary: COX enzymes participate in the formation of a wide range of eicosanoid mediators. The separate quantification of each of these fatty acid derivatives in biological samples is time consuming and has specific analytical requirements. More convenient assays that take advantage of the capacity of COX-derived peroxy intermediates to co-oxidise artificial reducing substrates, such as TMPD, are now being developed. By this approach suitable screening methods for potential COX inhibitors or alternate PUFA substrates have been identified.

References

1. Herschman HR (1996) Prostaglandin synthase 2. Biochim Biophys Acta 1299:125–140

2. Smith W, Garavito R, DeWitt D (1996) Prostaglandin endoperoxide H synthases (cyclooxygenases)-1 and 2. J Biol Chem 271:33157–33160

3. Kutchera W, Jones DA, Matsunami N, Groden J, McIntyre TM, Zimmerman GA, White RL, Prescott SM (1996) Prostaglandin H synthase-2 is expressed abnormally in human colon cancer: evidence for a transcriptional effect. Proc Natl Acad Sci U S A 93:4816–4820

4. Reddy ST, Herschman HR (1996) Transcellular prostaglandin production following mast cell activation is mediated by proximal secretory phospholipase A2 and distal prostaglandin synthase 1. J Biol Chem 271:186–191

5. Shao J, Sheng H, Inoue H, Morrow JD, DuBois RN (2000) Regulation of constitutive cyclooxygenase-2 expression in colon carcinoma cells. J Biol Chem 275: 33951–33956

6. Vane JR (1971) Inhibition of prostaglandin synthesis as a mechanism of action for aspirin-like drugs. Nature 231:232–235

7. Van der Ouderaa FJG, Buytenhek M (1982) Purification of PGH synthase from sheep vesicular glands. Methods Enzymol 86:60–68

8. Markey CM, Alward A, Weller PE, Marnett LJ (1987) Quantitative studies of hydroperoxide reduction by prostaglandin H synthase: reducing substrate specificity and the relationship of peroxidase to cyclooxygenase activities. J Biol Chem 262:6266–6279

9. Kulmacz RJ, Lands WEM (1983) Requirements for hydroperoxide by the cyclooxygenase and peroxidase activities of prostaglandin H synthase. Prostaglandins 25:531–540

10. Towne V, Will M, Oswald B, Zhao Q (2004) Complexities in horseradish peroxidase-catalyzed oxidation of dihydroxyphenoxazine derivatives: appropriate ranges for pH values and hydrogen peroxide concentrations in quantitative analysis. Anal Biochem 334: 290–296

11. Forghani F, Ouellet M, Keen S, Percival MD, Tagari P (1998) Analysis of prostaglandin G/H synthase-2 inhibition using peroxidase-induced luminol luminescence. Anal Biochem 264:216–221

12. Noreen Y, Ringbom T, Perera P, Danielson H, Bohlin L (1998) Development of a radiochemical cyclooxygenase-1 and -2 in vitro assay for identification of natural products as inhibitors of prostaglandin biosynthesis. J Nat Prod 61:2–7

13. Stenson WF (2001) Measurement of prostaglandins and other eicosanoids. Curr Protoc Immunol Chapter 7, Unit 7.33

14. Barrière G, Rabinovitch-Chable H, Cook-Moreau J, Faucher K, Rigaud M, Sturtz F (2004) PHGPx overexpression induces an increase in COX-2 activity in colon carcinoma cells. Anticancer Res 24:1387–1392

15. Brideau C, Kargman S, Liu S, Dallob AL, Ehrich EW, Rodger IW, Chan CC (1996) A human whole blood assay for clinical evaluation of biochemical efficacy of cyclooxygenase inhibitors. Inflamm Res 45:68–74

16. Guilemany JM, Roca-Ferrer J, Mullol J (2008) Cyclooxygenases and the pathogenesis of chronic rhinosinusitis and nasal polyposis. Curr Allergy Asthma Rep 8:219–226

17. Yang H, Chen C (2008) Cyclooxygenase-2 in synaptic signaling. Curr Pharm Des 14: 1443–1451

18. Ragel BT, Jensen RL, Couldwell WT (2007) Regulation of cyclooxygenase-2 expression by cyclic AMP. Biochim Biophys Acta 1773:1605–1618

19. Hsieh HL, Sun CC, Wang TS, Yang CM (2008) PKC-delta/c-Src-mediated EGF receptor transactivation regulates thrombin-induced COX-2 expression and PGE(2) production in rat vascular smooth muscle cells. Biochim Biophys Acta 1783:1563–1575

20. Wang L, Chen W, Xie X, He Y, Bai X (2008) Celecoxib inhibits tumor growth and angiogenesis in an orthotopic implantation tumor model of human colon cancer. Exp Oncol 30:42–51

21. Wong E, Bayly C, Waterman HL, Riendeau D, Mancini JA (1997) Conversion of prostaglandin G/H synthase-1 into an enzyme sensitive to PGHS-2-selective inhibitors by a double His513→Arg and Ile523→Val mutation. J Biol Chem 272:9280–9286

22. Van der Ouderaa FJ, Buytenhek M, Nugteren DH, Van Dorp DA (1977) Purification and characterisation of prostaglandin endoperoxide synthetase from sheep vesicular glands. Biochim Biophys Acta 487:315–331

23. Raz A, Needleman P (1990) Differential modification of cyclooxygenase and peroxidase activities of prostaglandin endoperoxide synthase by proteolytic digestion and hydroperoxides. Biochem J 269:603–607

24. Gierse JK, Hauser SD, Creely DP, Koboldt C, Rangwala SH, Isakson PC, Seibert K (1995) Expression and selective inhibition of the constitutive and inducible forms of human cyclooxygenase. Biochem J 305:479–484

25. Butler MJ, Lachance MA (1987) The use of N,N,N',N'-tetramethylphenylenediamine to detect peroxidase activity on polyacrylamide electrophoresis gels. Anal Biochem 162: 443–445

26. Tam SS, Lee DH, Wang EY, Munroe DG, Lau CY (1995) Tepoxalin, a novel dual inhibitor of the prostaglandin-H synthase cyclooxygenase and peroxidase activities. J Biol Chem 270:13948–13955

27. Ouellet M, Falgueyret JP, Percival MD (2004) Detergents profoundly affect inhibitor potencies against both cyclo-oxygenase isoforms. Biochem J 377:675–684

28. Copeland RA, Williams JM, Giannaras J, Nurnberg S, Covington M, Pinto D, Pick S, Trzaskos JM (1994) Mechanism of selective inhibition of the inducible isoform of prostaglandin G/H synthase. Proc Natl Acad Sci USA 91:11202–11206

29. Gierse JK, Koboldt CM, Walker MC, Seibert K, Isakson PC (1999) Kinetic basis for selective inhibition of cyclo-oxygenases. Biochem J 339:607–614

30. Szymczak M, Murray M, Petrovic N (2008) Modulation of angiogenesis by omega-3 polyunsaturated fatty acids is mediated by cyclooxygenases. Blood 111:3514–3521

31. Fischer S, Weber PC (1984) Prostaglandin I3 formed in vivo in man after dietary eicosapentaenoic acid. Nature 307: 165–168

32. Malkowski MG, Thuresson ED, Lakkides KM, Rieke CJ, Micielli R, Smith WL, Garavito RM (2001) Structure of eicosapentaenoic and linoleic acids in the cyclooxygenase site of prostaglandin endoperoxide H synthase-1. J Biol Chem 276:37547–37555

Chapter 10

Structural and Functional Changes in the Insulin Molecule Produced by Oxidative Stress

Rafael Medina-Navarro, Alberto M. Guzmán-Grenfell, Ivonne Olivares-Corichi, and Juan J. Hicks

Abstract

The change produced by oxidative stress on proteins (cross-links, backbone cleavage, amino acid modification) generates structural changes with a wide range of consequences such as increased propensity to the aggregation or proteolysis, altered immunogenicity and frequently enzymatic and binding inhibition. Insulin is particularly sensitive to conformational changes, aggregation and cross-linking; any change on insulin could impair its function. We have examined the biological activity of insulin modified by hydroxyl radical and exposed to acrolein in rats and adiposites. We found out important changes that we have shown as prototype of possible effect of oxidative stress on the structural and functional damage to insulin. Whereas, hydroxyl radical and acrolein both have diminished the hypoglycemic effect of insulin in vivo, and the effect of acrolein seems be to involved in carbonylation and not derived from inter-molecular cross-links formation or aggregates. The effect was highly stimulated at alkaline pH, concomitant with carbonyl formation and then probably aldolic condensation type reaction-dependent. Hydroxyls radical generates tyrosine derivative formation and introduces non aldehyde dependent carbonyls in the insulin molecule.

Key words: Oxidative stress, Insulin, Hydroxyl radical, Aldehydes, Acrolein, Protein oxidation, Carbonyls, Quinones

1. Introduction

During diabetes type 1 and 2, the oxidative stress associated to hyperglycemia and the formation of the reactive oxygen species produces in addition, dangerous compounds generated by the lipid peroxidation process (1). Some of these compounds include highly reactive aldehydes that can react with amino acids, peptides, proteins, carbohydrates, and nucleic acids, and therefore form molecular adducts that could modify the structural and

D. Armstrong (ed.), *Advanced Protocols in Oxidative Stress II*, Methods in Molecular Biology, vol. 594
DOI 10.1007/978-1-60761-411-1_10, © Humana Press, a part of Springer Science+Business Media, LLC 2010

functional properties of many of these biomolecules, and then generates adverse effects on specific biological processes (2, 3). It has been postulated that α,β-unsaturated aldehydes could be particularly toxic for the pancreatic beta cell (4) and, since there is evidence of a higher pancreas beta cell oxidative stress vulnerability (5, 6), these compounds could be associated to the beta cell decaying process and perhaps diabetes. Important levels of aldehydes like 4-hydroxy-2-nonenal (HNE) and others, have been detected in the pancreatic islets, and could produce inhibition of glucose stimulated insulin secretion, probably by an effect on glycolitic way and the tricarboxylic acid cycle (7, 8). The reason why these compounds exert that effect could be their interaction with sulfhydryl groups, as well as the formation of molecular adducts in Michael addition-type reactions (9). The protein thioethers formed from the former mechanism will contain aldehyde free groups, favoring the Schiff base formation and cross-linking (10). After, in an advanced state, cross-linkings can result in more stable proteins, and less vulnerable to proteolytic degradation (11). The strongest electrophile aldehyde derived from lipid peroxidation is acrolein, and therefore, posses the highest activity with the sulfhydryl groups of cysteine and glutathion, imidazole group of histidine and the amine groups of lysine (12, 13). Acrolein in vivo is formed from the oxidation of polyunsaturated fatty acids and generally in reactions catalyzed by transition metals.

It is possible to demonstrate the effect of an α,β-unsaturated aldehyde on the structure and function of a protein. In a recent study, we have demonstrated how the insulin-aldehyde interaction results in serious effects on insulin biological function, hypoglycemic activity and insulin-dependent glucose uptake (14).

As we discussed previously, the secondary product of oxidation of lipids can undergo important changes in proteins. However, there are primary modifications, produced by the metal-catalyzed oxidation, radiation, oxidants like ozone and hypochlorite, etc. Carbonyl groups may be introduced into proteins by the two kinds of oxidative mechanisms, primary or secondary, and the detection of carbonyls is a convenient evidence of protein oxidative modification.

In the present chapter, we show the effect of hydroxyl radical and acrolein on the most important functions of insulin, the effect on the hypoglycemic activity in vivo and the consequences on the glucose transportation in adipocytes (acrolein). Here we exemplify that, agents like free radicals and aldehydes have an important impact on the biological activity of insulin and that, this effect is closely associated with a reduction of glucose utilization. The results have shown that the effects could rather be associated to the protein carbonylation extension and probably the formation of intro-molecular cross-links and amino acid modifications.

2. Materials

2.1. Equipment

1. Glucose analyzer (Abbott Laboratories Inc., MediSense Products, Bedford, MA 01730 USA).
2. Beckman LS 6000SE β counter. 1. Beckman Instruments Inc., Fullerton, CA, USA.
3. Fume hood.
4. Micro-centrifuge filters (5 kDa cut off). (Microcon, Millipore, Bedford, MA, USA).
5. Incubator for cell culture with CO_2 supply.
6. 6-well cell culture clusters.
7. Spectrophotometer with possibilities of scanning.
8. Vortex mixer.

2.2. Reagents

1. Insulin from bovine pancreas, lyophilized powder, cell culture tested. Potency: ≥27 USP units per mg. Sigma Chemical Co., USA.
2. Acrolein Sigma Chemical Co., USA.
3. 2-deoxy-[2,6-^{3}H] glucose (1 µCi/mL). (Amersham Pharmacia Biotech., England).
4. Sodium Dodecylsulfate (SDS). Sigma Chemical Co., USA.
5. Triton X100. Sigma Chemical Co., USA.
6. β-mercaptoethanol. Sigma Chemical Co., USA.
7. Scintillation cocktail. 1. Beckman Instruments Inc., Fullerton, CA, USA.
8. Bovine serum albumin (BSA). Sigma Chemical Co., USA.
9. 10% calf serum (HyClone Laboratories, Inc. Logan, UT, USA).
10. 2,4-Dinitrophenylhydrazine (DNPH). Sigma Chemical Co., USA.
11. Tris (2-amino-2-hydroxymethylpropane-1,3-diol) Sigma Chemical Co., USA.
12. Sephadex G-50. Sigma Chemical Co., USA.
13. HPLC grade water.

2.3. Buffer and Solutions (see Note 1)

1. Insulin stock solution (10 mg/mL). Add 10 mL of acidified Milli Q water prepared by addition of 1 N HCL and agitate continuously until dissolution.
2. Insulin aliquots. Each aliquot of insulin containing 855 µg/mL insulin (150 µM) should be adjusted to pH 7.4 or 8.0 with 0.5 M Tris base.

3. Solution of hydrogen peroxide 50 mM. Take 57 uL of peroxide from bottle (30%, J.T. Baker) and dilute it with 10 mL of Milli Q water.

4. Solution of CuSO4·5H$_2$O 40 mM. Prepare stock solution with 0.998 g of CuSO4 and dilute it in 100 mL of deionized water.

5. Acrolein stock solution (*see* Note 2). Prepare stock solution aspirate approximately 50 µL of acrolein with Hamilton syringe from the compound flask and add water until the volume is adjusted to 50 mL. Compound bottle has a special stopper from which, it is possible to aspirate the aldehyde.

6. Krebs Ringer Phosphate (KRP). The buffer must contain 128 mM NaCl, 1.4 mM CaCl$_2$, 1.4 mM MgSO$_4$, 5.2 mM KCl, and 10 mM Na$_2$H PO$_4$, and the pH 7.4 (*see* Note 3).

7. Dulbeco's Modified Eagle Medium (DMEM). (Gibco BRL, Life Technologies. Grand Island, NY, USA) with 10% bovine serum albumin.

   Formulation: components (mg/L): Amino Acids: L-Arginine HCl 84.00, L-Cystine 2HCl 62.57, L-Glutamine 584.00, Glycine 30.00, L-Histidine HCl H$_2$O 42.00, L-Isoleucine 104.80, L-Leucine 104.80, L-Lysine HCl 146.20, L-Methionine 30.00, L-Phenylalanine 66.00, L-Serine 42.00, L-Threonine 95.20, L-Tryptophan 16.00, L-Tyrosine 2Na 2H$_2$O 103.79, L-Valine 94.00; Vitamins: Choline Chloride 4.00, d-Calcium Pantothenate 4.00, Folic Acid 4.00, myo-Inositol 7.20, Nicotinamide 4.00, Pyridoxine HCl 4.00, Riboflavin 0.40, Thiamine HCl 4.00; Inorganic Salts: Calcium Chloride (CaCl$_2$) 200.00, Ferric Nitrate (Fe(NO$_3$)$_3$·9H$_2$O) Nonahydrate 0.10, Magnesium Sulfate (MgSO$_4$) 97.70, Potassium Chloride (KCl) 400.00, Sodium Bicarbonate (NaHCO3) 3700.00, Sodium Chloride (NaCl) 6400.00, Sodium Phosphate Monobasic (NaH$_2$ PO$_4$ H$_2$O) Monohydrate 125.00; Other: Dextrose 4500.00, Phenol Red Sodium Salt 15.00, Sodium Pyruvate 110.00 (*see* Note 4).

8. Human recombinant insulin (Humulin® Lilly Lilly, México). Working solution for hydroxyl radical effect is prepared using 5 IU equivalent to 33.5 nmol or 194.55 µg dissolved in 1 mL of sterile water.

9. 10 mM 2,4-Dinitrophenylhydrazine (DNPH) in 2 M HCl.

10. 150 mM phosphate buffer (pH 7).

11. 100 mM Tris/HCl, pH 8.0.

12. 0.5 M Tris base.

3. Methods

3.1. Insulin–Aldehyde Complex Formation

To verify the aldehyde–insulin interaction, both components are placed together, with distinct conditions of incubation time and pH. Aliquots of insulin containing 855 µg/mL hormone (150 µM) were adjusted to pH 7.4 or pH 8.0 with 0.5 M Tris base and incubated for 2, 4, and 12 h at 37 C with or without 300 µM acrolein (*see* Note 2). At the end of the incubation, the pH of all of the samples was adjusted to 7.4, placed in micro-centrifuge filters (5 kDa cutoff) (*see* Subheading 2.2), and centrifuged at $5,000 \times g$ for 45 min. The filter residues were washed and dialyzed three times and dissolved in Krebs Ringer Phosphate (KRP).

3.2. Intraperitoneal Insulin Tolerance Test

With the objective of determining the functional modification of insulin's hypoglycaemic properties, an intraperitoneal insulin tolerance test can be performed. Male Sprague–Dawley rats, weighing 350 ± 5 g, maintained under controlled light–dark conditions at 20 C with food and water ad libitum has been used previously with good results. Animals are anesthetized with Phenobarbital (*see* Note 5), and native bovine insulin or treated insulin (complex) is administered intraperitoneally (0.35 mg in 100 µL/kg). Blood samples are taken from the tails at different times (0–90 min), and glucose concentration is measured by the glucose oxidase method using a glucose analyzer (*see* Subheading 2.2). The results of representative experiment are plotted in Fig. 10.1. For hydroxyl radical effect, native or treated insulin (incubated in Fenton Reaction, see Subheading 3.7) is injected intraperitoneally (1.0 UI/kg) and each 10 min is taken for a blood sample from the tails until 50 min are completed. In this case, the 100 % is considered the initial value (Figs. 10.1 and 10.8).

3.3. Cell Culture

Cell line 3T3-L1 (15) pre-adipocytes are used frequently for studies of glucose transport. In this case, the purpose of using the cell line is to know how modified insulin acts on glucose transport dynamics. For the culture, cells are growing in Dulbeco's modified Eagle's medium (DMEM) containing 25 mM of glucose, supplemented with 10% calf serum (*see* Subheading 2.2) in a 5% CO_2 atmosphere. The confluence was considered day 0, and the differentiation is induced with 200 nM insulin, 100 µM 1-methyl-3-isobutyl-xantine, 0.25 µM dexamethasone in DMEM supplied with 10% FCS for 2 days. Thereafter, medium DMEM supplied with FCS 10% is changed, every 2 days.

3.4. Glucose Transport

Glucose transport is measured in KRP buffer with the use of 2-Deoxy-[2,6-³H] glucose as described by Frost and Lane (16). Insulin adducts and untreated hormone (0, 0.1 nM, 1 nM,

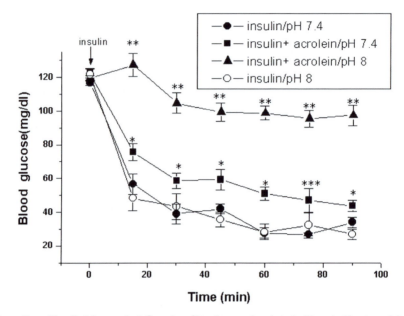

Fig. 10.1. Intraperitoneal insulin tolerance test. Samples of insulin were incubated with and without acrolein for 4 h at pH 7.4 and 8.0. After incubation, the pH was adjusted to 7.4 and processed as described. The samples (0.3 mg/Kg) were then injected into rats after an overnight fast. At time 0 and every 15 min, blood glucose concentration was measured by the glucose oxidase method. Data are means $\pm$ SD, ($n=5$) $*p<0.01$ vs. insulin, pH 7.4; $**p<0.001$ vs. insulin pH 7.4; $***p<0.05$ vs. insulin, pH 7.4 (From ref. (14), with permission).

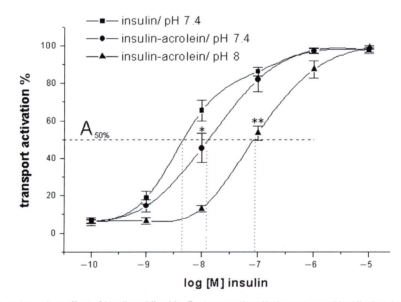

Fig. 10.2. Hypoglycemiant effect of insulin oxidized by Fenton reaction. Native or treated insulin (incubated in Fenton Reaction) was injected intraperitoneally (1.0 UI/kg) and each 10 min a blood sample from the tails until 50 min are taken. The glucose concentration is measured using a glucose analyzer. To calculate the inhibition percent of hypoglycemiant effect, the 100 % is considered the initial value. Data are expressed as mean $\pm$ SD (From ref. (20), with permission).

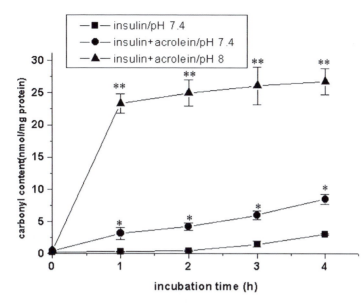

Fig. 10.3. Effect of modified insulin on glucose transport. Regular insulin and insulin incubated with acrolein were processed as described. The dotted lines intercepting x axis indicate the insulin concentration where the transport activation reaches a 50% (A50%) and corresponds to the followings values: 4.5, 12.3 and 94 nM (non modified pH 7.4, modified pH 7.4 and modified pH 8.0). Each point represents means ± S.D. of five experiments. *$p < 0.05$ vs. insulin pH 7.4; **$p < 0.001$ vs. insulin pH 7.4 (From ref. (14), with permission).

10 nM, 100 nM, and 1 µM) are added to the cells in 1 mL of KRP. After 20 min of incubation at 37 C in the dark and a second 20 min incubation with 50 µM 2-desoxy-(2,6-3H) glucose (1 µCi/mL) under the same conditions, the culture medium is aspirated, and the cells are washed four times with cold 1% BSA in KRP and then lysed with 1% Triton. The mixture obtained (1 mL) is added to 10 mL of scintillation cocktail and counted on a β counter (*see* Subheading 2.2). The results of a representative experiment are plotted in Fig. 10.3 (*see* Note 6).

3.5. Determination of Carbonyls Introduced into Insulin Molecule

The procedure to know how many aldehyde units have been incorporated to the insulin molecule, is to measure the protein carbonyl groups, after the separation of the free aldehyde and the components of the Fenton reaction. At the end of incubation period, the samples are filtered and re-dissolved as it was described before. The protein carbonyl content then can be determined as was described by Levine et al. (17). The protein is incubated for 45 min with 2,4-Dinitrophenylhydrazine (DNPH) 10 mM in 2 M HCL and occasionally mixed. After precipitation with 20 % trichloroacetic acid, precipitates are washed three times with ethanol/ethyl acetate (1:1, v/v). Pellets are dissolved in 6 M guanidine hydrochloride, 0.5 M potassium phosphate, pH 2.5, and the absorbance of the protein-2,4-Dinitrophenylhydrazone derivative is determined at 375 nm. For each sample, a blank without DNPH must be run, and their absorbencies are subtracted. Carbonyl content is reported as nmol/mg protein using a molar absorption coefficient of 22,000 M-1 cm-1. Protein concentration is calculated in the samples by determining the absorbance at 280 nm (molar absorption 5.31 for a 1 g/100 mL solution) (*see* Note 7). The results of a representative experiment are plotted in Figs. 10.3 and 10.7.

3.6. Molecular Weight Modification

Under several circumstances including the increase of concentration, insulin forms dimeric, tetrameric and hexameric species and loses bioactivity. The formation of these class of complexes is a possible mechanism for α,β-unsaturated aldehyde interaction with proteins and consequent with insulin and its loss of activity. To verify the possible formation of molecular complexes of insulin–insulin, the molecular weight modification can be verified with the use of electrophoresis, gel filtration chromatography and

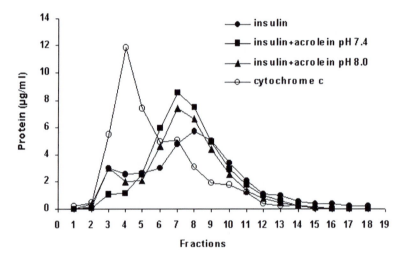

Fig. 10.4. Carbonyl introduction to insulin exposed to acrolein. Insulin (1 mg/mL) was incubated with and without acrolein at 37 C for a 4-h period, in deionized water at pH 7.4 and 8. At time 0 and before each hour, aliquots of the samples were collected, all pH's adjusted (7.4), and processed as described in Subheading 3. The insulin carbonyl content was determined using 2, 4-dinitrophenylhydrazine. Each point is the mean ± S.D. from three experiments. *$p < 0.01$ vs. insulin pH 7.4; **$p < 0.001$ vs. insulin pH 7.4 (From ref. (14), with permission).

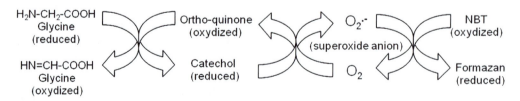

Fig. 10.5. Carbonyl introduction to insulin exposed to hydroxyl radical. The procedure was essentially such as Fig. 10.3. and described in Subheading 3. In this case aliquots of the samples were collected each 10 min. Osazones in graph correspond to protein-2,4-dinitrophenylhydrazone. Additional data are presented inside the graph (From ref. (20), with permission).

probably others. In the present chapter, we include the gel filtration chromatography as just one of the possibilities.

3.6.1. Gel Filtration Chromatography

A solution containing 1 mg of insulin/mL of deionized water is incubated for 12 h at 37 C in the presence or absence of 300 µM acrolein, and the pH is adjusted to 7.4 or 8.0 using 0.5 M Tris. After the incubation and before dialysis, samples containing 50 µg of protein must be dissolved in a buffer containing 0.5 M Tris at pH 6.8, 1% SDS, and 15 mM dithiothreitol to preclude spontaneous aggregation. Under these conditions, the protein is applied to a Sephadex G-50 column (10×3.0 cm), equilibrated, and eluted with 150 mM phosphate buffer (pH 7) at 20 C. Cytochrome c (1 mg/mL) must be applied to the column under the same conditions. The protein concentrations in the 1 mL fractions collected is measured by the method of Bradford (18). The results of the representative experiment are plotted in Fig. 10.4.

3.7. Insulin Oxidation by Fenton Reaction

1. Five units (IU) of human recombinant insulin equivalent to 194.55 µg (33.5 nmol) is diluted in 925 µL of deionized water, then 100 µL of $CuSO_4 \cdot 5H_2O$ (40 mM) is added to obtain a concentration of 4 mM. The incubation time starts after 100 µL of H_2O_2 (50 mM) addition, coincident with the hydroxyl radical generation.

2. During incubation time, each point (0–15 min) represents an independent reaction. The protein is precipitated adding 1 mL

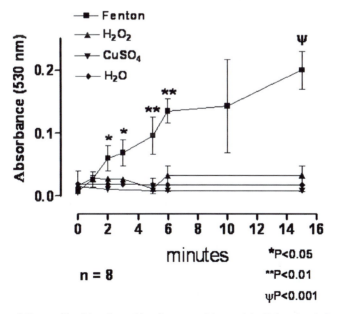

Fig. 10.6. Gel filtration elution profile of insulin and insulin exposed to acrolein. Before incubation (12 h), insulin and treated insulin were applied to a Sephadex G50 column (10× 3.0 cm) equilibrated and eluted using 150 mM phosphate buffer, pH 7. The profile of cytochrome c corresponds approximately to the void volume. Each fraction volume corresponds to 1 mL (From ref. (14), with permission).

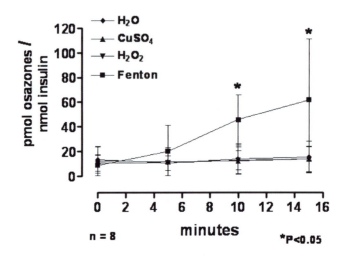

Fig. 10.7. The NBT/Glycinate redox cycle assay. The residues of phenylalanine and tyrosine are prone to hydroxylation reactions on the backbone of a protein (insulin). The products of the reaction correspond to 3,4-dihydroxyphenylalanine–protein (protein–DOPA) and the correspondent ortho-quinones.

of TCA 5% (w/v). The sample is centrifugated at $2{,}000 \times g$ for 10 min.

3. The precipitates are washed twice with 1 mL of TCA 2.5 % (w/v) and centrifuged at $2{,}000 \times g$ for 10 min to eliminate the residues of the Fenton reaction; at this point, the insulin is ready to be used in hydroxyl radical experiments.

4. Controls with each component of the Fenton reaction must be performed.

3.8. Insulin Quinone Formation

Quinoneproteins can be detected with the use of nitroblue tetrazolium (NBT). Proteins and aromatic amino acids previously exposed to hydroxyl radicals reduce NBT (Fig. 10.5). The reduction of nitro blue tetrazolium by radical-damaged protein is consistent with the generation of quinones in the protein (19). A simple procedure can be achieved as follows:

1. The insulin treated as previously described is dissolved in 1 mL of water and then 1 mL of NBT (0.28 mM in glycine 2 M pH 10) adding. The reaction is mixed and incubated at room temperature for 30 min in dark conditions.

2. After the incubation time, the reaction has to be measured immediately at 530 nm of wavelength in a spectrophotometer.

3. The values obtained are expressed as the change in absorbance with respect to the time. An example of the results is shown in the Fig. 10.6.

3.9. Data Analyses

The statistical significance of differences presented in the figures is evaluated by Student's t test, taking a probability of 0.05 % as the criterion of significance. All values reported are the mean $\pm$ S.D.

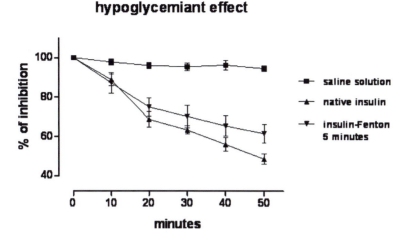

hypoglycemiant effect

Fig. 10.8. Reductive capacity of insulin on nitroblue tetrazolium (NBT). Changes in the absorbance at 530 nm correspond to the increased of quinone generation on insulin molecule (From ref. (20), with permission).

4. Notes

1. Water used for solution preparation is taken from a Milli-Q water purification unit (Millipore Filter Corp., Bedford, MA, USA). Buffers used for cell treatment are passed twice through a column of Chelex-100 resin prior to each experiment.

2. Acrolein is an extremely toxic and volatile compound. It must be used taken it in account. Open the container and flask in the extraction hood; use always gloves and store the flask at 4 C with a visible warning label. It is convenient to manipulate the containers cold and make a stock solution from the compound flask; it has a special stopper from which, it is possible to draw in the aldehyde with a Hamilton syringe.

3. All buffers and solutions should be prepared in water presented a resistivity of 18.2 $M\Omega \times cm$; when we use the reference water in the text, we are talking about these kind of water.

4. Buffer should be filtered through 0.45-μm filters prior to use. It should be stored in refrigerator and has to be discarded after 4 days.

5. The use of ketamine is not convenient, because the compound modifies the capillary glucose concentrations.

6. The transport activation is defined as the rate of glucose transport in the absence and presence of enough insulin to achieve the maximal response. On the experimental conditions described here, the non-modified insulin concentration in which the glucose transport activation results zero (basal glucose transport) is 0.1 nM, whereas, that necessary to reach the maximum glucose transport activation and 50% of them

($A_{50\%}$) is 1,000 nM and 4.5 nM, respectively (*see* Fig. 10.3). The concentration of acrolein modified insulin incubated at pH 7.4 is necessary to reach the $A_{50\%}$ results 12.3 nM, more than two fold of the $A_{50\%}$ normal insulin concentration. In case of acrolein-modified protein incubated at pH 8, the insulin concentration required for the same parameter is apparently more than 20-fold the normal insulin concentration (94 nM). In this case, the insulin concentration needed to provoke the initial glucose uptake is close to 10 nM, almost the concentration in which the non-modified insulin reaches the 50% of activity.

7. Because protein is lost in the washing steps, and the amount of protein has to be calculated from a bovine serum albumin standard curve.

Acknowledgments

We gratefully acknowledge the Consejo Nacional de Ciencia y Tecnología (CONACYT) for the economical support and for the realization of some experiments presented here.

References

1. Oberlay L (1988) Free radicals and diabetes. Free Radic Biol Med 5:113–124

2. Uchida K, Stadtman ER (1993) Covalent attachment of 4-hydroxynonenal to glyceraldehyde-3-phosphate dehydrogenase. A possible involvement of intra-and intermolecular cross-linking reaction. J Biol Chem 268: 6388–6393

3. Esterbauer H (1993) Cytotoxicity and genotoxicity of lipid-oxidation products. Am J Clin Nutr 57:7795–7865

4. Ihara Y, Toyokuni S, Uchida K et al (1999) Hyperglycaemia causes oxidative stress in pancreatic β-cells of GK rats, a model of type 2 diabetes. Diabetes 48:927–932

5. Cornelius JG, Luttge BG, Peck AB (1993) Antioxidant enzyme activities in IDD-prone and IDD-resistant mice. A comparative study. Free Radic Biol Med 14:409–420

6. Tiedge M, Lortz S, Drinkgern J, Lensen S (1999) Relation between antioxidant enzyme gene expression and antioxidative defence status of insulin-producing cells. Diabetes 46:1733–1742

7. Miwa I, Ichimura N, Sugiura M et al (2000) Inhibition of glucose-induced insulin secretion by 4-hydroxy-2-nonenal and other lipid peroxidation products. Endocrinology 141:2767–2772

8. Suarez-Pinzon WL, Strynadka K, Rabinovitch A (1996) Destruction of rat pancreatic Islet-cells by cytokines involve the production of cytotoxic aldehydes. Endocrinology 137:5290–5296

9. Uchida K, Stadtman ER (1992) Selective cleavage of thioether linkage in proteins modified with 4-hydroxynonenal. Proc Natl Acad Sci USA 89:5611–5615

10. Cohn JA, Tsai L, Friguet B et al (1996) Chemical characterization of protein-4-hydroxy-2-nonenal cross-link: immunochemical detection in mitochondria exposed to oxidative stress. Arch Biochem Biophys 328:158–164

11. Friguet B, Stadtman ER, Szweda LI (1994) Modification of glucose-6-phosphate dehydrogenase by 4-hydroxy-2-nonenal. Formation of cross-linked protein that inhibits the multicatalytic protease. J Biol Chem 269: 21639–21643

12. Esterbauer H, Schaur RJ, Zollner H (1991) Chemistry and biochemistry of 4-hydroxynonenal, malonaldehyde and relates aldehydes. Free Radic Biol Med 11:81–128

13. Uchida K, Kanematsu M, Sakai K, Matsuda T, Hattori N, Mizuno Y et al (1999) Protein-bound acrolein: potential markers for oxidative stress. Proc Natl Acad Sci U S A 95:4882–4887

14. Medina-Navarro R, Guzmán-Grenfell AM, Díaz-Flores M, Duran-Reyes G, Ortega-Camarillo C, Olivares-Corichi IM, Hicks JJ (2007) Formation of an adduct between insulin and the toxic lipoperoxidation product acrolein decreases both the hypoglycemic effect of the hormone in rat and glucose uptake in 3T3 adipocytes. Chem Res Toxicol 20:1477–1481

15. Green H, Kehinde O (1974) Sublines of mouse 3T3 cells that accumulate lipids. Cell (Cambridge) 1:113–116

16. Frost SC, Lane MD (1985) Evidence for the involvement of vicinal sulfhydryl groups in insulin-activated hexose transport by 3T3-LI adipocytes. J Biol Chem 260:2646–2652

17. Levine RL, Williams JA, Stadtman ER, Shacter E (1994) Carbonyl assays for determination of oxidatively modified proteins. Methods Enzymol 233:346–357

18. Bradford MM (1976) A rapid and sensitive method for the quantitation of microgram quantities of protein utilizing the principle of protein-dye binding. Anal Biochem 77: 248–254

19. Gieseg SP, Simpson JA, Charlton TS, Duncan MW, Dean RT (1993) Protein-bound 3, 4-dihydroxyphenylalanine is a major reductant formed during hydroxyl radical damage to proteins. Biochemistry 32:4780–4786

20. Olivares-Corichi IM, Ceballos G, Ortega-Camarillo C, Guzmán-Grenfell AM, Hicks JJ (2005) Reactive oxygen species (ROS) induce chemical and structural changes on human insulin in vitro, including alterations in its immunoreactivity. Front Biosci 10: 838–843

Chapter 11

Multiphoton Redox Ratio Imaging for Metabolic Monitoring In Vivo

Melissa Skala and Nirmala Ramanujam

Summary

Metabolic monitoring at the cellular level in live tissues is important for understanding cell function, disease processes, and potential therapies. Multiphoton imaging of the relative amounts of NADH and FAD (the primary electron donor and acceptor, respectively, in the electron transport chain) provides a noninvasive method for monitoring cellular metabolic activity with high resolution in three dimensions in vivo. NADH and FAD are endogenous tissue fluorophores, and thus this method does not require exogenous stains or tissue excision. We describe the principles and protocols of multiphoton redox ratio imaging in vivo.

Key words: Reduced nicotinamide adenine dinucleotide (NADH), Flavin adenine dinucleotide (FAD), Redox ratio, Metabolism, Mitochondria, Multiphoton microscopy

1. Introduction

The metabolic rate of a cell is an important marker for the diagnosis, staging, and treatment of diseases ranging from Alzheimer's to cancer, and can serve as a general marker of cell health. Optical imaging of endogenous tissue fluorophores provides a noninvasive, fast, and inexpensive method for evaluating the metabolic rate of cells in vivo. The electron transport chain is the primary means of energy production in the cell. The electron transport chain produces energy in the form of adenosine triphosphate (ATP) by transferring electrons to molecular oxygen. There are two endogenous (occurring naturally in the cell) fluorophores in tissue related to cellular metabolism in the electron transport chain. The first fluorophore is the reduced form of nicotinamide adenine dinucleotide (NADH), which transfers electrons to

D. Armstrong (ed.), *Advanced Protocols in Oxidative Stress II*, Methods in Molecular Biology, vol. 594
DOI 10.1007/978-1-60761-411-1_11, © Humana Press, a part of Springer Science+Business Media, LLC 2010

molecular oxygen. NADH has fluorescence excitation and emission maxima at 350 nm and 460 nm, respectively (1). The second fluorophore is flavin adenine dinucleotide (FAD), which is an electron acceptor. FAD has fluorescence excitation and emission maxima at 450 and 535 nm, respectively (1). An approximation of the oxidation–reduction ratio of the mitochondrial matrix space can be determined from the "redox ratio," which is the fluorescence intensity of FAD divided by the fluorescence intensity of NADH (2). This optical redox ratio provides relative changes in the oxidation–reduction state in the cell without the use of exogenous stains or dyes, and can thus be measured in vivo in both human and animal studies. This advantage is important because it eliminates possible artifacts in metabolic measurements that can be introduced by tissue excision, processing, or staining. The redox ratio is sensitive to changes in the cellular metabolic rate and vascular oxygen supply (2–5). A decrease in the redox ratio usually indicates increased cellular metabolic activity (6).

Multiphoton microscopy is an attractive method for imaging the redox ratio in vivo, because it provides high resolution (~400 nm) three-dimensional images deep within living tissue (~1 mm). Multiphoton excitation occurs when a fluorophore is excited simultaneously by two photons of half the absorption energy of the fluorophore (or by three photons of one-third the absorption energy of the fluorophore, etc.), and probes the same biological fluorophores as single-photon fluorescence (7). Multiphoton excitation of NADH and FAD occurs in the near infrared (NIR) wavelength region (8), and these wavelengths of non-ionizing radiation are relatively benign (9). The NIR wavelength range between 650 and 900 nm is called the "optical window" where light can penetrate deep into tissue, due to reduced tissue scattering and minimal absorption from water and hemoglobin. Thus, multiphoton excitation also allows for increased imaging depth compared to single photon excitation.

A typical multiphoton microscope includes a titanium sapphire (Ti–Sapphire) laser, a raster scan unit, a dichroic mirror, a microscope objective, and a photomultiplier tube (PMT). The most common excitation sources are femtosecond titanium sapphire (Ti–Sapphire) lasers that generate 100 fs pulses at a repetition rate of about 80 MHz. This allows for sufficient two-photon excitation without excessive heat or photodamage to the sample. The tuning range of Ti–Sapphire systems are 700–1,000 nm, sufficient to excite both NADH and FAD. The raster scan unit scans the excitation beam across the x–y plane so that two-dimensional images can be created at each image depth. The focal point of the objective can be moved in the z- (depth) direction by a z-stage motor. A short pass dichroic mirror reflects the longer-wavelength IR light onto the sample, and transmits the shorter-wavelength fluorescence (usually in the visible wavelength range) to the detector.

High NA (numerical aperture) microscope objectives are used to maximize the excitation efficiency. PMTs are a popular choice for detector because they are robust, low cost, and relatively sensitive. Multiphoton microscopes can be custom built or purchased from most microscope companies. Multiphoton endoscopes (10–12) are required for redox ratio measurements in organ sites that are not accessible with a microscope.

In vivo metabolic measurements require that anesthesia minimally perturbs the metabolic state of cells, or at least, that the anesthesia has the same effect on experimental and control groups. A previous study found that Isoflurane (1.5%) produced constant muscle pO_2 and blood perfusion in mice, and thus Isoflurane is a good choice for metabolic imaging using the redox ratio (13). Baudelet et al also found that the pO_2 of both tumors and normal muscle tissue decreased approximately the same percentage with ketamine/xylazine anesthesia in mice (13).

2. Materials

The advantage of multiphoton redox imaging is that no sample processing is necessary. The required components include a multiphoton microscope, a live sample, and a computer for analysis.

2.1. Equipment

1. Multiphoton microscope can be custom built (14) or purchased from vendors including Lavision Biotech (Pittsford, NY), Zeiss (Thornwood, NY), Prairie Ultima (Middleton, WI), Olympus (Center Valley, PA), and Leica (Bannockburn, IL).

2. Titanium–sapphire lasers can be purchased from vendors including PicoQuant (Berlin, Germany) and Newport (Mountain View, CA) if they do not come with your microscope.

3. x–y translator for the microscope stage.

4. Anesthesia machine or injectable anesthesia.

2.2. Reagents and Supplies

1. 5.8×10^{-5} g/ml Rhodamine B in distilled water.

2. Standard coverslips.

3. Methods

3.1. Image Collection

1. Select appropriate excitation wavelength(s) and emission filters. Huang et al. found that NADH and FAD fluorescence is isolated at 750 nm and 900 nm excitation, respectively (8). Optical filters, which only allow wavelengths in a selected range to pass, can also be used with one or more detectors to

further isolate NADH and FAD fluorescence. Huang et al. found that a 410–490 bandpass filter isolated NADH emission and a 510–560 nm bandpass filter isolated FAD fluorescence at 800 nm excitation (8). Mitochondrial uncouplers and inhibitors can be used to further verify the isolation of NADH and FAD fluorescence (8).

2. Collect calibration standard images at the NADH excitation wavelength (*see* Note 1). Daily calibration standards must be measured to account for fluctuations in the throughput and excitation efficiency of the system. A standard calibration sample is Rhodamine B, and previous in vivo multiphoton studies have used a concentration of 5.8×10^{-5} g/ml Rhodamine B in distilled water (15). For added precision, the mean of three intensity images of the Rhodamine standard can be used to correct for system variations. The Rhodamine standard images should be collected under conditions identical to those of the in vivo NADH data from that day. Alternatively, an internal microscope reference that reports the two-photon excitation efficiency can be used to correct the measured fluorescence intensities for incident power (16).

3. Place the anesthetized animal on the imaging stage. Anesthesia that minimally alters the metabolic rate of the tissue should be chosen. The tissue of interest should be secured flush with a coverslip to ensure the deepest possible imaging depth. Bulk motion can be avoided by properly securing the tissue on the microscope stage. However, it is important that normal circulation be maintained so as not to alter the metabolic rate of the cells of interest.

4. Collect multiphoton image stack number one at the NADH excitation wavelength (*see* Notes 2 and 3). Use the NADH emission filter, if necessary. The resolution, field of view and imaging depth can be optimized with the appropriate objective. Resolution and field of view are determined from the numerical aperture (NA) of the objective, and the imaging depth is determined from the working distance of the objective. Water immersion objectives allow for deeper imaging depths than oil immersion objectives because of index matching with the tissue. The number of slices in the stack is determined by the z-stack slice separation, which can be as small as the axial resolution of the system. Typical z-stack slice separations range from 2 to 10 μm.

5. Collect additional image stacks at the NADH excitation wavelength, noting the position of each image stack on an $x–y$-translator.

6. Collect multiphoton image stack number one at the FAD excitation wavelength (use the FAD emission filter, if necessary).

Collect additional image stacks at the FAD excitation wavelength from the remaining x–y positions.

7. Collect calibration standard images at the FAD excitation wavelength under conditions identical to those of the in vivo FAD data from that day.

3.2. Image Analysis   ImageJ software is sufficient for all image analysis steps and is available free online at http://rsbweb.nih.gov/ij/.

1. Calculate the redox image. Choose your region of interest and divide the FAD image by its corresponding NADH image. Multiply the resulting image by a scalar value that accounts for the Rhodamine standard measured for the NADH and FAD images.

$$[\text{Redox}] = \frac{[\text{FAD}]}{[\text{NADH}]} \cdot \frac{R_{\text{NADH}}}{R_{\text{FAD}}} \qquad (11.1)$$

In Eq 11.1, [Redox] is the redox ratio image, [FAD] is the FAD intensity image, [NADH] is the corresponding NADH intensity image, and R_{FAD} and R_{NADH} are the mean Rhodamine intensity values measured under identical experimental conditions as [FAD] and [NADH], respectively. Note that the ratio of R_{FAD} and R_{NADH} is a scalar value, and the ratio of [FAD] and [NADH] is a matrix value. The division of [FAD] and [NADH] should be done pixel-by-pixel.

An example of relative redox ratio images from the normal and precancerous hamster cheek pouch in vivo are shown in Fig. 11.1. Three-dimensional redox ratio images can be useful for quantifying changes in the distribution of the redox ratio within cells, and with depth in tissue, as well as volume-averaged changes in the bulk tissue (*see* Note 4). Statistical analysis on multiple images from the study shown in Fig. 11.1 (15) indicates increased heterogeneity of the redox ratio within precancerous cells compared to normal cells, a decrease in the redox ratio with depth within normal tissues and no change in the redox ratio with depth within precancerous tissues. There was no change in the volume-averaged redox ratio with precancer compared to normal tissues in this study, which indicates the importance of high resolution, depth resolved imaging of the redox ratio in vivo (*see* Note 5).

4. Notes

1. If the incident power and detector gain settings are identical for all measurements, a Rhodamine calibration is not necessary. In cases in which the dynamic range of the NADH and FAD fluorescence intensities across all samples is large, the

+z (µm)

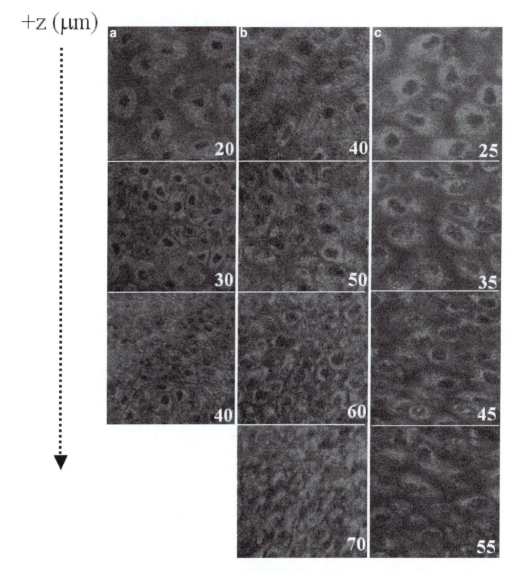

Fig. 11.1. Representative in vivo three-dimensional multiphoton images of the redox ratio (fluorescence intensity of FAD/NADH) from tissues diagnosed as normal (**a**), low-grade precancer (**b**), and high-grade precancer (**c**) in the 7,12-dimethylbenz(a)anthracene (DMBA)-treated hamster cheek pouch model of oral cancer. The numbers in the corner of each image indicate the depth below the tissue surface in microns, and each image is 100 × 100 µm. From ref. 15, © 2007, National Academy of Sciences, U S A.

incident power and detector gain settings must change between samples, and a Rhodamine calibration is necessary.

2. Monitor photon count rates while collecting multiphoton images, to ensure that photobleaching does not occur. Alternatively, after all images in the multiphoton image stack have been collected, collect the first image in the stack again to make sure that the image intensity has not diminished because of photobleaching.

3. Previous multiphoton measurements of tissue NADH at 780 nm excitation used an average power, peak power, and the focused fluence incident on the sample of approximately 15 mW, 300 GW/cm^2, and 3.7 MW/cm^2, respectively, with a pixel dwell time of 11.5 μs (16). Incident powers and integration times such as these should give sufficient signal for in vivo redox measurements using multiphoton microscopy.

4. This in vivo optical redox ratio is a relative measurement, so careful consideration of the appropriate controls must be made. All changes in the redox ratio must be considered relative to these controls.

5. These methods can be used to monitor metabolic changes in cell culture and excised tissue. However, these measurements may not reflect the true in vivo metabolic state, which is largely determined by oxygen delivery from the vasculature.

Acknowledgments

This work was supported by the NIH (R01 EB000184). M.S. acknowledges individual fellowship support from the DOD (W81XWH-04-1-0330) and the NIH (F32 CA130309).

References

1. Ramanujam N (2000) Fluorescence spectroscopy of neoplastic and non-neoplastic tissues. Neoplasia 2:89–117

2. Chance B, Schoener B, Oshino R, Itshak F, Nakase Y (1979) Oxidation–reduction ratio studies of mitochondria in freeze-trapped samples. NADH and flavoprotein fluorescence signals. J Biol Chem 254:4764–4771

3. Drezek R, Brookner C, Pavlova I, Boiko I, Malpica A, Lotan R, Follen M, Richards-Kortum R (2001) Autofluorescence microscopy of fresh cervical-tissue sections reveals alterations in tissue biochemistry with dysplasia. Photochem Photobiol 73:636–641

4. Gulledge CJ, Dewhirst MW (1996) Tumor oxygenation: a matter of supply and demand. Anticancer Res 16:741–749

5. Ramanujam N, Kortum RR, Thomsen S, Jansen AM, Follen M, Chance B (2001) Low temperature fluorescence imaging of freeze-trapped human cervical tissues. Opt Express 8:335–343

6. Chance B (1989) Metabolic heterogeneities in rapidly metabolizing tissues. J Appl Cardiol 4:207–221

7. Denk W, Strickler JH, Webb WW (1990) Two-photon laser scanning fluorescence microscopy. Science 248:73–76

8. Huang S, Heikal AA, Webb WW (2002) Two-photon fluorescence spectroscopy and microscopy of NAD(P)H and flavoprotein. Biophys J 82:2811–2825

9. Squirrell JM, Wokosin DL, White JG, Bavister BD (1999) Long-term two-photon fluorescence imaging of mammalian embryos without compromising viability. Nat Biotechnol 17:763–767

10. Bird D, Gu M (2003) Two-photon fluorescence endoscopy with a micro-optic scanning head. Opt Lett 28:1552–1554

11. Helmchen F (2002) Miniaturization of fluorescence microscopes using fibre optics. Exp Physiol 87:737–745

12. Jung JC, Schnitzer MJ (2003) Multiphoton endoscopy. Opt Lett 28:902–904

13. Baudelet C, Gallez B (2004) Effect of anesthesia on the signal intensity in tumors using BOLD-MRI: comparison with flow measurements by Laser Doppler flowmetry and oxygen measurements by luminescence-based probes. Magn Reson Imaging 22:905–912

14. Wokosin DL, Squirrell JM, Eliceiri KW, White JG (2003) An optical workstation with concurrent independent multiphoton imaging and experimental laser microbeam capabilities. Rev Sci Instrum 74:193–201

15. Skala MC, Riching KM, Gendron-Fitzpatrick A, Eickhoff J, Eliceiri KW, White JG, Ramanujam N (2007) In vivo multiphoton microscopy of NADH and FAD redox states,

fluorescence lifetimes, and cellular morphology in precancerous epithelia. Proc Natl Acad Sci U S A 104:19494–19499

16. Skala MC, Squirrell JM, Vrotsos KM, Eickhoff JC, Gendron-Fitzpatrick A, Eliceiri KW, Ramanujam N (2005) Multiphoton microscopy of endogenous fluorescence differentiates normal, precancerous, and cancerous squamous epithelial tissues. Cancer Res 65:1180–1186

Chapter 12

Using Fluorescence-Activated Flow Cytometry to Determine Reactive Oxygen Species Formation and Membrane Lipid Peroxidation in Viable Boar Spermatozoa

H. David Guthrie and Glenn R. Welch

Abstract

Fluorescence-activated flow cytometry analyses were developed for determination of reactive oxygen species (ROS) formation and membrane lipid peroxidation in live spermatozoa loaded with, respectively, hydroethidine (HE) or the lipophilic probe 4,4-difluoro-5-(4-phenyl-1,3-butadienyl)-4-bora-3a,4a-diaza-s-indacene-3-undecanoic acid, C_{11}BODIPY[581/591] (BODIPY). ROS was detected by red fluorescence emission from oxidization of HE and membrane lipid peroxidation was detected by green fluorescence emission from oxidation of BODIPY in individual live sperm. Of the reactive oxygen species generators tested, BODIPY oxidation was specific for FeSo4/ascorbate (FeAc), because menadione and H_2O_2 had little or no effect. The oxidization of hydroethidine to ethidium was specific for menadione and H_2O_2; FeAc had no effect. The incidence of basal or spontaneous ROS formation and membrane lipid peroxidation were low in boar sperm (<1% of live sperm) in fresh semen or after low temperature storage; however the sperm were quite susceptible to treatment-induced ROS formation and membrane lipid peroxidation.

Key words: C11-BODIPY[581/591], Lipid peroxidation, Hydroethidine, Flow cytometry, Motility

1. Introduction

The fertility of boar sperm after freeze-thawing or long-term hypothermic liquid storage is less than that of fresh liquid semen (1, 2). Part of this reduction in sperm fertility may be due to oxidative damage from inappropriate formation of reactive oxygen species (ROS) in the plasma membrane, cytoplasm, or nucleus. Sperm isolated from fresh (3), hypothermic liquid-stored (4), and thawed (5) boar semen are susceptible to the combination of

D. Armstrong (ed.), *Advanced Protocols in Oxidative Stress II*, Methods in Molecular Biology, vol. 594
DOI 10.1007/978-1-60761-411-1_12, © Humana Press, a part of Springer Science+Business Media, LLC 2010

FeSO$_4$ and ascorbate (FeAc) catalyzed lipid peroxidation as measured by malondialdehyde (MDA) formation. Chemiluminescence measurements of ROS formation, such as accumulation of superoxide and hydrogen peroxide (H$_2$O$_2$), are problematic because of poor specificity (6). As an alternative approach for boar spermatozoa, we modified a flow cytometric procedure that used hydroethidine (HE) to identify ROS in leukocytes (7, 8) by addition of the fluorochrome Yo Pro-1 (9), which is excluded from viable cells.

Measurement of MDA, an oxidation breakdown product derived from arachodonic acid (10), provides an estimate of lipid peroxidation in a population of cells (11), but has the disadvantages of poor sensitivity, inability to distinguish between viable and dead cells, and inability to provide an estimate of the proportion of the viable population that contains membrane lipid peroxidation. We modified a flow cytometric analysis of membrane lipid peroxidation based on oxidation of a fluorescent fatty acid conjugate 4,4-difluoro-5-(4-phenyl-1,3-butadienyl)-4-bora-3a,4a-diaza-s-indacene-3-undecanoic acid, C11-BODIPY[581/591] (BODIPY) by the addition of propidium iodide (PI) which is excluded from viable cells (12), similar to a procedure for bovine sperm (13). BODIPY is a membrane probe whose fluorescence changes irreversibly from red to green upon exposure to ROS (13, 14). Due to the high quantum yield of BODIPY-based probes and the high sensitivity of fluorescence techniques, the appearance of green fluorescence in BODIPY-labeled cells can be regarded as an indication of physiological relevant exposure of phospholipids to ROS (15, 16). We found that basal ROS formation and membrane lipid peroxidation in spermatozoa of fresh and frozen-thawed boar semen was low (present in less than 1% of viable cells). However, the spermatozoa were quite susceptible to treatment-induced ROS formation and membrane lipid peroxidation.

2. Materials

2.1. Equipment

1. SpermaCue micro-cuvette (catalog 12300/1100, Minitube of America)
2. Photometer (SpermaCue™, catalog 12300/0500, Minitube of America)
3. Epics XL-MCL Analyzer (Beckman-Coulter)
4. Hamilton Thorne IVOS version-12 motion analysis system (Hamilton Thorne Biosciences)
5. 20 µm four-chamber glass counting slide (SC 20-01 FA, Leja Products)

2.2. Reagents

2.2.1. Boar Sperm Extender

2.2.2. Percoll and Suspending Media

Beltsville Thawing Solution (BTS) (Minitube of America)

1. Percoll (P-1644, Sigma-Aldrich)

2. "10× saline-Hepes medium" consists of NaCl, 1.37 M; Hepes, 200 mM; glucose, 100 mM; and KOH, 25 mM (*see* Note 1).

3. "1× saline-Hepes medium" consists of one volume of 10× saline-Hepes medium diluted with 9 volumes of deionized water (approximately 290 mOsmol/kg, pH 7.4)

4. Prepare "Percoll stock suspension" by mixing 56.3 mL of Percoll with 5 mL of 10× saline-Hepes medium (final osmolality 311 mOsm/kg).

5. The 70% Percoll suspension is prepared by adding 13.16 mL of 1× saline-Hepes medium to 30.65 mL of the Percoll stock suspension (final osmolality typically 310 mOsm/kg).

6. The 35% Percoll suspension is prepared by adding 56.92 mL of 1× saline-Hepes medium to 30.65 mL of Percoll stock solution (final osmolality typically 300 mOsm/kg). Kanamycin is added to each Percoll suspension (50 µg/mL).

7. 4mL of 35% Percoll and 2 mL of 70% Percoll are transferred into 15 mL conical and 12×75 mm tubes, respectively, and stored at 4°C until use.

2.2.3. Modified Tyrode's Sperm Medium (mTYR)

1. Noncapacitating basal medium excluded bicarbonate and consists of 116 mM NaCl, 3.1 mM KCl, 0.4 mM $MgSO_4$, 0.3 mM NaH_2PO_4, 5 mM glucose, 21.7 mM sodium lactate, 1 mM sodium pyruvate, 20 mM Hepes, 0.1% polyvinyl chloride, and 50 µg kanamycin/mL (*see* note 2).

2. Medium pH is adjusted to 7.3–7.4 at 37°C

3. Osmolarity is adjusted to 300 mOsm/kg with NaCl.

2.2.4. Working Solution for Menadione

Menadione (M5625, Sigma Chemical Co.) is dissolved in dimethyl sulfoxide, 200 mM.

2.2.5. Working Solution for H_2O_2

H_2O_2 (216763, Sigma-Aldrich) is received as a 30% solution in deionized water

2.2.6. Working Solution for HE

HE (D-1168, Molecular Probes) is dissolved in dimethyl sulfoxide, 20 mM

2.2.7. Working Solution for Yo Pro-1

Yo Pro-1, (Y-3603 Molecular Probes) purchased in dimethyl sulfoxide (1 mM), is diluted in dimethyl sulfoxide, 0.1 mM.

2.2.8. Working Solution for FeAc

$FeSO_4$ (Iron (II) sulfate heptahydrate: 215422, Sigma-Aldrich, St. Louis) and Na ascorbate (sodium L-ascorbate: A7631 Sigma-Aldrich) in combination are dissolved in water (2 and 60 µM, respectively).

2.2.9. Working Solution for BODIPY^{581/591}	BODIPY$^{581/591}$ (3861, Molecular Probes, Inc., Eugene, OR) is dissolved in ethanol, 2 mM
2.2.10. Working Solution for PI	PI (P-4170, Sigma Chemical Co.) is dissolved in deionized water, 2.4 mM

3. Methods

3.1. Determination of Sperm Concentration

1. A 20 µL aliquot of either fresh boar semen diluted with one volume of BTS, or a washed sperm suspension ($150–400 \times 10^6$ sperm/mL) in mTYR or BTS is pipetted into a SpermaCue micro-cuvette.

2. The micro-cuvette is placed into the instrument's micro-cuvette holder and inserted into a photometer and analyzed for optical density calibrated to record concentration as 10^6 sperm/mL.

3.2. Removal of Seminal Plasma and Semen Extenders by 2-Step Discontinuous Percoll Gradient Centrifugation

1. Sperm are extended to a concentration of 150×10^6/mL in 3 mL of BTS or mTYR.

2. On day of use, the tubes containing 35 and 70% Percoll are warmed to room temperature and 70% Percoll (2 mL) is underlaid beneath 4 mL of 35% Percoll in 15 mL conical centrifuge tube.

3. The sperm suspension (3 mL) is layered on top of the 35% Percoll layer and then sperm are isolated from semen constituents and extenders by centrifugation at room temperature for 5 min at 200 g followed by 15 min at 1,000 g.

4. The medium is aspirated from the loosely pelleted sperm cells and washed once in mTYR at 350 g for 10 min.

5. Sperm in the pellet of each sample are resuspended, counted, and diluted to 60×10^6/mL in mTYR.

3.3. Computer-Assisted Motion Analysis

1. Sperm are incubated at a concentration of 30×10^6/mL at 37°C in a temperature-controlled water bath and aliquots are removed at specific times for motion analysis using an IVOS version-12 motion analysis system (12).

3.4. Menadione Induction of Reactive Oxygen Species

1. A total of 30×10^6 spermatozoa in 0.5 mL mTYR (pre-warmed in a 37°C water bath for 5 min) is added to a series of 17×100 mm polypropylene tubes containing 0 or 60 µM of menadione in 0.5 mL of mTYR (pre-warmed in a 37°C water bath for 5 min), which was prepared from the 200 mM working solution (*see* note 3).

2. Cells are incubated aerobically at 37°C for sampling at various times up to 120 min for motion analysis and flow cytometry.

3.5. H_2O_2 Treatment

1. A total of 30×10^6 spermatozoa 0.5 mL mTYR (pre-warmed in a 37°C water bath for 5 min) is added to a series of 17×100 mm polypropylene tubes containing 0 or 600 µM of H_2O_2 in 0.5 mL of mTYR (pre-warmed in a 37°C water bath for 5 min), which was prepared from the 30% working solution.

2. Cells are incubated aerobically at 37°C for sampling at various times up to 120 min for motion analysis and flow cytometry.

3.6. $FeSO_4$/Ascorbate Treatment

1. A total of 30×10^6 spermatozoa 0.5 mL mTYR (pre-warmed in a 37°C water bath for 5 min) is added to a series of 17×100 mm polypropylene tubes containing 0 µM of $FeSO_4$ and Na ascorbate or 0.5 mL of the $FeSO_4$/ascorbate working solution containing 2 and 60 µM, respectively (pre-warmed in a 37°C water bath for 5 min).

2. Cells are aerobically incubated at 37°C for various periods of time for up to 120 min for motion analysis and flow cytometry.

3.7. Loading of Spermatozoa with HE and Yo Pro-1 Staining

1. After the incubation treatments, aliquots of 1×10^6 sperm (33 µL) were transferred to 12×75-mm polypropylene tubes containing 467 µL of mTYR for a final concentration of 4 µM HE and 0.05 µM Yo Pro-1.

2. Cells were incubated for 40 min at 25°C before flow cytometric analysis.

3.8. Loading of Spermatozoa with BODIPY and PI Staining

1. BODIPY working solution is added to Percoll washed spermatozoa at a concentration of 60×10^6 sperm in 1 mL in mTYR to a final concentration of 2 µM from a working solution of 2 mM in ethanol.

2. After aerobic incubation at 37°C for 30 min, excess probe is removed by washing the cells once with mTYR (7 min × 300 g).

3. Experiments are then conducted in vitro using BODIPY-loaded spermatozoa as described in items 3.4, 3.5, and 3.6.

4. Just prior to cytometry, 1×10^6 spermatozoa/mL are transferred to 12×75 mm polypropylene tubes containing 0.5 mL of mTYR and are stained with PI at 9.6 µM from the working solution, and are incubated for 10 min at 25°C.

3.9. Fluorescence-Activated Flow Cytometry of HE Oxidation

1. A Beckman-Coulter Epics XL-MCL Analyzer (Beckman-Coulter, Hialeah, FL) equipped with a single 488 nm excitation source was used for all flow cytometric analyses.

2. After linear amplification, forward and side light scatter parameters are displayed for gating on the major population of single cells, excluding debris and aggregates.

3. Fluorescence from ethidium- and Yo Pro-1-stained sperm are collected in fluorescence detector 3 with a 620 nm band pass (BP) filter and fluorescence detector 1 with a 525 nm BP filter, respectively.

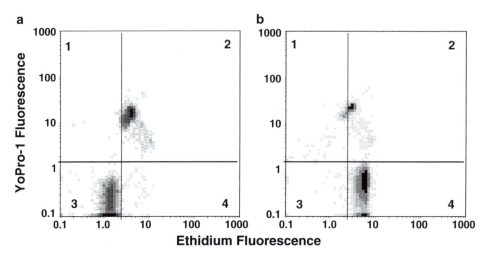

Fig. 12.1. Flow cytometric two-dimensional dot-plots of fluorescence (FL) intensities of 10,000 individual boar sperm from fresh semen stained with hydroethidine (HE) and Yo Pro-1. Sperm were incubated for 30 min at 37°C as controls, in the absence of menadione (panel **a**) and in the presence of 30 μM menadione (panel **b**). Events in quadrants 3 and 4 represent viable sperm with a low, basal level of Yo Pro-1 FL. Only 1.4% of the control population (panel **a**, quadrant 4) contained sufficient reactive oxygen species to oxidize HE to ethidium compared with 94.4% of the sperm incubated with menadione (panel **b**, quadrant 4).

4. Statistical regions were set manually to enumerate the total viable sperm population (Yo Pro-1-negative), as shown in Fig. 12.1 (quadrants 3 and 4 in panels A and B), and the viable subpopulation containing ROS, designated ethidium-positive, as shown in quadrant 4 of panels A and B.

3.10. Fluorescence Activated Flow Cytometry BODIPY Oxidation

1. A Beckman-Coulter Epics XL-MCL Analyzer is used as described in Subheading 3.9.

2. Green fluorescence from oxidation of BODIPY in individual sperm was collected in fluorescence detector 1 with a 525 nm band pass (BP) filter and red fluorescence from PI was collected in fluorescence detector 4 with a 675 nm BP filter.

3. Statistical regions (Beckman Coulter System II Software, Hialeah, FL) were drawn to delineate the viable sperm population (low-intensity PI fluorescence) and were used to determine the percentage of viable sperm, the percentage of viable sperm with lipid peroxidation (green BODIPY fluorescence) as shown in quadrants 1 and 3 of Fig. 12.2, and the mean BODIPY fluorescence intensity/cell (mean channel fluorescence number) as shown in quadrant 1 of Fig. 12.2.

3.11. Effects of ROS Generators

3.11.1. Specificity of BODIPY and HE Oxidation

1. Boar semen ejaculates were divided into 4 portions for aerobic incubations at 37°C with four treatments: control (no ROS generators), 30 μM MEN, 300 μM H_2O_2, and 1/30 μM FeAc (Tables 12.1 and 12.2).

2. In control spermatozoa, <1% of the sperm population contained oxidized HE (Table 12.1) or oxidized BODIPY

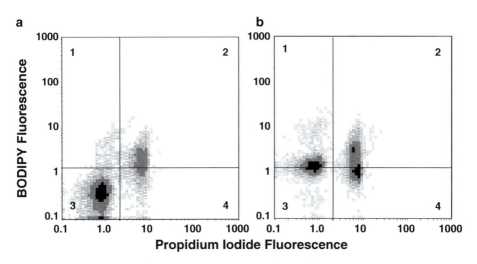

Fig. 12.2. Flow cytometric two-dimensional dot-plots of fluorescence intensities of 10,000 individual boar sperm from thawed semen loaded with C_{11}-BODIPY[591/691] (BODIPY) and stained with propidium iodide. Sperm were incubated for 30 min at 37°C with a control treatment (no ROS generator) (panel **a**) and with $FeSO_4$-Na ascorbate (FeAc) 1 and 30 µM, respectively (panel **b**). Propidium iodide was added to distinguish between membrane intact and permeant cells. Data in quadrants 1 and 3 represent viable sperm with basal, low-intensity propidium iodide fluorescence. In control spermatozoa (panel **a**, quadrant 1) only 4% of viable thawed sperm contained oxidized BODIPY (quadrant 1). The presence FeAc (panel **b**, quadrant 1) increased the percentage of sperm containing oxidized BODIPY to 78% of the viable sperm.

Table 12.1
Comparison of the effects of H_2O_2, menadione (MEN), and $FeSO_4$-Na ascorbate (FeAc) on hydroethidium (HE) oxidation in viable sperm of boars

H_2O_2 (µM)	MEN (µM)	FeAc (µM)[a]	Sperm with oxidized HE (%)	Ethidium fluorescence intensity/sperm[b]	Sperm motility (%)
0	0	0/0	0.6 ± 0.3^c	1.51 ± 0.15^c	72 ± 4^c
0	30	0/0	99.5 ± 0.1^c	6.69 ± 0.13^c	20 ± 7^c
300	0	0/0	93.7 ± 5.9^c	5.30 ± 0.66^c	4 ± 4^c
0	0	1/30	0.5 ± 0.2^c	1.32 ± 0.14^c	40 ± 8^c

Values are means ± SEM for spermatozoa from 3 boars aerobically incubated for 120 min at 37°C
[a]FeAc = a combination of $FeSO_4$ and Na ascorbate, expressed as µM/µM
[b]Mean channel fluorescence number, fluorescence detector 3 output
[c]Means, within a column with no superscript letter in common, differ ($P < 0.05$)

(Table 12.2); this level of oxidation did not change between 30 and 120 min of incubation at 37°C (data not shown).

3. Compared with control spermatozoa, incubation with 300 µM H_2O_2 or 30 µM menadione increased ($P < 0.05$) the percentage of sperm with HE oxidation to greater than 90% and increased ($P < 0.05$) ethidium fluorescence intensity/cell three to fourfold (Table 12.1). By contrast, in comparison to the control spermatozoa, FeAc did not oxidize additional HE.

Table 12.2

Comparison of the effects of H_2O_2, menadione (MEN), and FeSO$_4$-Na ascorbate (FeAc) on C_{11}-BODIPY[581/591] (BODIPY) oxidation in viable sperm of boars

H_2O_2 (μM)	MEN (μM)	FeAc (μM)[a]	Sperm with oxidized BODIPY (%)	BODIPY fluorescence intensity/sperm[b]
0	0	0/0	0.2 ± 0.1^c	0.38 ± 0.01^c
0	30	0/0	0.2 ± 0.1^c	0.36 ± 0.02^c
300	0	0/0	2.0 ± 0.7^c	0.48 ± 0.04^c
0	0	1/30	99.0 ± 0.4^c	3.66 ± 0.80^c

Values are means ± SEM for spermatozoa from 4 boars aerobically incubated for 120 min at 37°C
[a]FeAc = a combination of FeSO$_4$ and Na ascorbate, expressed as $\mu M/\mu M$
[b]Mean channel fluorescence number, fluorescence detector 1 output
[c]Means, within a column with no superscript letter in common, differ ($P < 0.05$)

4. The FeAc treatment increased ($P < 0.01$) the mean percentage of sperm containing oxidized BODIPY to 99% and increased ($P < 0.05$) BODIPY fluorescence intensity/cell ninefold, compared to the control group (Table 12.2). Compared with control spermatozoa, incubation with menadione and H_2O_2 had no effect on the percentage of sperm containing oxidized BODIPY or on the cells' BODIPY fluorescence intensity.

3.12. Effects of ROS Generators on Sperm Viability and Motility

1. Sperm viability, as measured by the exclusion of PI (for BODIPY loaded cells) or Yo-Pro-1 (for HE loaded cells) from the spermatozoa, averaged 92% (data not shown) and was not affected by presence of ROS generators ($P = 0.84$ and $P = 0.88$, respectively) or by incubation time up to 2 h ($P = 0.07$ and 0.30, respectively). These methodologies are sensitive enough to investigate the effects of ROS formation and membrane lipid peroxidation on sperm functions such as motility and mitochondrial function in an early stage of the processes before the cells have suffered enough damage to cause death.

2. Sperm motility was 72% in control spermatozoa (Table 12.1) following a 2 h incubation; the presence of ROS generators decreased motility during the same period, especially for H_2O_2.

4. Notes

1. The 10× saline-Hepes can be conveniently stored frozen 20 mL aliquots as 10 mL is needed for the 1× buffer and 5 mL for the base Percoll solution.

2. Sodium pyruvate should be added to the mTYR on day of use.

3. Transfer of menadione from the reactant tube containing mTYR needs to be done while vigorously mixing on vortex mixer to keep menadione in solution.

References

1. Johnson LA (1985) Fertility results using frozen boar spermatozoa: 1970–1985. In: Johnson LA, Larsson K (eds) Deep freezing of Boar semen. Swedish Univ. Agric Sci, Uppsala, pp 199–222

2. Waberski D, Weitze KF, Gleumes T, Schwarz M, Willmen T, Petzoldt R (1994) Effect of time of insemination relative to ovulation on fertility with liquid and frozen boar semen. Theriogenology 42:831–840

3. Roca J, Gil MA, Hernandez M, Parrilla I, Vazquez JM, Martinez EA (2004) Survival and fertility of boar spermatozoa after freeze-thawing in extender supplemented with butylated hydroxytoluene. J Androl 25:397–405

4. Cerolini S, Maldjian A, Surai P, Noble R (2000) Viability, susceptibility to peroxidation and fatty acid composition of boar semen during liquid storage. Anim Reprod Sci 58:99–111

5. Breininger E, Beorlegui NB, O'Flaherty CM, Beconi MT (2005) Alpha-tocopherol improves biochemical and dynamic variables in cryopreserved boar semen. Theriogenology 63:2126–2135

6. Fridovich I (2003) Editorial commentary on "Superoxide reacts with hydroethidine but forms a fluorescent product that is distinctly different from ethidium: Potential implications in intracellular fluorescence detection of superoxide" by H. Zhao et al. Free Radic Biol Med 34:1357–1358

7. Bass DA, Parce JW, DeChatelet LR, Szejda P, Seeds MC, Thomas M (1983) Flow cytometric studies of oxidative product formation by neutrophils: a graded response to membrane stimulation. J Immunol 130:1910–1917

8. Carter WO, Narayanan PK, Robinson JP (1994) Intracellular hydrogen peroxide and superoxide anion detection in endothelial cells. J Leukoc Biol 55:253–258

9. Guthrie HD, Welch GR (2006) Determination of intracellular reactive oxygen species and high mitochondrial membrane potential in percoll-treated viable boar sperm using fluorescence-activated flow cytometry. J Anim Sci 84: 2089–2100

10. Spiteller G (2006) Peroxyl radicals: inductors of neurodegenerative and other inflammatory diseases. Their origin and how they transform cholesterol, phospholipids, plasmalogens, polyunsaturated fatty acids, sugars, and proteins into deleterious products. Free Radic Biol Med 41:362–387

11. Storey BT (1997) Biochemistry of the induction and prevention of lipoperoxidative damage in human spermatozoa. Mol Hum Reprod 3:203–213

12. Guthrie HD, Welch GR (2007) Use of fluorescence-activated flow cytometry to determine membrane lipid peroxidation during hypothermic liquid storage and freeze-thawing of viable boar sperm loaded with4, 4-difluoro-5-(4-phenyl-1,3-butadienyl)-4-bora-3a,4a-diaza-s-indacene-3-undecanoic acid. J Anim Sci 85: 1402–1411

13. Brouwers JF, Gadella BM (2003) In situ detection and localization of lipid peroxidation in individual bovine sperm cells. Free Radic Biol Med 35:1382–1391

14. Pap EH, Drummen GP, Winter VJ, Kooij TW, Rijken P, Wirtz KW, Op den Kamp JA, Hage WJ, Post JA (1999) Ratio-fluorescence microscopy of lipid oxidation in living cells using C11-BODIPY(581/591). FEBS Lett 453:278–282

15. Drummen GPC, van Liebergen LCM, Op den Kamp JAF, Post JA (2002) C11-BODIPY[581-591], an oxidation-sensitive fluorescent lipid peroxidation probe: (micro) spectroscopic characterization and validation of methodology. Free Radic Biol Med 33:473–490

16. Haugland RP (2005) The handbook, 10th edn. Molecular Probes Inc., Eugene

Chapter 13

Lipofuscin: Detection and Quantification by Microscopic Techniques

Tobias Jung, Annika Höhn, and Tilman Grune

Abstract

Since lipofuscin, the so-called "aging pigment", turned out to play a fundamental role in the aging process, particularly in the postmitotic senescence of muscle or neuronal cells, it became a focus of aging and stress research. During normal aging, lipofuscin accumulates in a nearly linear way, whereas its rate of formation can increase in the final stages of senescence or in the progress of several pathologic processes.

Thus, both in senescence and pathologic processes, lipofuscin can be used as a detectable "marker" to estimate the remaining lifetime of single cells, the amount of long-term oxidative stress cells were subjected to or to quantify and qualify a pathologic progress *in vivo* or *in vitro*. To enable this, a quick and easy applicable method of detection and quantification of lipofuscin has to be used, as is provided by fluorescence microscopy determining the autofluorescence via of the "aging pigment".

In this review, we take a look at different methods of detection and quantification of lipofuscin in single cells by using its physical or chemical features.

Key words: Lipofuscin, Fluorescence microscopy, Confocal laser scanning microscopy, Autofluorescence, Protein oxidation, Immunocytochemistry

1. Introduction

1.1. Lipofuscin

Lipofuscin (1-4), in the literature sometimes named "ceroid" (5-9), "age fluorophore" or "age pigment" (10, 11) is a strongly oxidized material build of covalently cross-linked proteins (12, 13), lipids (14, 15) and even (oligo)saccharides (16). Moreover, it contains up to 2% transition metals such as iron, zinc, manganese, and copper (17). Though these metals can cause a redox active surface, they can catalyze radical releasing processes like the Fenton reaction. This is discussed in this chapter.

In transmission light microscopy, lipofuscin is visible as a perinuclear-distributed material in the cell, while mostly being included

D. Armstrong (ed.), *Advanced Protocols in Oxidative Stress II*, Methods in Molecular Biology, vol. 594
DOI 10.1007/978-1-60761-411-1_13, © Humana Press, a part of Springer Science + Business Media, LLC 2010

in the lysosomal compartment. Sometimes lipofuscin is termed as "age pigment", since it shows a nearly linear accumulation in cells, tissues, and whole organs over time during aging and is negatively correlated with the remaining life expectancy (18, 19).

This feature and the fact that increased lipofuscin formation and accumulation are playing a key role in several pathologies, such as Batten's disease (20-23), other ceroid lipofuscinoses (24), or age-related macular degeneration (AMD) (25-27), made the mechanisms of its origin an interesting object of investigation, since it has been discovered by Hannover in 1842 (28). The interactions of lipofuscin with cellular functions are able to increase its rate of formation, resulting in a vicious cycle, finally causing cellular malfunction and death.

The intracellular lipofuscin formation is a very complex network of compartmentalized biochemical and chemical reactions located in lysosomes, the cytosol and the mitochondria.

1.2. A Model of Lipofuscin Formation

One of the presently existing hypothesis of lipofuscin formation is the so-called "mitochondrial-lysosomal axis theory of postmitotic cellular aging", formulated by Brunk and Terman in 2002 (29). This model is on principle depicted in Fig. 13.1.

Mitochondria are the accredited main sources of free radicals in mammalian cells, producing the primary radical superoxide ($O_2^{\bullet-}$) by electron leakage from the respiratory chain, which generates further radicals or highly reactive particles in following reactions. Though mitochondria are the organelles most susceptible to oxidative stress, and increase their radical production when being oxidatively damaged, oxidative damage might inhibit mitochondrial division, resulting in the formation of so-called "giant" or "enlarged mitochondria" by incorporation of additional proteins and lipids needed for duplication into the organelle (29). Since these damaged mitochondria show an increased radical release, a vicious cycle starts that is usually ended by incorporation and degradation of the mitochondria by autophagic endosomes and lysosomal system. Within such autophagosomes, numerous hydrolases are able to degrade mitochondrial structures. However, if intramitochondrially covalently cross-linked proteins are formed, these are resistant to lysosomal hydrolases. Often such cross-linked proteins are formed due to cross-linking reactions by products of lipid peroxidation such as malondialdehyde (MDA) (30, 31) or 2-hydroxy-4-*trans*-nonenal (HNE) (32). Those non-degradable structures might be the "initiation cores" of growing lipofuscin clusters that accumulate material driven by its hydrophobic surface and covalently linked by further oxidation. The fact that mitochondria contain large amounts of iron, which are able to catalyze the Fenton reaction, can be an explanation for increased protein oxidation and lipid peroxidation that is necessary for the genesis lipofuscin.

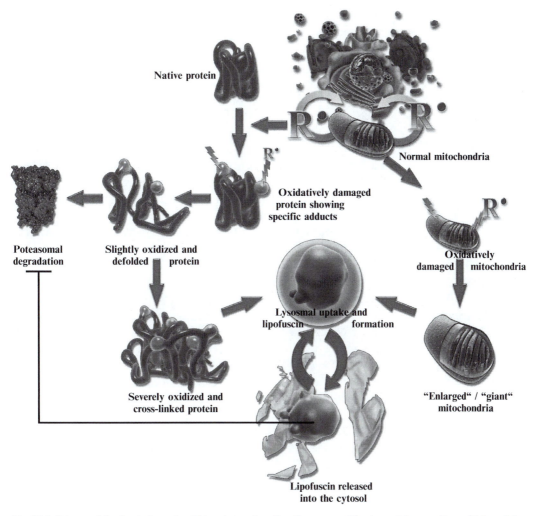

Fig. 13.1. Scheme of lipofuscin formation. This scheme describes the proposed fundamental connections of intracellular lipofuscin formation and the major other involved compartments and enzymatic processes.

Taking up multiple-damaged mitochondria and other cellular oxidized structures, with time the lipofuscin in the lysosomes is growing and perhaps impaired in its function. Eventually such a lysosome is so heavily loaded, that it ruptures, spilling the lipofuscin into the cytosol (33). Such material is taken up again by the endosomal-lysosomal compartment. Moreover, it has been shown that lipofuscin, even though it is only found in low amounts free in the cytosol, is able to inhibit the proteasome competitively (34). The result is an increased amount of oxidized proteins that are not degraded (35, 36) but further oxidized representing preliminary stages of lipofuscin. There are two arguments for this hypothesis: the fact that about 50% of some of the *in vivo* occurring lipofuscin are built of the remains from subunit c of the mitochondrial ATP synthase in CNS neurons (37, 38) and that lysosomal membranes contain a very high amount of α-tocopherol (39) in comparison to other cell membranes (up to 37-fold increase) (40), fulfilling the function of an antioxidant. Even though the source of the redox elements, which are essential for transition metal catalyzed radical genesis, still has to be figured out, a connection of vitamin E deficiency and an increased formation of lipofuscin has already been shown (41). Other data suggest a correlation of damaged mitochondria and a deficiency of vitamin E (42).

Tough work has been published showing a lack of oxidatively damaged mitochondria in the same time with a decrease of protein oxidation in aged skeletal muscle cells, but showing still linear increase of the amount of lipofuscin during aging (43).

Obviously, the biochemical interactions resulting in lipofuscin formation are very complex and an efficient method of intracellular lipofuscin detection might help to enlighten this system.

1.3. Methods of Lipofuscin Detection and Quantification

Particularly since there is no available antibody specific for lipofuscin, its autofluorescence has become the most important tool in detection using fluorescence microscopy. Despite of its autofluorescence, other histochemical methods are available such as techniques for lipid staining and lipophile agents: Sudan Black, Berlin Blue, Nile Blue, Fontana-Masson, osmic acid, hematoxilin, ferric ferrycianide, Ziehl-Neelson or Eosin. Transition light microscopy without staining shows lipofuscin as a transparent or yellow-brownish material (44).

A representative spectrum of *in vivo* lipofuscin shows an absorption maximum at 366 nm and an emission maximum in the range of 570–605 nm. Probably these spectral characteristics are caused by reacting carbonyls and amino groups forming fluorescent Schiff bases such as 1,4-dihydropyridines or 2-hydroxy-3-imino-1,2-dihydropyrrol derivates (45) (Fig. 13.2), which present a spectrum of autofluorescence that is comparable to that of lipofuscin. One of the very few identified fluorophores of lipofuscin is A2E (46) (Fig. 13.2), a pyridinium bis-retinoid found in retinal pigment cells showing a Schiff base structure (N-retinyl-N-retinylidene ethanolamine), resulting from reacting retinal-derivates and supposed to be contributing to the phototoxicity of lipofuscin resulting in retinal atrophy in AMD (47).

All these molecules have aromatic ring structures, containing delocalized π-electrons, the physical precondition to show fluorescence. Therefore, autofluorescence of lipofuscin has become the primary tool to detect lipofuscin. This can be done by fluorescence spectroscopy, by flow cytometry and by fluorescence microscopy, the methods we are focussing on here.

Lipofuscin, its autofluorescene, particularly, is detectable in virtually all kind of animal cells, especially in cells that have reached a postmitotic state. Fast dividing cells have the ability to "dilute" lipofuscin by mitosis to an almost nondetectable amount.

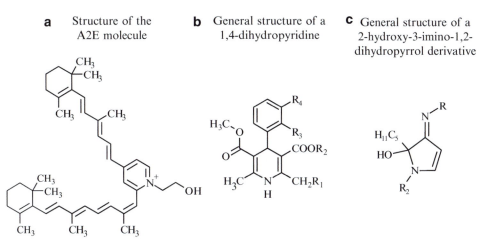

a Structure of the A2E molecule

b General structure of a 1,4-dihydropyridine

c General structure of a 2-hydroxy-3-imino-1,2-dihydropyrrol derivative

Fig. 13.2. Chemical structures of the fluorescent compounds found in lipofuscin. (**a**) The structure of the pyridinium bis-retinoid A2E, a blue light-phototoxic (67) compound found in retinal lipofuscin (68) a lipid peroxidation product (69), showing autofluorescence in CLSM if its excited using an Argon laser (458 nm) (70). (**b**) The general structure of a 1,4-dihydropyridine resulting from reactions of MDA and glycine (71-73). (**c**) The general structure of a 2-hydroxy-3-imino-1,2-dihydropyrrol derivative, supposed to be a results from an oxidative cyclization of a lysine-HNE Michael adduct Schiff-base cross-link, before cyclization not showing autofluorescence (45).

Investigations showed a perinuclear intracellular distribution, whereas the lipofuscin exists in globular structures showing diameters in a range from 0.1 to 5.0 μm. In most cases, lipofuscin seemed to be localized near the Gogli apparatus or the nucleus of the cell mostly colocalized with the lysosomal compartment.

1.4. Technical Approaches to Detect Lipofuscin by Fluorescence Microscopy

At present, two different kinds of fluorescence microscopy are used: the common fluorescence microscopy (FM) and the more recent confocal laser scanning microscopy (CLSM) (48, 49), suggested by Marvin Minsky 1955. These two variations of fluorescence microscopy are virtually used in every corresponding investigation. A third and very interesting new method varying common confocal laser scanning microscopy is the so-called Stimulated Emission Depletion (STED)-microscopy (50), which is actually able to evade Ernst Abbe's diffraction limit, resulting in magnifications up to 13-fold higher than known from FM/CLSM and thus able to resolve structures of about 15 nm (lateral). This common fluorescence microscopic method, developed by the group of Stefan Hell from Göttingen (Germany), is still highly experimental and the first commercial available microscopes are still expensive.

The generally known FM, just like the CLSM method, uses the characteristics of certain molecular structures (fluorophores) to absorb photons in a certain range of energy, resulting in a very short-lived (10^{-9}–10^{-5} s) intramolecular electron shift from a lower (S_0) to a higher (S_2) state of energy. If the electron falls back from S_2 to a lower state of energy S_1 and finally to S_0, it emits a photon

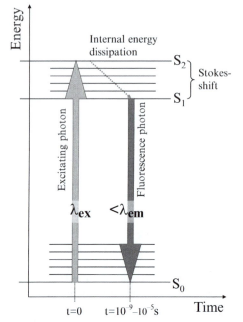

Fig. 13.3. The Jablonski-diagram. This graph shows the energy transition of an electron from the S_0 state to the S_2 state induced by the absorption of an exciting photon. The transition from S_2 to S_1 happens without radiation and the energy difference is released in the form of thermic vibrations. The transition from S_1 to S_0 after 10^{-9} to 10^{-5} s is accompanied by the emission of a single photon that is detected as fluorescence emission light. The energy difference between S_2 and S_1 is manifest in the so-called Stokes-shift.

with a lower energy (higher wavelength) than the previous absorbed one. This process is termed "fluorescence" and depicted in the Jablonski-diagram (Fig. 13.3).

The energy difference (Stokes-shift) between the absorbed and the emitted photons, resulting in an increase of the wavelength of about 20–80 nm, enables to distinguish between excitation and emission light. In FM/CLSM, this is realized by the use of different filter-sets.

The binding of such a fluorochrome to a highly specific antibody enables the fluorescence labeling of a cell, respectively cellular structures or compartments (fluorescence immunohistochemistry). Combinations of antibodies with different specificity and different attached fluorochromes make a colocalization of various cellular structures at the same time possible.

The main difference between FM and CLSM is found in the according depth of focus, which means the thickness of the focal xy-layer in the direction of the z-axis of the investigated sample (in this case, a single cell).

Using common FM, the whole investigated sample is exited by a high photon density; in this case also, a large amount of scattered light from the whole sample enters the objective and thus light from sample volumes above and beyond the focal plane. This scattered light decreases the contrast of the mapped image, and the illumination of the whole sample results sometimes in a quick bleaching of the permanently excited fluorochromes. FM is only able to yield a sharply defined image if the thickness of the sample is smaller or equal to the optical depth of focus of the used objective. Using an objective with 40-fold magnification and an optical aperture of 0.75 that thickness is about 1.5 μm, using oil immersion and a 63/1.4-objective it can be decreased to ≈1 μm. Considering that an adherent cell is about 5 μm high, common FM is not able to focus the whole object at the same time.

The main advantage of CLSM in contrast to FM is its ability to fade out scattered light from nonconfocal layers. CLSM enables the observation of a single highly defined xy-layer along the z-axis of the sample. The thickness of this confocal layer is regulated by the so-called "pinhole". Decreasing the diameter of the pinhole results in a thinning of the confocal plane, even though at the same time, the amount of detectable fluorescence light is reduced. A pinhole at maximal diameter, on the other hand, turns CLSM into common FM. The low amount of scattered light entering the microscopic objective increases the contrast of the visible image. The thickness of the visible layer can be lower than 0.5 μm (Fig. 13.4); the lateral optic resolution in the *xy*-layer of a confocal system is not increased in comparison to FM until the pinhole reaches extreme low diameters. In the optimum, the lateral resolution can be increased up to 1.4-fold (theoretical value) (51, 52). Moreover, the CLSM resolution is only limited by the excitation wavelength and not by the emission wavelength of the fluorochrome as in the case of FM.

Fluorescence excitation in CLSM is performed punctiform by a focused laser beam, rasterizing the sample line-by-line, causing only a low bleaching of the fluorochromes outside of the confocal layer by a lowered photon density, and at the same time minimizing the amount of scattered light. In this way, it is possible to dissect the sample following the z-axis in thin slices, showing a cross section of the object (optical slicing). Modern software enables slices following various object layers and a three-dimensional reconstruction from the obtained data sets. Thus, in CLSM investigations of the spatial shape of (intra)cellular structures and in living cells, even the exploration of their dynamics are possible with precise intracellular localization of structures that would be impossible using common FM (Fig. 13.5).

To illustrate this, (Fig. 13.6) the same cell is depicted using common FM and CLSM. CLSM reveals that the nucleus of the cell is almost completely free of oxidative protein modification, whereas it is impossible in common FM to discern if the fluorescence is emitted from the nucleus or the layer of cytosol above.

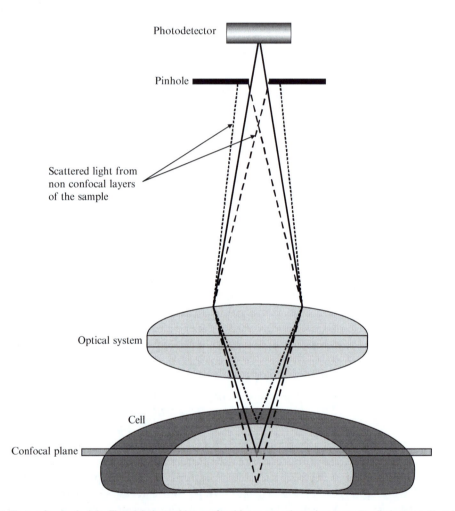

Fig. 13.4. The confocal principle. The pinhole enables confocal laser scanning microscopy to eliminate scattered light from non confocal layers (*a dashed lines*) of the sample. Since only light from the confocal layer (continuous line) is detected by the photosensor, the contrast of the image is increased compared to common fluorescence microscopy; without the pinhole CLSM becomes common FM.

Fig. 13.6. (continued) oxidative protein modification (own data). The six CLSM images beyond show different equidistant confocal *xy*-planes of this cell along its *z*-axis. The schematic drawing beyond shows a cross section of this cell as a simplified model. The position of this cross section is depicted in the top right image showing the cell using CLSM as white dashed line. The numbers from 1 to 6 show the positions of the corresponding confocal slices following the cell's *z*-axis. The CLSM images 5 and 6 clearly show the "dome" of cytosol above the cells nucleus. Beyond the nucleus in attached cells is virtually no cytosol detectable.

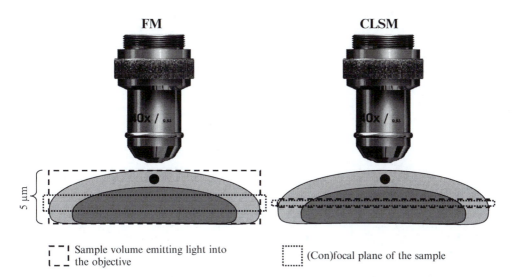

Fig. 13.5. Fluorescence and confocal laser scanning microscopy. This image depicts the main difference between fluorescence (FM, *left side*) and confocal laser scanning microscopy (CLSM, *right side*). In FM fluorescence light from the whole cell, respectively the immunolabeled cellular structure enters the objective (*dashed line*), decreasing the contrast of the focal plane (*dotted line*). In CLSM only light from the confocal plane is visible, resulting in a very contrasty image. The *black dot* represents an object located in the cytosol of the cell. In contrast to FM, CLSM allows an exact determination of its intracellular position and to distinguish between a nuclear and cytosolic localization. So in the here presented FM example this structure would be visible, whereas in the CLSM not.

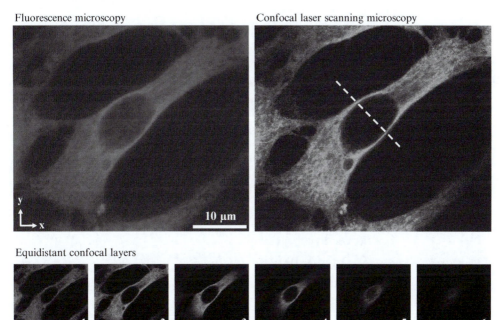

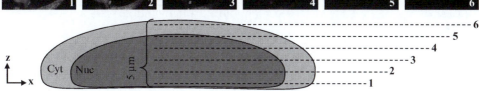

Fig. 13.6. Fluorescence and confocal laser scanning microscopy compared. The two upper images show the same cell viewed using FM (*left image*) and CLSM (*right image*), with immunocytochemical labeling of protein carbonyls, an

2. Materials

2.1. Equipment

1. 4/−20°C storage
2. Cell culture incubator
3. Pipettes
4. Exhauster
5. Vortexer
6. Centrifuge (≥12,000 g, capable for 50 mL vials)
7. Fluorescence/confocal laser scanning microscope

2.2. Reagents and Supplies

2.2.1. Preparation of Cell Culture Materials

1. Cell culture plastic ware
2. Poly-D-lysine (sterile)
3. Collagen (sterile)
4. Sterile and de-ionized water
5. Sterile filters (≤0.2 μm pore size)
6. Phosphate-buffered saline (PBS, 1× PBS: 8.0 g NaCl, 0.2 g KCl, 1.44 g Na_2HPO_4, 0.24 g KH_2PO_4, adjusted to pH = 7.4 using HCl, sterile for cell culture)
7. Acetic acid (sterile)
8. Parafilm

2.2.2. Cell Preparation for FM/CLSM

1. Cell culture dishes/glass bottom dishes
2. Trypsin (sterile)
3. EDTA (sterile)
4. PBS (see above)

2.2.3. Combination with Immunocytochemistry

1. Primary antibodies, matching the used cell line
2. FITC/TRITC-labeled secondary antibodies matching the used primary antibody
3. Ethanol (≥96%)
4. Diethylether (≥99%)
5. Triton X-100
6. Fetal calf serum (FCS, sterile and heat inactivated for cell culture)

2.2.4. Fluorescence/ Confocal Laser Scanning Microscopy of the Sample

1. DAPI (4′,6-diamidino-2-phenylindole)

3. Methods

3.1. Selection of the Microscopic Method

Before the cells are prepared, the appropriate method, i.e., FM or CSLM has to be chosen according to the object to be investigated. If both methods are applicable, the time-saving method should be used, that is, in most cases, common FM. Advantages and disadvantages of FM and CLSM are listed in Table 13.1.

3.2. Sample Preparation

Several cell lines detach during the procedure of several washings. Therefore, either poly-D-lysine or collagen coating can be used to keep the cells attached.

3.2.1. Poly-D-Lysine Coating of Culture Dishes

3.2.1.1. Poly-D-Lysine Solution

1. Dissolve 5 mg of sterile poly-D-lysine (Sigma-Aldrich P-6407) in 100 mL of sterile H_2O (50 µg/mL final concentration); if no sterile material is available, sterilize the solution after preparation using a sterile filter ≤0.2 µm pore size

2. Vortex until complete dissolution (long-term storage of the solution at –20°C, short-term storage at 4°C)

3.2.1.2. Preparation of Poly-D-Lysine Coated Culture Dishes

1. Cover the whole surface of the sterile culture dish with poly-d-lysine solution

2. Incubate at room temperature (25°C) for 12 h

3. Exhaust solution and wash the culture dish gently using sterile 1× PBS

4. Air dry the culture dish and use immediately or store at 4°C (sealed with parafilm)

3.2.2. Collagen Coating of Culture Dishes

1. Prepare a sterile collagen (Sigma-Aldrich) solution with a final concentration of 2 mg/mL in 0.2% acetic acid in H_2O (if sterile filter solution is needed as described above, store the stock solution at 4°C)

2. Cover your dishes with a 1:5 diluted solution in sterile H_2O

Table 13.1
Selection of the adequate microscopic method

Investigative objective	FM	CLSM
Quantification of fluorescence intensity	++	+
Intracellular distributions	+	++
Colocalization of different structures	+	++
(Intracellular) distributions even in tissues	––	++
3D reconstruction from microscopical data	––	++

3. Incubate dishes at 37°C for 1 h

4. Exhaust collagen solution and wash dishes gently using sterile 1× PBS

5. Air dry culture dishes and use immediately or store at 4°C (parafilm-sealed, not longer than 2 weeks)

3.3. Cell Preparation for FM/CLSM

It is best suitable for FM/CLSM that investigations are slowly dividing or postmitotic and adherent cells. Fast dividing and cells that have to be attached by collagen or poly-D-lysine to an object slide are less suitable, i.e., they produce lower quality of microscopical images. From the authors' experience, very good results were obtained using human fibroblasts accumulating large amounts of autofluorescent material in late stages of mitotic and postmitotic aging.

Cell culture is done using normal culture flasks, according to the used cell line. Cells are to be plated out in Petri dishes before microscopic investigation. For inverse FM/CLSM, the use of sterile glass bottom dishes is suggested. In every case, the cell density should be low enough to ensure the microscopy of single nonoverlapping cells, simplifying the evaluation of the resulting data. While seeding, the division rate of the used cell line has to be considered too, in order to receive optimal conditions for microscopy.

If the autofluorescence of lipofuscin has to be colocalized with another cellular structure using immunocytochemistry, an antibody with a nonoverlapping emission spectra has to be chosen; in this case, fluorescein-5-isothiocyanat (FITC), another fluorescent antibody showing an absorption maximum at 490 nm and an emission maximum at 520 nm is recommended (*see* Note 1).

Self-attaching cells can be singularized using trypsination and plated out for FM/CLSM:

1. Prepare a solution containing 0.05% trypsin (500 µg/mL) and 0.02% EDTA (200 µg/mL) dissolved in 1× PBS (without Ca^{2+} and Mg^{2+})

2. Remove cell medium from culture flask

3. Incubate cells with trypsin/EDTA solution at 37°C for max. 2–3 min (not more than 3 mL for a T75 flask)

4. If cells are oversensitive to trypsin, substitute that step by scraping

5. Pipette cells several times to singularize them effectively

6. Transfer cell-trypsin-suspension into a falcon tube, add the five-fold amount of 1× PBS to dilute trypsin and centrifuge cells down ($\approx$12,000 g)

7. Discard supernatant and resuspend carefully cell pellet in a proper amount of full medium to adjust desired cell density

8. Transfer the amount of cell suspension needed to guarantee low cell density into a (coated if necessary) culture dish (consider rate of cell division) appropriate for FM/CLSM

9. Keep cells under growth conditions in full medium for 16–24 h to ensure cell attachment

If fixation of the cells is needed, in the case of additional immunocytochemistry, poly-D-lysine coated dishes should be used to avoid excessive cell loss during the following preparation steps.

3.4. Combination with Immunocytochemistry

If the autofluorescence of lipofuscin has to be colocalized with another cellular structure using immunocytochemistry, an antibody with nonoverlapping emission spectra has to be chosen. It is recommended to use fluorescein-5-isothiocyanat (FITC), another fluorescent antibody showing an absorption maximum at 490 nm and an emission maximum at 520 nm.

Fixation of the cells for immunocytochemistry

1. Plated cells are washed exhaustively using 1× PBS

2. Fixation using a mixture of ethanol:diethylether (1:1) at –20°C for not more than 10 min

3. Cells are washed three times with 1× PBS

4. Optional step: to increase antibody-access samples are incubated at room temperature for maximal 5 min using a 0.5% solution of Triton X-100 in 1× PBS

5. Cells are washed again and are incubated in 1% fetal calf serum (FCS) in 1× PBS in order to block unspecific binding sites

6. After washing, cells are incubated for at least 1 h at 4°C/room temperature, using the according primary antibody following the producers' advice (range of dilution is usually in the range of 1:80–1:200 in 1× PBS containing 1% FCS)

7. After another exhaustive washing using 1× PBS, the secondary antibody, showing a maximum in its emission spectra >500 nm, is applied (1 h at 4°C/room temperature, protected from light)

8. Samples are washed again followed by immediate microscopy

3.5. Fluorescence Microscopy of the Sample

The autofluorescence (*see* Note 2) of the sample is detected using the DAPI-excitation wavelength (320–430 showing a maximum at 366 nm) and an emission maximum ranging from 460–630 nm (peaks from 570–605 nm) (53). The fact that the absorption and the emission peaks show a larger distance than expected from simple Stokes-shift, suggests several intramolecular energy-transfers between different electron systems, before the absorbed energy is finally emitted in form of a "multi-Stokes-shifted" fluorescence photon. This is also the explanation for the broad emission and absorption spectra displayed by lipofuscin.

Technically, there might be a problem in common fluorescence microscopy to excite the sample at 366 nm while detecting the emission at ≥570 nm, so an adequate filter set has to be chosen, that matches representative characteristics of the lipofuscin spectrum.

In common FM, there are usually three filter sets available, that are not overlapping and clearly to discern from each other: 400/460 nm (I_{ex}/I_{em}), 495/530 nm and 570/610 nm. This is the so-called "triple bandpass", sometimes represented by the fluorescences of the three often used and clearly separable fluorochromes DAPI/FITC/Cy3 using the adequate filter sets. Though, the exact wavelengths of each single bandpass can vary from different fluorescence microscopes.

If only the autofluorescence of the cells has to be detected, the samples don't have to be fixed before microscopy, but should be stored under growth conditions to the very end before investigation. pH-indicator containing cell medium should be replaced by 1× PBS in order to minimize fluorescence absorption. For sensitive cells, a pH-stable (CO_2-independent) medium can be used. In every case, the liquid layer covering the cells should be as thin as possible, even if inverse microscopy is used. Although lipofuscin is exclusively found in the cytosolic compartment, FM should be done without DAPI-labeling of the nuclear DNA since both spectra overlap.

Principally, neither the objective nor the exposure time should be changed during FM measurement if a quantification of the lipofuscin (*see* Note 1) amount compared to a control group is required. In cases of intracellular localization of the autofluorescent material is the only required information the exposure time and the objective should be chosen respecting optimal image quality.

The exposure time should be adjusted using the darkest and the brightest sample (if known) in order to avoid overexposure (Fig. 13.7).

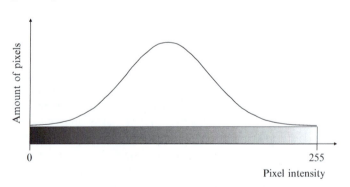

Fig. 13.7. Ideal distribution of pixel intensities for FM/CLSM. Exposure time of the sample should be adjusted in this way that the average intensity is in the middle of the detection range. The intensities at both ends of the resulting Gaussian function should be cut as little as possible by the chosen sensitivity.

Though it is possible to produce a calibration curve to correct the fluorescence intensities resulting from different exposure times, this excessive labor can be avoided using 10-bit instead of 8-bit color depth, spreading the amount of detectable different gray levels from 256 to 1,024. In this case, the compatibility of the program used for image analysis has to be considered. If needed, 10-bit images can be downscaled to 8-bit, even though this is accompanied with loss of information.

Evaluation of the resulting images is done on principle by determination of the background intensity that is due to scattered light, "unspecific (auto)fluorescence" or noise of the used camera. That background intensity has to be detracted from the each measured value. Even in modern programs an automatic recognition of structures is at best implemented rudimentary, so an area of the obtained image has to be defined as "background", in the same way, cells have to be marked manual. It is possible to define the intensity above a certain threshold as "fluorescent structure," but this automated method has to be verified per eye. In this case, a well known open source freeware (54-58) might be very useful, since many licence free plugins for scientific evaluation are available, including automatic cell counting and a simple threshold or color driven structure recognition. For manual selection of fluorescent structures from an image, a graphic tablet may be recommended.

3.6. Confocal Laser Scanning Microscopy of the Sample

Using highly sophisticated confocal laser scanning microscopy, the combination of different excitation/emission channels is alleviated since many different laser wavelengths are available and powerful algorithms allow an unravelling of even overlapping spectres. For investigating the intracellular distribution of lipofuscin, a fluorescence excitation using 405 nm and long pass filtering, excluding wavelengths shorter than 420 nm is recommended in order to maximize the yield of detectable autofluorescence (*see* Note 3).

Since CLSM is usually used for exact determination of intracellular distributions, the experiments should be performed considering three dimensional reconstructions from the data sets. With modern software, it is not a problem to avoid an overlapping or gaping of two adjacent confocal planes in z-direction, if the settings are correctly chosen (Fig. 13.8).

Commercial high-end software packages, primarily specialized in 3D reconstruction, are able to import raw CLSM-data and to compute 3D structures (59) from these sets with a minimum of additional user settings. The resulting structures can be exported as three dimensional meshes and further processed by commercial 3D packages for enhanced visualisation of intracellular distributions (Fig. 13.9) (*see* Note 4) or even time dependent changes.

Position of two adjacent confocal layers

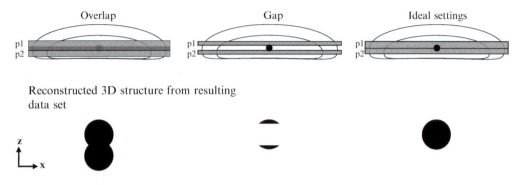

Reconstructed 3D structure from resulting
data set

Fig. 13.8. Aberrations in 3D reconstruction of intracellular structures. If thickness and position of the confocal layers are not properly set dependent from the pinhole diameter and the dimensions of the investigated structure, the 3D reconstruction from the resulting data will show errors like the partial doubling of structures (*left image*) or loss of data (*middle image*). Only if the settings define the confocal layers attaching without overlaps, the reconstruction will be flawless.

4. Results

A representative image, showing the intracellular distribution of lipofuscin is shown in Fig. 13.9. The channels "lipofuscin" (showing CLSM image of the cell) and "Fluorescence microscopy" (showing the same cell via FM) displays the typical perinuclear cytosolic distribution of lipofuscin detected by its autofluorescence.

Using immunochemical labeling of the 20S proteasome (60-63), the major intracellular proteolytic system, for 20S-lipofuscin colocalization (as displayed in the image "merge"), reveals no colocalization. In fact, the visible lipofuscin aggregates appear as dark spots in the 20S-channel. Since it is known, that lipofuscin is mostly found in the lysosomal compartment, thus a colocalization with the cytosolic 20S is not to be expected. Furthermore, no lipofuscin is found in the nuclear compartment and thus here is no colocalization found, too. In order to discriminate between nucleus and cytosol, the nuclear DNA has been marked via DAPI, in the same way showing no colocalization with lipofuscin.

The detected distributions can be visualized very impressive by a three dimensional reconstruction of the cell by using the data sets of every single channel as depicted in panel B of Fig. 13.9.

5. Notes

1. During the formation of lipofuscin, several fluorescent intermediates can occur, like pyridine-dialdehydes, pyralines, or pentosidin complexes (64), resulting from lipid peroxidation

a   CLSM   **b   Fluorescence Microscopy**

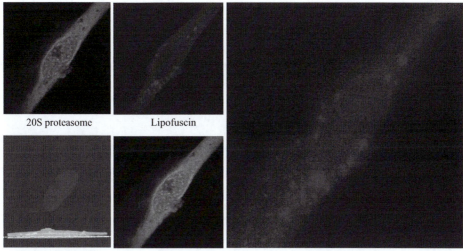

c   3D reconstruction

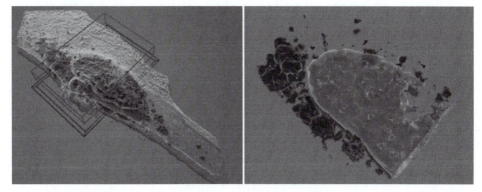

Fig. 13.9. CMLS data and the 3D reconstruction of intracellular distributions from CLSM data sets. The four pictures of panel (**a**) show CLSM images from an aged human fibroblast (own data). The images show immunolabeled 20S proteasome (*top left*, polyclonal anti-20S-proteasome antibody), the autofluorescence of lipofuscin (top right, $I_{ex} = 488$ nm), a DAPI-labeling of the nuclear DNA (*bottom left*) and a merge image of the three channels (*bottom right*). Furthermore, in the DAPI-image a side view of the investigated cell is depicted, the visible confocal layer marked by a white dotted line. Panel (**b**) demonstrates the autofluorescence of the same cell using common fluorescence microscopy. The different signal intensities and the differences in image quality are clearly visible. A 3D reconstruction of the same cell from the CLSM data set is demonstrated in panel (**c**). The cell dimensions are defined by the distribution of the 20S proteasome, the nucleus via the DAPI fluorescence, and the lipofuscin using its autofluorescence. The left image shows the reconstruction of the whole cell, the right image shows details of the nuclear volume (frame in the *left image*) and some of the perinuclear distributed lipofuscin without the 20S proteasome.

or advanced glycation endproduct (AGE)-formation (65, 66). These structures may be preliminary stages of lipofuscin, but at this time they cause only a diffuse autofluorescence and are not accumulated in larger aggregates.

2. Considering the mentioned autofluorescence of aromatic structures caused by delocalized π-electron systems, it is difficult to distinguish between normal cellular autofluorescence shown by native (not oxidatively modified) proteins, mainly caused by aromatic amino acids. For validation, it would be necessary to compare the whole spectra of autofluorescence with the spectra of lipofuscin. This is very difficult in single cells but perhaps possible using a white light laser and the whole filter set in high end CLSM.

   Thus even fast dividing tumor cells, containing almost no detectable amount of lipofuscin, show an autofluorescence, as found in cellular structures that are – even in postmitotic aging cells – usually free of lipofuscin like the nucleus. Even an aged fibroblast (Fig. 13.9) shows a certain amount of autofluorescence in the nuclear compartment that can not just be defined as lipofuscin.

3. Since only an extreme low amount of lipofuscin is found to be free in the cytosol (about 0.3–0.5%, unpublished data), verified via colocalization of the autofluorescence with LysoTracker® blue, this colocalization can be used for confirmation in FM/CLSM. Most of the cellular lipofuscin is found in aggregates showing diameters ranging from 0.1–5.0 μm, these structures should be clearly visible in FM (at least in CLSM). Thus, a microscopic resolvable volume can be another criterion for the validation of lipofuscin detection.

4. Until now no considerable structure/particle showing a lipofuscin-like autofluorescence in the nucleus of a cell could be found by us, and so at this time it can be concluded that lipofuscin aggregation is a cytosolic/lysosomal process. As mentioned, there is an established model of lipofuscin formation but the subtleties of this very complex network are still under investigation. Furthermore, it is still discussed, if "lipofuscin" (caused by aging) and "ceroid" (caused by pathologic processes) have to be considered as two different materials (53). Moreover, it can be supposed, that the composition of lipofuscin differs in cells of different organs (as the bis-retinoid A2E may play a secondary role apart from the retina), in organs of different animals, and possibly in different cells of the same organ.

References

1. Jung T, Bader N, Grune T (2007) Lipofuscin: formation, distribution, and metabolic consequences. Ann N Y Acad Sci 1119:97–111

2. Brunk UT, Terman A (2002) Lipofuscin: mechanisms of age-related accumulation and influence on cell function. Free Radic Biol Med 33:611–619

3. Terman A, Brunk UT (2004) Lipofuscin. Int J Biochem Cell Biol 36:1400–1404

4. Terman A, Gustafsson B, Brunk UT (2007) Autophagy, organelles and ageing. J Pathol 211:134–143

5. Levine AS, Lemieux B, Brunning R, White JG, Sharp HL, Stadlan E, Krivit W (1968) Ceroid

accumulation in a patient with progressive neurological disease. Pediatrics 42:583–591

6. Terman A, Dalen H, Brunk UT (1999) Ceroid/lipofuscin-loaded human fibroblasts show decreased survival time and diminished autophagocytosis during amino acid starvation. Exp Gerontol 34:943–957

7. Terman A, Abrahamsson N, Brunk UT (1999) Ceroid/lipofuscin-loaded human fibroblasts show increased susceptibility to oxidative stress. Exp Gerontol 34:755–770

8. Jolly RD, Dalefield RR, Palmer DN (1993) Ceroid, lipofuscin and the ceroid-lipofuscinoses (Batten disease). J Inherit Metab Dis 16:280–283

9. Terman A, Brunk UT (1998) Ceroid/lipofuscin formation in cultured human fibroblasts: the role of oxidative stress and lysosomal proteolysis. Mech Ageing Dev 104:277–291

10. Koistinaho J, Hartikainen K, Hatanpaa K, Hervonen A (1989) Age pigments in different populations of peripheral neurons in vivo and in vitro. Adv Exp Med Biol 266:49–59

11. Gutteridge JM (1984) Age pigments: role of iron and copper salts in the formation of fluorescent lipid complexes. Mech Ageing Dev 25:205–214

12. Yim MB, Kang SO, Chock PB (2000) Enzyme-like activity of glycated cross-linked proteins in free radical generation. Ann N Y Acad Sci 899:168–181

13. Friguet B, Szweda LI (1997) Inhibition of the multicatalytic proteinase (proteasome) by 4-hydroxy-2-nonenal cross-linked protein. FEBS Lett 405:21–25

14. Bourre JM, Haltia M, Daudu O, Monge M, Baumann N (1979) Infantile form of so-called neuronal ceroid lipofuscinosis: lipid biochemical studies, fatty acid analysis of cerebroside sulfatides and sphingomyelin, myelin density profile and lipid composition. Eur Neurol 18:312–321

15. Granier LA, Langley K, Leray C, Sarlieve LL (2000) Phospholipid composition in late infantile neuronal ceroid lipofuscinosis. Eur J Clin Invest 30:1011–1017

16. Benavides SH, Monserrat AJ, Farina S, Porta EA (2002) Sequential histochemical studies of neuronal lipofuscin in human cerebral cortex from the first to the ninth decade of life. Arch Gerontol Geriatr 34:219–231

17. Jolly RD, Douglas BV, Davey PM, Roiri JE (1995) Lipofuscin in bovine muscle and brain: a model for studying age pigment. Gerontology 41(Suppl 2):283–295

18. Sohal RS, Donato H Jr (1979) Effect of experimental prolongation of life span on lipofuscin content and lysosomal enzyme activity in the brain of the housefly, Musca domestica. J Gerontol 34:489–496

19. Brunk UT, Jones CB, Sohal RS (1992) A novel hypothesis of lipofuscinogenesis and cellular aging based on interactions between oxidative stress and autophagocytosis. Mutat Res 275:395–403

20. Dawson G, Cho S (2000) Batten's disease: clues to neuronal protein catabolism in lysosomes. J Neurosci Res 60:133–140

21. Collins J, Holder GE, Herbert H, Adams GG (2006) Batten disease: features to facilitate early diagnosis. Br J Ophthalmol 90:1119–1124

22. Luiro K, Kopra O, Blom T, Gentile M, Mitchison HM, Hovatta I, Tornquist K, Jalanko A (2006) Batten disease (JNCL) is linked to disturbances in mitochondrial, cytoskeletal, and synaptic compartments. J Neurosci Res 84:1124–1138

23. Kristensen K, Lou HC (1983) Central nervous system dysfunction as early sign of neuronal ceroid lipofuscinosis (Batten's disease). Dev Med Child Neurol 25:588–590

24. Katz ML, Eldred GE, Siakotos AN, Koppang N (1988) Characterization of disease-specific brain fluorophores in ceroid-lipofuscinosis. Am J Med Genet Suppl 5:253–264

25. Ben-Shabat S, Itagaki Y, Jockusch S, Sparrow JR, Turro NJ, Nakanishi K (2002) Formation of a nonaoxirane from A2E, a lipofuscin fluorophore related to macular degeneration, and evidence of singlet oxygen involvement. Angew Chem Int Ed Engl 41:814–817

26. Hammer M, Konigsdorffer E, Liebermann C, Framme C, Schuch G, Schweitzer D, Strobel J (2008) Ocular fundus auto-fluorescence observations at different wavelengths in patients with age-related macular degeneration and diabetic retinopathy. Graefes Arch Clin Exp Ophthalmol 246:105–114

27. Holz FG, Bindewald-Wittich A, Fleckenstein M, Dreyhaupt J, Scholl HP, Schmitz-Valckenberg S (2007) Progression of geographic atrophy and impact of fundus autofluorescence patterns in age-related macular degeneration. Am J Ophthalmol 143:463–472

28. Hannover A (1842) Mikroskopiske undersögelser af nervesystemet. Kgl.Danske Vidensk. Kabernes Selkobs Naturv.Math.Afh.Copenhagen 10, 1–112. Ref Type: Magazine Article

29. Brunk UT, Terman A (2002) The mitochondrial-lysosomal axis theory of aging: accumulation of damaged mitochondria as a result of imperfect autophagocytosis. Eur J Biochem 269:1996–2002

30. Bichile LS (1994) Malondialadehyde: a marker of lipid peroxidation. J Assoc Physicians India 42:769

31. Gutteridge JM, Quinlan GJ (1983) Malondialdehyde formation from lipid peroxides in the thiobarbituric acid test: the role of lipid radicals, iron salts, and metal chelators. J Appl Biochem 5:293–299

32. Esterbauer H, Koller E, Slee RG, Koster JF (1986) Possible involvement of the lipid-peroxidation product 4-hydroxynonenal in the formation of fluorescent chromolipids. Biochem J 239:405–409

33. Sitte N, Huber M, Grune T, Ladhoff A, Doecke WD, von ZT, Davies KJ (2000) Proteasome inhibition by lipofuscin/ceroid during postmitotic aging of fibroblasts. FASEB J 14:1490–1498

34. Grune T, Jung T, Merker K, Davies KJ (2004) Decreased proteolysis caused by protein aggregates, inclusion bodies, plaques, lipofuscin, ceroid, and 'aggresomes' during oxidative stress, aging, and disease. Int J Biochem Cell Biol 36:2519–2530

35. Sitte N, Merker K, von ZT, Grune T (2000) Protein oxidation and degradation during proliferative senescence of human MRC-5 fibroblasts. Free Radic Biol Med 28:701–708

36. Sitte N, Merker K, Grune T (1998) Proteasome-dependent degradation of oxidized proteins in MRC-5 fibroblasts. FEBS Lett 440:399–402

37. Elleder M, Sokolova J, Hrebicek M (1997) Follow-up study of subunit c of mitochondrial ATP synthase (SCMAS) in Batten disease and in unrelated lysosomal disorders. Acta Neuropathol 93:379–390

38. Kida E, Wisniewski KE, Golabek AA (1993) Increased expression of subunit c of mitochondrial ATP synthase in brain tissue from neuronal ceroid lipofuscinoses and mucopolysaccharidosis cases but not in long-term fibroblast cultures. Neurosci Lett 164:121–124

39. Wang X, Quinn PJ (1999) Vitamin E and its function in membranes. Prog Lipid Res 38:309–336

40. Rupar CA, Albo S, Whitehall JD (1992) Rat liver lysosome membranes are enriched in alpha-tocopherol. Biochem Cell Biol 70:486–488

41. Fattoretti P, Bertoni-Freddari C, Casoli T, Di SG, Solazzi M, Corvi E (2002) Morphometry of age pigment (lipofuscin) and of ceroid pigment deposits associated with vitamin E deficiency. Arch Gerontol Geriatr 34:263–268

42. Bertoni-Freddari C, Fattoretti P, Casoli T, Di SG, Solazzi M, Corvi E (2002) Morphometric investigations of the mitochondrial damage in ceroid lipopigment accumulation due to vitamin E deficiency. Arch Gerontol Geriatr 34:269–274

43. Hutter E, Skovbro M, Lener B, Prats C, Rabol R, Dela F, Jansen-Durr P (2007) Oxidative stress and mitochondrial impairment can be separated from lipofuscin accumulation in aged human skeletal muscle. Aging Cell 6:245–256

44. Porta EA (2002) Pigments in aging: an overview. Ann N Y Acad Sci 959:57–65

45. Tsai L, Szweda PA, Vinogradova O, Szweda LI (1998) Structural characterization and immunochemical detection of a fluorophore derived from 4-hydroxy-2-nonenal and lysine. Proc Natl Acad Sci USA 95:7975–7980

46. Eldred GE, Lasky MR (1993) Retinal age pigments generated by self-assembling lysosomotropic detergents. Nature 361:724–726

47. Boulton M, Rozanowska M, Rozanowski B (2001) Retinal photodamage. J Photochem Photobiol B 64:144–161

48. Onetti MA (2000) Confocal laser scanning microscopy. Adv Clin Path 4:235–239

49. Roderfeld M, Matern S, Roeb E (2003) Confocal laser scanning microscopy: a deep look into the cell. Dtsch Med Wochenschr 128:2539–2542

50. Willig KI, Harke B, Medda R, Hell SW (2007) STED microscopy with continuous wave beams. Nat Methods 4:915–918

51. Martini J, Andresen V, Anselmetti D (2007) Scattering suppression and confocal detection in multifocal multiphoton microscopy. J Biomed Opt 12:034010

52. Shao ZF, Baumann O, Somlyo AP (1991) Axial resolution of confocal microscopes with parallel-beam detection. J Microsc 164:13–19

53. Seehafer SS, Pearce DA (2006) You say lipofuscin, we say ceroid: defining autofluorescent storage material. Neurobiol Aging 27:576–588

54. Girish V, Vijayalakshmi A (2004) Affordable image analysis using NIH Image/ImageJ. Indian J Cancer 41:47

55. Papadopulos F, Spinelli M, Valente S, Foroni L, Orrico C, Alviano F, Pasquinelli G (2007) Common tasks in microscopic and ultrastructural image analysis using ImageJ. Ultrastruct Pathol 31:401–407

56. Hachet-Haas M, Converset N, Marchal O, Matthes H, Gioria S, Galzi JL, Lecat S (2006) FRET and colocalization analyzer – a method to validate measurements of sensitized emission FRET acquired by confocal microscopy and available as an ImageJ Plug-in. Microsc Res Tech 69:941–956

57. Collins TJ (2007) ImageJ Microsc. Biotechniques 43:25–30

58. Feige JN, Sage D, Wahli W, Desvergne B, Gelman L (2005) PixFRET, an ImageJ plug-in for FRET calculation that can accommodate variations in spectral bleed-throughs. Microsc Res Tech 68:51–58

59. Anderson JR, Wilcox MJ, Wade PR, Barrett SF (2003) Segmentation and 3D reconstruction of biological cells from serial slice images. Biomed Sci Instrum 39:117–122

60. Jung T, Grune T (2008) The proteasome and its role in the degradation of oxidized proteins. IUBMB Life 60:743–752

61. Bader N, Jung T, Grune T (2007) The proteasome and its role in nuclear protein maintenance. Exp Gerontol 42:864–870

62. Chondrogianni N, Stratford FL, Trougakos IP, Friguet B, Rivett AJ, Gonos ES (2003) Central role of the proteasome in senescence and survival of human fibroblasts: induction of a senescence-like phenotype upon its inhibition and resistance to stress upon its activation. J Biol Chem 278:28026–28037

63. Friguet B (2002) Aging of proteins and the proteasome. Prog Mol Subcell Biol 29:17–33

64. Nilsson E, Yin D (1997) Preparation of artificial ceroid/lipofuscin by UV-oxidation of subcellular organelles. Mech Ageing Dev 99:61–78

65. Jobst K, Lakatos A (1996) The liver cell histones of diabetic patients contain glycation endproducts (AGEs) which may be lipofuscin components. Clin Chim Acta 256:203–204

66. Ahmed N (2005) Advanced glycation endproducts – role in pathology of diabetic complications. Diabetes Res Clin Pract 67: 3–21

67. Pawlak A, Rozanowska M, Zareba M, Lamb LE, Simon JD, Sarna T (2002) Action spectra for the photoconsumption of oxygen by human ocular lipofuscin and lipofuscin extracts. Arch Biochem Biophys 403:59–62

68. Framme C, Schule G, Birngruber R, Roider J, Schutt F, Kopitz J, Holz FG, Brinkmann R (2004) Temperature dependent fluorescence of A2-E, the main fluorescent lipofuscin component in the RPE. Curr Eye Res 29:287–291

69. Hammer M, Richter S, Kobuch K, Mata N, Schweitzer D (2008) Intrinsic tissue fluorescence in an organotypic perfusion culture of the porcine ocular fundus exposed to blue light and free radicals. Graefes Arch Clin Exp Ophthalmol 246:979–988

70. Marmorstein AD, Marmorstein LY, Sakaguchi H, Hollyfield JG (2002) Spectral profiling of autofluorescence associated with lipofuscin, Bruch's Membrane, and sub-RPE deposits in normal and AMD eyes. Invest Ophthalmol Vis Sci 43:2435–2441

71. Fang C, Peng M, Li G, Tian J, Yin D (2007) New functions of glucosamine as a scavenger of the lipid peroxidation product malondialdehyde. Chem Res Toxicol 20:947–953

72. Yin DZ, Brunk UT (1991) Microfluorometric and fluorometric lipofuscin spectral discrepancies: a concentration-dependent metachromatic effect? Mech Ageing Dev 59:95–109

73. Li L, Li G, Sheng S, Yin D (2005) Substantial reaction between histamine and malondialdehyde: a new observation of carbonyl stress. Neuro Endocrinol Lett 26:799–805

Part II

Antioxidant Technology and Application

Chapter 14

OXY-SCORE: A Global Index to Improve Evaluation of Oxidative Stress by Combining Pro- and Antioxidant Markers

Fabrizio Veglia, Viviana Cavalca, and Elena Tremoli

Abstract

Our group recently proposed OXY-SCORE, a summary index of oxidative stress, computed by combining plasma free and total malondialdehyde (F- and T-MDA), glutathione in disulphide/reduced forms (GSSG/GSH), α- and γ-tocopherol (TH), urine isoprostanes (iPF$_{2\alpha}$-III) levels, and plasma individual antioxidant capacity. Here, we describe the methods for the determination of the analytes and for the computation of the scores. We also report the results of two studies testing the performances of OXY-SCORE, and showing its value in assessing the oxidative status of cardiovascular patients and healthy subjects.

Keywords: Oxidative stress, Malondialdehyde, Isoprostanes, Glutathione, Vitamin E, Coronary artery disease

1. Introduction

Oxidative stress occurs in cells when the production of free radicals overwhelms the antioxidant defence system, thus inducing oxidant damage to macromolecules (1, 2). Consequently, alterations in gene expression, protein structure and function, and lipid and membrane integrity take place. Reactive oxygen species (ROS) are continuously generated during the normal processes of cell metabolism and may be induced by external sources, such as ultraviolet light radiation, smoke, ozone, chemotherapeutic agents, and toxins. Under physiological conditions, the negative effects of an overproduction of ROS are neutralized by endogenous antioxidant systems and free-radical scavengers.

D. Armstrong (ed.), *Advanced Protocols in Oxidative Stress II*, Methods in Molecular Biology, vol. 594
DOI 10.1007/978-1-60761-411-1_14, © Humana Press, a part of Springer Science+Business Media, LLC 2010

Although a number of indexes to measure prooxidant or antioxidant status have been proposed, up to now, no single parameter has been recognized as a gold standard to assess oxidative stress or changes in oxidative status. *In vivo* and *in vitro* investigations have been so far performed by estimating, individually, the consumption of a specific antioxidant defense (*i.e.* vitamin E, glutathione, etc), the compensatory increase of an antioxidant enzyme mass or activity (superoxide dismutase, catalase, glutathione peroxidase, etc), or the production of metabolites derived from oxidative damage to biological molecules (malondialdehyde, isoprostanes, nitrotyrosine, 8-OH-guanosine, etc).

Our group has recently proposed a comprehensive index derived from the computation of different relevant parameters of the oxidative balance, defined OXY-SCORE (3). OXY-SCORE brings together several measures of both oxidative damage and antioxidant defences, in order to account for the multi-factorial nature of oxidative stress. Four common markers of oxidative damage and four accepted antioxidant factors have been selected to compute OXY-SCORE. Specifically, the free (F-) and total (T-) forms of malondialdehyde (MDA) are indexes of lipoperoxidation, indicative of chronic or recent oxidative injury (4); isoprostanes (8-iso-PGF2α) also represent a well-established marker of in vivo oxidation (5); the ratio of disulphide/reduced forms of glutathione (GSSG/GSH) is indicative of a cellular unbalance toward oxidation. Among markers of antioxidant status, GSH and tocopherol (TH), in both its α and γ isoforms, are representative of hydrophilic and lipophilic compartments of the body, respectively. Furthermore, GSH plays a pivotal role in the regeneration of other antioxidants, and tocopherols act as both potent peroxyl radical scavengers and chain-breaking antioxidants, regulating enzyme activity and membrane fluidity. An indirect index of the individual plasma antioxidant capacity (IAC) was also included in the computation of OXY-SCORE.

In this chapter, we show that OXY-SCORE reflects subjects characteristics, such as age and gender, reported to affect oxidative stress and pathological conditions, such as coronary artery disease, thus discriminating between clinical settings characterized by high or low oxidative status.

2. Materials

2.1. Equipment

2.1.1. Isoprostane Determination

Accela High Speed Liquid Chromatography (Thermo Scientific Inc. Waltham, MA 02454) equipped with TSQ Quantum triple quadrupole system with electrospray ionization source (ESI) (Thermo Fisher Scientific Inc.).

2.1.2. Plasma Malondialdehyde (MDA) Assay	HPLC is constituted by two pumps ESA® mod 582 (ESA Biosciences, Chelmsford, MA, USA) equipped with an autosampler ESA® mod 542 and a fluorimetric detector (Jasko FP-1520, Japan).
2.1.3. GSH and GSSG Measurements	HPLC is constituted by two pumps ESA® mod 582 (ESA Biosciences, Chelmsford, MA, USA) equipped with an autosampler ESA® mod 542 and a fluorimetric detector (Jasko FP-1520, Japan)
2.1.4. Vitamin E Measurement	HPLC is constituted by two pumps ESA® mod 582 (ESA Biosciences, Chelmsford, MA, USA) equipped with an autosampler ESA® mod 542 and an electrochemical detector (ESA CoulArray mod. 5600A).

2.2. Reagents and Supplies

All organic solvents are HPLC grade and purchased from Sigma (Sigma-Aldrich, St Louis, MO, USA) if not otherwise specified

2.2.1. Isoprostane Determination (see Note 1)

1. Standards (*see* Note 2): 8-iso-PGF-2α-D$_4$ 10 pg/μL (Cayman Chemical Co, Ann Arbor, MA, USA) used as internal standard; 8-iso-PGF-2α as standard solution to make the calibration curve from 25 to 0.780 pg/μL.

2. Solid phase extraction: Sep-Pak 3 mL (500 mg) cartridge (Waters, Milford, MA, USA).

3. Chromatographic columns: precolumn Hypersil gold C18 (2.1×20 mm, 3 μm) (Thermo Scientific Inc. Waltham, MA 02454), column Hypersil gold C18 (2.1×100 mm, 3 μm) (Thermo Scientific Inc. Waltham, MA 02454).

4. Software: Excalibur 2.0.7 (Thermo Scientific Inc. Waltham, MA 02454).

5. Mobile phases: (A) Ammonium acetate (CH$_3$COONH$_4$) 2.5 mmol/L; (B) Acetonitrile; (C) Methanol.

2.2.2. Plasma Malondialdehyde (MDA) Assay

1. Standards (*see* Note 3): 1,1,3,3-tetraethoxy-propane is used for calibration curve.

2. Alkaline hydrolysis: NaOH 10 mol/L.

3. Proteins precipitation (*see* Note 4): perchloric acid (HClO$_4$) (5.85 mol/L).

4. Tiobarbituric Acid (TBA) reagent: 4,6-dihydroxy 2-thiopyrimidine is prepared as a 10 g/L solution in 50 mmol/L phosphate buffer, pH 7, stirring on a hot plate (50–55°C). Stored at room temperature, this reagent is stable for 4 weeks.

5. Chromatographic columns: XBridge C18, (4.6×150 mm, 5 μm) (Waters, Milford, MA, USA)

6. Software: EZStart (Scientific software group, Sandy, Utah, USA).

7. Mobile phases: (A) Methanol (20% v/v) in phosphate buffer (KH_2PO_4) 20 mmol/L, pH 7; (B) Methanol (60% v/v) in phosphate buffer (KH_2PO_4) 20 mmol/L, pH 7.

2.3. GSH and GSSG Measurements

1. Standards: GSH and GSSG 1 mmol/L solutions are prepared in chloridric acid (HCl) 0.1 mol/L and stored at –80°C until use.

   (a) Working solutions (GSH 160 μmol/L and GSSG 40 μmol/L) are prepared just before the analysis.

   (b) Standard solution is obtained combining GSH and GSSG working solutions (1:1).

2. Chromatographic columns: Synergy Hydro-RP C18 (4.6 × 150 mm, 5 μm) (Phenomenex, Torrance, CA, USA).

3. Software: ESA CoulArray for Windows (ESA Inc. Biosciences, Chelmsford, MA, USA).

4. Mobile phases: Acetonitrile (1% v/v) in phosphate buffer (KH_2PO_4) 20 mmol/L, pH 2.7.

2.4. Vitamin E Measurement

1. Standards (*see* Note 5): α-Tocopherol and γ-Tocopherol (TH) 10 mg/mL solutions are prepared in ethanol and stored at –80°C until use. Working solutions (α-TH 20 μg/mL and γ-TH 2 μg/mL) are prepared just before the analysis.

2. Deproteinization and organic extraction: Ethanol, Hexane.

3. Chromatographic columns: Discovery® C18 (4.6 × 250 mm, 3.5 μm) (Supelco, Bellefonte, PA, USA).

4. Software: EZStart (Scientific software group, Sandy, Utah, USA).

5. Mobile phases: Methanol (100%).

3. Methods

3.1. Sample Collection

3.1.1. Whole Blood

Peripheral blood samples are collected on ice from fasting subjects into ethylenediamine tetra-acetic acid (EDTA)-containing tubes (9.3 mmol/L; Vacutainer Systems, Becton, Dickinson Co., Franklin Lake, NJ, USA), immediately precipitated with 10% trichloroacetic acid (TCA) in 1 mmol/L EDTA solution, and stored at –80°C until analysis.

3.1.2. Plasma

Plasma is obtained from peripheral blood anticoagulated with EDTA after centrifugation (3,000 × g for 10 min at 4°C) within 30 min. Samples are stored at –80°C until analysis.

3.1.3. Urine

Overnight urine samples are added with the antioxidant 4-hydroxy-tempo (1 mmol/L; Sigma-Aldrich Chemical Co., St Louis, MO, USA) and stored at −80°C before extraction.

3.2. Isoprostane Determination

1. Isoprostanes are widely considered as one of the most reliable indexes of lipid peroxidation in vivo (6, 7). Their determination in urine samples offers some advantages over its plasma counterpart in which isoprostanes are present in minor concentration. Urinary isoprostanes have been shown to be stable for up to 5 days at room temperature (8), and there are no special requirements for collection or preservation (9) as these metabolites are not formed *ex vivo* by auto-oxidation. Various analytical methods have been used to measure iPF2α-III, including Enzyme Immunoassay (EIA) (9-11), radioimmunoassay (RIA) (10), gas-chromatography mass spectrometry (GC-MS) (12-17), and liquid chromatography mass spectrometry (LC-MS) (11). The analytical method in which HPLC is coupled to electrospray ionization tandem mass spectrometers (LC-MS/MS) can avoid the derivatization step and greatly reduce the likelihood of unrelated contaminants, which arises using single-ion monitoring with GC-MS (12).

2. Urinary isoprostanes iPF2α-III are purified using a solid phase extraction protocol onto SPE cartridge and determined by LC-MS equipped with ESI. One millilitre of urine, centrifuged at $3,000 \times g$ for 10 min, is added with 8-iso-PGF-2α-D$_4$ (200 μL) and loaded onto a Sep-Pak cartridge, which is previously equilibrated with methanol (5 mL) and water (5 mL). After washing with water (2.5 mL), methanol 95% (2.5 mL), and hexane (3 mL), the sample is eluted with ethyl acetate 100% (4 mL). After complete drying by SpeedVac system, the sample is resuspended in acetonitrile (10% v/v in water).

3. Ten microliters of the sample is chromatographed at 30°C onto C18 column (Hypersil gold C18, 2.1×100 mm, 3 μm) with a flow rate of 200 μL/min and a linear gradient with 3% C beginning with 7% B in A and programmed to 37% B at 9 min and 92% at 12, with the column being flushed for another 5 min at 92% B prior to returning to 7% B in 1 min and holding for 15 min before initiating the next injection in the sequence. Mass spectrometry is performed using a Thermo Scientific TSQ Quantum triple quadrupole equipped with a standard ESI source outfitted with a deactivated fused silica capillary. Nitrogen is used for both the sheath and the auxiliary gases. The sheath and auxiliary gases are set to 45 and 35 (arbitrary units), respectively. The mass spectrometer is operated in the negative ion mode, and the electrospray needle is maintained at 3,700 V. The ion-transfer tube is operated at 35 V and 270°C. The tube lens voltage is set to 86 V.

4. Typical elution times for 8-iso-PGF2α and 8-iso-PGF2α-D$_4$ are both 9.50 ± 0.04 min. 8-iso-PGF2α and 8-iso-PGF2α- D$_4$ are monitored in the multiple-reaction monitoring (MRM) mode using the following transitions: m/z 353.1 to 192.87 (8-iso-PGF2α) and m/z 357.05 to 197.1 (8-iso-PGF2α-D$_4$). A seven-point linear calibration curve is established using both internal and external standards over a range of 0–25 pg/µL ($r = 0.9996$). The intra- and inter-assay CVs are 0.3 and 5.2%, respectively, with quantification (LOQ) and detection (LOD) limits of 0.38 and 0.13 pg/µL.

3.3. Plasmatic Malondialdehyde (MDA) Assay

1. Plasma malondialdehyde, free and total, is detected by high-performance liquid chromatography with a method modified from Carbonneau et al.(13). Two hundred microliters of plasma are acidified at pH 1 with HClO$_4$ (50 µL). After centrifuging at $13,000 \times g$ for 15 min, the supernatant is treated with TBA reagent (50 µL). The color reaction is activated by heating at 100°C for 1 h.

2. To determine total MDA, a previous alkaline hydrolysis is performed. Plasma samples (200 µL) are adjusted to pH 13 using 20 µL NaOH 10 mol/L and incubated at 60°C for 60 min. The acidification of the samples is obtained by adding 100 µL of HClO$_4$, before TBA treatment. After cooling, the samples are centrifuged at $13,000 \times g$ for 10 min.

3. Supernatants (20 µL) are injected into XBridge C18 reversed-phase column and eluted at 30°C with a flow rate of 1.1 mL/min. A linear gradient is applied beginning with 10% B in A and programmed to 100% B at 10 min with the column being flushed for another 8 min at 100% B prior to returning to 10% B in 1 min and holding for 16 min before initiating the next injection in the sequence. A fluorescent detector settled at $\lambda_{exc} = 515$ nm, $\lambda_{em} = 553$ nm is used to detect the analytes.

4. The concentrations of individual samples are calculated from calibration curves using 1,1,3,3-tetraethoxypropane as standard. Calibration of the analytical procedure gave a linear signal over the malondialdehyde range of 0.06–2 µmol/L ($r = 0.9992$), with a quantification and detection limits of 0.03 and 0.01 µmol/L, respectively. The intra- and inter-assay coefficients of variation are 5.8 and 5.9%.

3.4. GSH and GSSG Measurements

1. Whole blood GSH and GSSG levels are measured by isocratic HPLC method. Whole blood samples are centrifuged at $14,000 \times g$ for 1 min, and GSH and GSSG (20 µL injections) are separated onto C18 column eluted at 30°C with phosphate buffer at a flow rate of 1 mL/min. Analysis is carried out using the electrochemical detector with electrodes set at 400, 700, 750 and 800 mV. The sum of peak heights is

integrated using commercial software. Under these conditions, GSH and GSSG elute at 5.08 and 9.30 min, respectively.

2. The sample concentrations are calculated from calibration curves using standard GSH and GSSG solutions. Calibration of the analytical procedure gave a linear signal over the GSH range of 2.5–80.0 μmol/L ($r=0.9999$) and the GSSG range of 0.625–20.0 μmol/L ($r=0.9998$), with a limit of quantitation (LOQ) of 0.265 and 0.426 μmol/L, respectively. The intra- and inter-assay CVs are 1.4 and 5.9% for GSH (20 μmol/L), and 6.7 and 7.1% for GSSG (5 μmol/L), respectively.

3.5. Vitamin E Measurement

1. α- and γ-Tocopherol concentrations are detected by HPLC after organic extraction. One hundred microliters of plasma or α- or γ-TH working solutions are diluted 1:2 with water and added with ethanol (500 μL) and hexane (1 mL). After vortexing, the centrifugation is performed ($3,000 \times g$ for 10 min) and the organic phase evaporated to dryness under vacuum.

2. The residue is dissolved in isopropyl alcohol (0.2 mL) and injected (25 μL) into a C18 reversed-phase column eluted at room temperature with methanol (100%) at a flow rate of 1 mL/min. The retention times of α– and γ-TH are 4.82 and 4.34 min, respectively, as determined by a fluorescent detector ($\lambda_{exc} = 292$ nm, $\lambda_{em} = 335$ nm). Commercial software is used for the chromatograms integration.

3. Data are obtained after comparison with calibration curves using working standard solutions, ranging from 0.312–10 μg/mL for α-TH ($r=0.9997$), and 0.03–1 μg/mL for γ-TH ($r=0.9998$). The intra-assay and the inter-assay CVs for plasma α- and γ-TH are 2.9, 3.9 and 3.3, 4.7, respectively, with a limit of quantitation (LOQ) of 0.27 and 0.03 μmol/L for α- and γ-TH, respectively.

3.6. Assay of Individual Antioxidant Capacity (IAC)

Plasma IAC, an index of the overall protection against oxidative damage, is assessed using a commercially available spectrophotometric assay, which measures the plasma capability to neutralize the massive oxidative action of hypochlorous acid (HClO) on alkyl-substituted aromatic amine in a chromogenic mixture (N,N-diethyl- paraphenilendiamine). The decrement of absorbance is detected at 505 nm, and the antioxidant capacity is expressed in μmol HClO/mL sample. The intra- and inter-assay CVs are 2.2 and 6.3%, respectively.

3.7. Computation of the Oxidative Stress Scores

The OXY-SCORE reflects the two main components of the oxidative stress, *i.e.* the accumulation of oxidative damage and the reduction of anti-oxidant defences. To calculate OXY-SCORE, two 'partial' scores are first determined: the Damage-Score (DS), which summarizes the damage component, and the Protection-Score

(PS), which summarizes the anti-oxidant defences. Thus, the global OXY-SCORE is computed by subtracting the PS from the DS, and it generates a single number, which reflects the balance between the oxidant-antioxidant systems.

The following section describes in details all the steps for the computation of the above described scores.

3.7.1. Computation of the Scores Indices of Oxidative Damage and of Anti-Oxidant Defence

First, variables with skewed distributions (F- and T-MDA, iPF_{2a}-III, GSSG, α-TH and γ-TH) are log-transformed; only GSH and IAC do not need log-transformation. The log-transformation renders the distributions with a positive skewness (a long 'tail' to the right) more symmetric; moreover, it limits the effects of outliers and enhances small differences near the bottom of the scale. Subsequently, to get rid of the problem of the different measurement units and of the different variability of the analytes, pro- and antioxidant variables are standardized according to the formula:

$$z_{ij} = \frac{(x_{ij} - m_j)}{s_j}$$

where z_{ij}: standardized value of variable j for subject i; x_{ij}: raw measure (possibly log transformed) of variable j for subject i; m_j: mean of variable j; s_j: standard deviation of variable j.

This method of standardization is used to fully preserve the quantitative nature of the original measurements. An alternative method is to dichotomize the single variables by assigning a value of 1 for measurements above the median and a value of 0 for measurements below. The score is then computed by summing up or averaging the dichotomized variables.

The damage score (DS) is computed as the average of standardized F-MDA (log transformed), T-MDA (log transformed) $iPF_{2\alpha}$-III (log transformed), and GSSG/GSH (log transformed).

The protection score (PS) is computed in a similar way by averaging the standardized IAC, GSH, α-TH (log transformed), and γ-TH (log transformed).

An example of DS and PS computation on a representative subject is reported in Table 14.1. Here the means and standard deviations were obtained from a group of 82 healthy subjects, but any reference population would have suited just as well. Thus, if an individual has a DS equal to 0, the average of his prooxidant variables is equal to the mean of the healthy population. Likewise, a PS value of 0 reflects an average level of antioxidant defences equivalent to the mean of healthy individuals.

Incidentally, computing the scores as means instead of sums allows to adjust for a mild proportion of missing data. In fact, if one individual has a missing value in one of the analytes, this procedure allows to automatically impute the missing value, which is assumed to have the same standardized value as the average of the

Table 14.1
Example of scores computation

Variable	Healthy sample $N=82$		Representative subject	
	Mean	SD	Raw values	Standardized values
Prooxidant				
F-MDA (log)	−0.42	0.22	−0.25	+0.77
T-MDA (log)	0.39	0.21	0.55	+0.76
iPF$_{2\alpha}$-III (log)	2.3	0.33	2.6	+0.91
GSSG/GSH (log)	−1.29	0.35	−1.36	−0.20
Damage score				+0.56
Antioxidant				
IAC	303.9	36.6	297	−0.19
GSH	69.1	17.2	75	+0.34
α-TH (log)	1.14	0.15	0.95	−1.27
γ-TH (log)	−0.43	0.2	−0.6	−0.85
Protection score				−0.49

Means and standard deviations (SD) refer to the healthy sample. The raw values of the representative subject are standardized and averaged to compute Damage and Protection scores

analytes with nonmissing values. For instance, if the representative subject of Table 14.1 had a missing value for F-MDA, its DS, computed on the remaining prooxidant analytes, would be $(0.76+0.91-0.2)/3=0.49$. This procedure assumes for the missing F-MDA a standardized value of 0.49, corresponding to a row value of −0.31 (which, in this case is not far from the measured log F-MDA). This approach is largely arbitrary. However, the analysis of pros and cons of the different missing-data imputation procedures goes beyond the purposes of this chapter.

3.7.2. Computation of OXY-SCORE

The two specific scores are then combined in order to obtain a single global score. OXY-SCORE is computed by subtracting the PS from the DS. Again, a value of zero is obtained when, on average, both prooxidant biomarkers and antioxidant defences are equal to the means of the control population. Conversely, an OXY-SCORE above or below zero indicates an unbalance of the two components.

3.8. Results

The performance of OXY-SCORE has been tested in two studies. In the first (3), we compared the ability of the scores and of the single analytes to discriminate the effect of age and sex differences on the oxidative stress status of healthy individuals. In the same study, we also compared patients with coronary artery disease (CAD) to healthy controls. In the second study, we tested the performance of OXY-SCORE, and of its individual components, in patients undergoing coronary artery bypass surgery, with or without employment of cardiopulmonary bypass, two clinical circumstances with established differences in the variations of the oxidative balance (14, 15)

3.8.1. Assessment of the Modulation of Oxidative Stress Status According Age and Sex in Healthy Individuals

Eighty-two healthy subjects (51 men and 31 women) were enrolled within the hospital staff. All subjects were asymptomatic, and their healthy status was confirmed by clinical assessment and laboratory analyses. Subjects receiving drugs, vitamin, antioxidant supplements, and synthetic estrogens as contraceptives or hormone replacement therapy were excluded.

As prooxidant biomarkers plasma T- and F-MDA, GSSG/GSH ratio and urine $iPF_{2\alpha}$-III were considered, whereas the antioxidant defences were plasma α- and γ-TH, GSH, and IAC.

The associations with age and sex were assessed by analysis of covariance (ANCOVA). Due to their skewed distributions, T- and F-MDA, GSSG/GSH, $iPF_{2\alpha}$-III, and α- and γ-TH were log-transformed before analysis. All analyses were performed using SAS statistical package v.8, SAS Institute, Cary, NC, USA.

Table 14.2 reports the means and SD of the single analytes and of the three scores, stratified by age class (panel a) or sex (panel b). Data were compared between groups by ANCOVA, adjusting for sex (panel a) or age (panel b).

A significant (and positive) association with age was found for T-MDA and $iPF_{2\alpha}$-III, a significant sex difference was found for GSH and tocopherols (higher in females), and a nearly significant difference in GSSG/GSH (higher in males). None of the analytes was significantly associated with both age and sex.

DS showed positive associations with both male sex and age, this latter only reaching statistical significance; PS was significantly and negatively associated with age and was higher in women. Finally, the association of OXY-SCORE with both age and sex was highly significant, indicating a potential unbalance of the proantioxidant substances toward a prooxidant status in men and in older subjects. The interaction between age and sex, as assessed by ANCOVA, was not significant ($F = 0.19$, $p = 0.66$).

3.8.2. Comparison of CAD Patients and Controls

Twenty stable CAD patients, whose pathological status had been documented by angiography, were enrolled in the study; all of them signed an informed consent to a protocol approved by the Institutional Review Board of Centro Cardiologico Monzino. Patients with a history of myocardial infarction in the previous six weeks or suffering from unstable angina were excluded. None of them was treated with vitamins or antioxidant supplements. The patients were compared with a random subgroup of 40 subjects

Table 14.2
Association of the individual analytes with age (panel A) and gender (panel B)

Panel A	Age 24–50 yr ($n=42$)	Age 51–77 yr ($n=40$)	
Analyte	Mean ± S.D.	Mean ± S.D.	P value
Prooxidant			
T-MDA (µmol/L)	3.27 ± 1.13	4.76 ± 1.99	0.04
F-MDA (µmol/L)	0.5 ± 0.21	0.61 ± 0.32	0.34
iPF$_{2\alpha}$-III (pmol/mmol creatinine)	174 ± 110	301.7 ± 143.1	0.04
GSSG/GSH	0.06 ± 0.04	0.089 ± 0.09	0.09
Antioxidant			
GSH (µmol/g Hb)	70.5 ± 18.3	66 ± 15.8	0.18
IAC (µmol HClO/mL)	305 ± 36	303.1 ± 38.6	0.55
α-TH (µg/mL)	14.3 ± 4	15 ± 7.4	0.36
γ-TH (µg/mL)	0.41 ± 0.13	0.42 ± 0.36	0.95
DS	−0.06 ± 0.72	0.38 ± 0.85	0.002
PS	−0.003 ± 0.58	−0.12 ± 0.71	0.03
OXY–SCORE	−0.054 ± 1.03	0.37 ± 1.13	0.004
Panel B			
	Women (n=31)	Men (n=51)	
Analyte	Mean ± S.D.	Mean ± S.D.	P value
Prooxidant			
T-MDA (µmol/L)	3.38 ± 1.62	3.62 ± 1.24	0.68
F-MDA (µmol/L)	0.48 ± 0.22	0.55 ± 0.23	0.39
iPF$_{2\alpha}$-III (pmol/mmol creatinine)	271 ± 145	209 ± 139	0.34
GSSG/GSH	0.05 ± 0.04	0.08 ± 0.08	0.06
Antioxidant			
GSH (µmol/gr Hb)	77.3 ± 17.7	63.7 ± 15.2	0.002
IAC (µmol HClO/mL)	297 ± 30	308 ± 41	0.25
α-TH (µg/mL)	17.8 ± 7.7	12.9 ± 3.7	0.0004
γ-TH (µg/mL)	0.49 ± 0.26	0.38 ± 0.17	0.04
DS	−0.1 ± 0.75	0.25 ± 0.81	0.1
PS	0.13 ± 0.63	−0.18 ± 0.63	0.04
OXY–SCORE	−0.21 ± 1.02	0.37 ± 1.09	0.03

MDA malondialdehyde, *iPF$_{2\alpha}$-III* isoprostanes, *GSSG* disulphide glutathione, *GSH* reduced glutathione, *IAC* individual antioxidant capacity, *TH* tocopherol

P values are adjusted for age, gender, and smoking status by ANCOVA

extracted from our healthy population, frequency matched for age and sex.

The ability of the scores to discriminate between CAD patients and healthy subjects was assessed by ROC curve analysis (16). This analysis accounts for both sensitivity and specificity, and a single number, the area under the curve (AUC), provides a measure of the discriminating ability of a variable: an AUC of 1.0 denotes a complete prediction, whereas an AUC of 0.5 is indicative of a null discriminating power. Excellent predictors yield AUCs above 0.9, whereas an AUC of 0.7 or above is usually considered as satisfactory.

After adjusting for age, sex, BMI, HDL-cholesterol, and smoking status, OXY-SCORE was significantly higher in CAD patients than in controls (2.70 ± 0.41 vs. 0.66 ± 0.29, $p < 0.001$). The two partial scores also differed between the two groups, only DS reaching statistical significance (DS: 1.94 ± 0.27 vs. 0.65 ± 0.27, $p < 0.007$; PS: -0.77 ± 0.24 vs. -0.22 ± 0.18, $p = 0.12$).

When the capacity of each score to discriminate CAD patients from matching controls was tested by ROC curve analysis, the AUC was 0.78 for PS, 0.94 for DS, and 0.96 for OXY-SCORE. Thus, the discriminating ability of OXY-SCORE in distinguishing CAD patients from controls is very high.

3.8.3. Oxidative Stress in Patients Undergoing Coronary Artery Bypass Surgery, with or Without Cardiopulmonary Bypass

As a further example of the application of OXY-SCORE, we analyzed the data of a study undertaken to assess oxidative stress in patients undergoing coronary artery bypass surgery (CABG) (14). Forty-seven patients scheduled for CABG were randomized to two different procedures: cardiopulmonary bypass (CPB) or off-pump coronary artery bypass (OPCAB). The two procedures are characterized by different production of oxidative stressors (14, 15), thus providing an ideal setting to test the relative performances of OXY-SCORE *vs.* its individual components.

Blood samples were collected at 7 time points: before induction of anesthesia (t1, baseline), after sternotomy (t2), 30 min after aortic cross-clamp in CABG or 30 min after the start of the first distal anastomosis in OPCAB (t3), after protamine administration (t4), at the end of surgery (t5, about 1.5 h from t3), 4 h after the arrival to the intensive care unit (t6), and 24 h after surgery (t7). The ability of OXY-SCORE and of individual biomarkers to discriminate between the two clinical situations was assessed by ROC curve analysis. Patients assigned to CPB or OPCAB were comparable for age, sex, body mass index, and risk factors.

Figure 14.1 compares the time course of OXY-SCORE during the two procedures: OXY-SCORE increased markedly in CPB, peaking during the procedure and with a sharp drop at the end of surgery (t5), followed by a progressive decline, like most of the prooxidant factors but remaining well above baseline values even at 24 h. In OPCAB, a minor, but significant

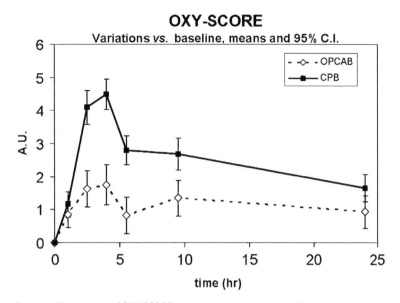

Fig. 14.1. Time course of OXY-SCORE in coronary artery bypass graft surgery, either with cardiopulmonary bypass (CPB) or without (OPCAB).

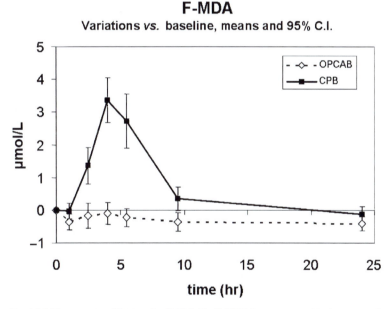

Fig. 14.2. Time course of free malondialdehyde (F-MDA) in coronary artery bypass graft surgery, either with cardiopulmonary bypass (CPB) or without (OPCAB).

increase in OXY-SCORE was observed with values above base-line up to 24 h.

The time course of F-MDA, as an example of prooxidant factors, is reported in Fig. 14.2. The variations in F-MDA values in the CPB group are similar to those of OXY-SCORE, except that the curve reaches baseline at 24 h. In the OPCAB group, no appreciable increase in F-MDA values occurred, but a slight decrease at 24 h.

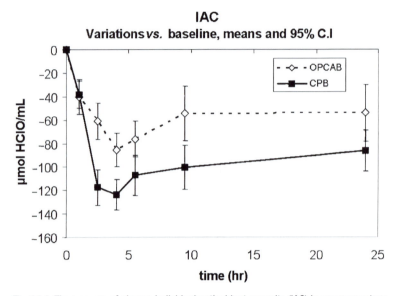

Fig. 14.3. Time course of plasma individual antioxidant capacity (IAC) in coronary artery bypass graft surgery, either with cardiopulmonary bypass (CPB) or without (OPCAB).

The time course of IAC is shown in Fig. 14.3, as an example of antioxidant defenses. Both procedures are associated with a significant reduction of IAC, more marked in the CPB group, and both show a partial recovery after surgery.

As compared with MDA, the picture provided by OXY-SCORE seems to be more sensitive to the minor variations occurring during surgery and to the apparent persistence of an oxidative stress in the postoperative phase. Both of these features are well represented in the variations of IAC, revealing a lesser but significant decrease also in the OPCAB group, and a persistent deprivation in both groups, lasting in the postsurgery phase. On the other hand, the variations of IAC only partially reveal the sharp increase of oxidative damage occurring during surgery in the CPB group. Thus, OXY-SCORE seems to be more sensitive than both damage and antioxidant defenses biomarkers, by accounting for all the oxidative balance modifications partially described by the single biomarkers.

3.8.4. ROC Analysis

The ROC analysis was applied considering two different phases of the surgical procedures: the acute phase during surgery, where the peak variations are located (points T3 to T5), and the recovery phase at 24 h after surgery (point T7). Table 14.3 reports the AUCs obtained using OXY-SCORE and its individual components. During surgery, OXY-SCORE well discriminates the two groups, with an AUC of 0.93, although both forms of MDA performed slightly better. OXY-SCORE had a good discriminating power even at 24 h (AUC = 0.70), where its performance was better then that of all the other biomarkers.

Table 14.3
Area under the ROC curve

Variable	AUC (95% C.I.)	
	During surgery (points T3–T5)	At 24 h (T7)
OXY-SCORE	0.93	0.70
F-MDA	0.99	0.64
T-MDA	0.98	0.59
$iPF_{2\alpha}$-III	0.79	0.67
GSSG/GSH	0.51	0.53
α-TH	0.69	0.40
γ-TH	0.57	0.44
GSH	0.55	0.53
IAC	0.81	0.67

3.9. Conclusions

Our data show that the comprehensive index OXY-SCORE is able to keep the entire information carried by different biomarkers, and it appears to maintain the advantages of both antioxidant defence indexes and oxidative damage markers. In a healthy population, it is sensitive to both age and gender, in contrast with all individual biomarkers, and with high sensitivity and specificity, it is able to discriminate between CAD patients and healthy subjects. In a clinical setting of two procedures characterized by different levels of oxidative stress (CABG with or without CPB), it is able to describe the parallel time courses of the two groups with high accuracy, still maintaining the ability to distinguish the high from the low grades of oxidative stress. In contrast with most of the individual analytes, characterized by very skewed distributions, OXY-SCORE, which is computed as a mean value, exhibits advantageous statistical properties, such as nearly normal distribution and a limited variability.

A few other approaches to achieve a global index of oxidative balance have been proposed. In one of these (17, 18), the integrated score was based on measured exposures to oxidant or antioxidant environmental and dietary factors, but no objective plasma or urine biomarkers were included. In another interesting approach, Kunt et al. (19) used an integrated oxidative stress index (OSI), summarizing total peroxide concentration and Total Antioxidant Capacity (analogous to IAC) to assess changes throughout surgical coronary revascularization.

Obviously, the present OXY-SCORE algorithm is to be considered as a starting point for the future development of a validated global index of oxidative stress, based on the integration of different approaches. Its precision and completeness are expected to increase with the number of parameters included, but a reasonable compromise should be found between power and feasibility. Among the new candidate parameters, markers of DNA and protein damage (e.g. 8-hydroxyguanosine, dityrosine, etc.) might be considered.

4. Notes

1. Use only water LC-MS to make dilutions

2. Standards are firstly dissolved in ethanol (10 ng/μL), and working solutions are made in CH_3CN (10%)

3. Standards are firstly dissolved in ethanol (200 mmol/L), and working solutions are made in $HClO_4$ (1.17 mol/L)

4. Pay attention to avoid contaminations.

5. Keep in dark

Acknowledgments

We thank Dr. Isabella Squellerio and Dr. Fabiana Minardi for their kind assistance in preparing the manuscript.

References

1. Droge W (2002) Free radicals in the physiological control of cell function. Physiol Rev 82:47–95

2. Valko M, Leibfritz D, Moncol J, Cronin MT, Mazur M, Telser J (2007) Free radicals and antioxidants in normal physiological functions and human disease. Int J Biochem Cell Biol 39:44–84

3. Veglia F, Cighetti G, De Franceschi M, Zingaro L, Boccotti L, Tremoli E, Cavalca V (2006) Age- and gender-related oxidative status determined in healthy subjects by means of OXY-SCORE, a potential new comprehensive index. Biomarkers 11:562–573

4. Cighetti G, Debiasi S, Paroni R, Allevi P (1999) Free and total malondialdehyde assessment in biological matrices by gas chromatography-mass spectrometry: what is needed for an accurate detection. Anal Biochem 266:222–229

5. Lawson JA, Rokach J, FitzGerald GA (1999) Isoprostanes: formation, analysis and use as indices of lipid peroxidation in vivo. J Biol Chem 274:24441–24444

6. Roberts LJ, Morrow JD (2000) Measurement of F(2)-isoprostanes as an index of oxidative stress in vivo. Free Radic Biol Med 28:505–513

7. Taylor AW, Bruno RS, Frei B, Traber MG (2006) Benefits of prolonged gradient separation for high-performance liquid chromatography-tandem mass spectrometry quantitation of plasma total 15-series F-isoprostanes. Anal Biochem 350:41–51

8. Morrow JD, Roberts LJ 2nd (1994) Mass spectrometry of prostanoids: F2-isoprostanes produced by non-cyclooxygenase free radical-catalyzed mechanism. Methods Enzymol 233:163–174

9. Saenger AK, Laha TJ, Edenfield MJ, Sadrzadeh SM (2007) Quantification of urinary 8-iso-PGF2alpha using liquid chromatography-tandem mass spectrometry and association with elevated troponin levels. Clin Biochem 40:1297–1304

10. Helmersson J, Basu S (1999) F2-isoprostane excretion rate and diurnal variation in human urine. Prostaglandins Leukot Essent Fatty Acids 61:203–205

11. Li H, Lawson JA, Reilly M, Adiyaman M, Hwang SW, Rokach J, FitzGerald GA (1999) Quantitative high performance liquid chromatography/tandem mass spectrometric analysis of the four classes of F(2)-isoprostanes in human urine. Proc Natl Acad Sci USA 96:13381–13386

12. Davies SS, Zackert W, Luo Y, Cunningham CC, Frisard M, Roberts LJ 2nd (2006) Quantification of dinor, dihydro metabolites of F2-isoprostanes in urine by liquid chromatography/tandem mass spectrometry. Anal Biochem 348:185–191

13. Carbonneau MA, Peuchant E, Sess D, Canioni P, Clerc M (1991) Free and bound malondialdehyde measured as thiobarbituric acid adduct by HPLC in serum and plasma. Clin Chem 37:1423–1429

14. Cavalca V, Sisillo E, Veglia F, Tremoli E, Cighetti G, Salvi L, Sola A, Mussoni L, Biglioli P, Folco G, Sala A, Parolari A (2006) Isoprostanes and oxidative stress in off-pump and on-pump coronary bypass surgery. Ann Thorac Surg 81:562–567

15. Gerritsen WB, van Boven WJ, Driessen AH, Haas FJ, Aarts LP (2001) Off-pump versus on-pump coronary artery bypass grafting: oxidative stress and renal function. Eur J Cardiothorac Surg 20:923–929

16. Zou KH, O'Malley AJ, Mauri L (2007) Receiver-operating characteristic analysis for evaluating diagnostic tests and predictive models. Circulation 115:654–657

17. Goodman M, Bostick RM, Dash C, Flanders WD, Mandel JS (2007) Hypothesis: oxidative stress score as a combined measure of pro-oxidant and antioxidant exposures. Ann Epidemiol 17:394–399

18. Wright ME, Mayne ST, Stolzenberg-Solomon RZ, Li Z, Pietinen P, Taylor PR, Virtamo J, Albanes D (2004) Development of a comprehensive dietary antioxidant index and application to lung cancer risk in a cohort of male smokers. Am J Epidemiol 160:68–76

19. Kunt AS, Selek S, Celik H, Demir D, Erel O, Andac MH (2006) Decrease of total antioxidant capacity during coronary artery bypass surgery. Mt Sinai J Med 73:777–783

Chapter 15

Cupric Ion Reducing Antioxidant Capacity Assay for Antioxidants in Human Serum and for Hydroxyl Radical Scavengers

Reşat Apak, Kubilay Güçlü, Mustafa Özyürek,
Burcu Bektaşoğlu, and Mustafa Bener

Abstract

Tests measuring the combined antioxidant effect of the nonenzymatic defenses in biological fluids may be useful in providing an index of the organism's capability to counteract reactive species known as pro-oxidants, resist oxidative damage, and combat oxidative stress-related diseases. The selected chromogenic redox reagent for the assay of human serum should be easily accessible, stable, selective, and respond to all types of biologically important antioxidants such as ascorbic acid, α-tocopherol, β-carotene, reduced glutathione (GSH), uric acid, and bilirubin, regardless of chemical type or hydrophilicity. Our recently developed cupric reducing antioxidant capacity (CUPRAC) spectrophotometric method for a number of polyphenols and flavonoids using the copper(II)-neocuproine reagent in ammonium acetate buffer is now applied to a complete series of plasma antioxidants for the assay of total antioxidant capacity of serum, and the resulting absorbance at 450 nm is recorded either directly (e.g., for ascorbic acid, α-tocopherol, and glutathione) or after incubation at 50°C for 20 min (e.g., for uric acid, bilirubin, and albumin), quantitation being made by means of a calibration curve. The lipophilic antioxidants, α-tocopherol and β-carotene, are assayed in dichloromethane. Lipophilic antioxidants of serum are extracted with n-hexane from an ethanolic solution of serum subjected to centrifugation. Hydrophilic antioxidants of serum are assayed in the centrifugate after perchloric acid precipitation of proteins. The CUPRAC molar absorptivities, linear ranges, and TEAC (trolox equivalent antioxidant capacity) coefficients of the serum antioxidants are established, and the results are evaluated in comparison with the findings of the ABTS/TEAC reference method. The intra- and inter-assay coefficients of variation (CVs) are 0.7 and 1.5%, respectively, for serum. The CUPRAC assay proved to be efficient for glutathione and thiol-type antioxidants, for which the FRAP (ferric reducing antioxidant potency) test is basically nonresponsive. The additivity of absorbances of all the tested antioxidants confirmed that antioxidants in the CUPRAC test do not chemically interact among each other so as to cause an intensification or quenching of the theoretically expected absorbance, and that a total antioxidant capacity (TAC) assay of serum is possible. As a distinct advantage over other electron-transfer based assays (e.g., Folin, FRAP, ABTS, DPPH), CUPRAC is superior in regard to its realistic pH close to the physiological pH, favorable redox potential, accessibility and stability of reagents, and applicability to lipophilic antioxidants as well as hydrophilic ones. The CUPRAC procedure can also assay hydroxyl radicals, being the most reactive oxygen species (ROS). As a more convenient, efficient, and less costly alternative to HPLC/electrochemical detection techniques and to the nonspecific, low-yield TBARS test, we use p-aminobenzoate, 2,4- and 3,5-dimethoxybenzoate probes for detecting hydroxyl radicals generated from an equivalent mixture of

D. Armstrong (ed.), *Advanced Protocols in Oxidative Stress II*, Methods in Molecular Biology, vol. 594
DOI 10.1007/978-1-60761-411-1_15, © Humana Press, a part of Springer Science+Business Media, LLC 2010

[Fe(II)+EDTA] with hydrogen peroxide. The produced hydroxyl radicals attack both the probe and the water-soluble antioxidants in 37°C-incubated solutions for 2 h. The CUPRAC absorbance of the ethylacetate extract due to the reduction of Cu(II)-neocuproine reagent by the hydroxylated probe decreases in the presence of ·OH scavengers, the difference being proportional to the scavenging ability of the tested compound. The developed method is less lengthy, more specific, and of a higher yield than the classical TBARS assay.

Key words: CUPRAC antioxidant capacity assay, Human serum, Plasma antioxidants, Uric acid, Bilirubin, Ascorbic acid, α-Tocopherol, β-Carotene, Glutathione, Hydroxyl radical scavengers

1. Introduction

When natural defenses of the organism (of enzymatic, nonenzymatic or dietary origin) are overwhelmed by an excessive generation of reactive oxygen species (ROS), a situation of oxidative stress occurs, in which cellular and extracellular macromolecules (proteins, lipids and nucleic acids) can suffer oxidative damage, causing tissue injury (1, 2). Living organisms have developed complex antioxidant systems to counteract reactive species and to reduce their oxidative damage (3). These antioxidant systems include enzymes such as superoxide dismutase, catalase, and glutathione peroxidase (4); macromolecules such as albumin, ceruloplasmin, and ferritin; and an array of small molecules, including ascorbic acid, α-tocopherol, β-carotene, ubiquinol-10, glutathione (GSH), methionine, uric acid, and bilirubin (5).

Blood has a central role in transporting and redistributing antioxidants to the body parts. Although enzymatic antioxidants constitute an important part of ROS scavengers in intracellular fluids, nonenzymatic low molecular weight antioxidants more easily assayed by total antioxidant capacity (TAC) tests are predominant in serum (6). Naturally, each antioxidant compound of serum having a specific reactivity and redox potential would show a different affinity towards the TAC test reagent. Thus, it is predictable that different TAC tests should yield different results for a given serum sample. The TAC of serum is a combination of the effects of all chain-breaking antioxidants, including protein thiols (probably constituting a large portion of the "dark" TAC, as these constituents cannot be properly identified) and uric acid. Since uric acid has the largest contribution to serum TAC, and its increasing concentrations may also indicate clinical conditions (e.g., kidney failure and metabolic disorders) associated with oxidative stress (6), measurement of absolute TAC values may only have a tentative meaning, and the relative changes of these values systematically measured in serum at regular intervals with definite tests may be more useful as potential indicators of human health.

Total antioxidant capacity (TAC) of serum is a reliable marker to detect changes of in vivo oxidative stress, which may not be detectable through the measurement of single specific antioxidants. Thus, serum TAC as an integrated parameter represents the cumulative action of all antioxidants in the sample (6). This cumulative action is important from the standpoint of providing greater protection against reactive O, N-species, because synergistic interactions among antioxidants may exist, such as regeneration of ascorbate by glutathione (7) and of α-tocopherol by ascorbate (8).

A quick look at various TAC assays applicable to serum antioxidants reveals that a majority of those suffer from the difficulties encountered in the formation and stability of colored radicals (9) such as ABTS (10) and DPPH (11). Re et al. developed an improved ABTS radical cation decolorization assay using persulfate as the oxidant, and thus compensated for the weaknesses of the original ferryl myoglobulin/ABTS assay (12). The total radical trapping parameter (TRAP) assay of Wayner et al. (13) was the most widely used method of measuring total antioxidant capacity of plasma or serum during the last decade; however, it suffered from the major drawback of oxygen electrode end-point in that the electrode would not maintain its stability over the required time period (14). The major limitation of the oxygen radical absorbance capacity, or $ORAC_{PE}$ (ORAC test based on β-phycoerythrin: β-PE) was reported as the inconsistency of β-PE in varying production lots, and lack of its photostability. The ferric reducing antioxidant power (FRAP) assay of antioxidants (15), which is based on ferric -to- ferrous reduction in the presence of the Fe(II)-stabilizing ligand: tripyridyltriazine (TPTZ), is both unrealistic (i.e., the colored complex is formed at a definitely acidic pH such as pH = 3.6, much lower than the physiological pH) and insufficiently reactive to thiol- type (i.e., –SH containing) antioxidants such as glutathione (16) (see Note 1).

Current literature states that there is no "total antioxidant" as a nutritional index available for food labeling because of the lack of standard quantitation methods (17). This is also valid for serum where different tests yield different results that do not correlate well. For example, Cao and Prior observed a weak linear correlation between serum ORAC and serum FRAP, but no correlation either between serum ORAC and serum TEAC, or between serum FRAP and serum TEAC (3). Total antioxidant capacity (TAC) assays measure the capacity of biological samples only under defined conditions prescribed by the given method using different oxidants or radicalic reagents in each case. Some assays measure only the hydrophilic antioxidants, without considering the lipophilic ones. Not all methods measure protein-thiols, or smaller molecule -SH compounds of different origin (such as GSH, with FRAP). Thus, the first aim of this work is to apply the cupric reducing antioxidant capacity (CUPRAC) assay (18), previously developed for food antioxidants, to the measurement of antioxidants in human serum (19). This would compensate for the deficiencies of other methods, as CUPRAC is employed at nearly physiological pH, is capable to react with thiol antioxidants, and is applicable to both hydrophilic and lipophilic antioxidants. Since the optimal pH of CUPRAC is close to the physiological one, there would be no risk of underestimation (under acidic conditions) or overestimation (under basic conditions) of TAC, due to either proton association or dissociation of phenolic compounds, respectively (20) (*see* Note 1).

A second aim is to report the use of a modified CUPRAC assay for detecting hydroxyl radicals, the most reactive ROS capable of producing cytotoxic effects in aerobic biological systems (1, 21, 22). The conventional colorimetric method of ·OH measurement is the thiobarbituric acid (TBA)-reactive substances (TBARS) assay. Oxidative attack of hydroxyl radicals generated from a Fenton reaction (i.e., $Fe(II)+H_2O_2$) on deoxyribose produces malondialdehyde (MDA) and similar substances that are colorimetrically (or fluorometrically) reactive toward (TBA) (based on the formation of colored TBA-MDA adducts), forming the essence of ·OH detection (23, 24). When ·OH radicals generated by iron-EDTA+H_2O_2 in the presence of ascorbic acid oxidize deoxyribose and the reaction products yield a pink chromogen absorbing at 532 nm upon heating with thiobarbituric acid, hydroxyl radical scavengers added to the medium compete with the deoxyribose probe and diminish chromogen formation, enabling the calculation of second-order rate constants of ·OH scavenging (25). However, there are also criticisms to the classical TBARS method in that it is rather nonspecific (e.g., also used for testing lipid peroxidation (26)), is of low yield (i.e., only a small percentage of the carbohydrate deoxyribose is converted to TBA-reactive substances), and cannot properly assay the ·OH scavenging power of phenolic antioxidants which may show pro-oxidant activity in the Fenton reaction system via iron recycling (27). Thus, a "test tube assay" similar to that of Halliwell et al. (25), but one having greater efficiency and selectivity, is developed using a modified CUPRAC assay. Aromatic probes such as *p*-aminobenzoate, 2,4-dimethoxybenzoate, and 3,5-dimethoxybenzoate are used for ·OH radical scavenging assay of a number of important water-soluble compounds. This assay makes use of competition kinetics to simultaneously incubate the probe with the scavenger under the attack of hydroxyl radicals generated in a Fenton system, and measures the difference in CUPRAC absorbance of the probe (extracted into ethylacetate at the end of the incubation period) in the absence and presence of the scavenger (i.e., the hydroxylation product of the probe would show a higher CUPRAC absorbance when alone) (28).

2. Materials

2.1. Equipment

1. Varian (Australia) CARY 1E UV-Vis spectrophotometer using a pair of matched quartz cuvettes of 1 cm thickness is used for the absorbance measurements.

2. MSE Mistral 2000 (MSE, UK) centrifuge apparatus using 10 cm-tubes of 1.5 cm diameter is used for serum centrifugation.

3. E512 Metrohm-Herisau (Switzerland) pH-meter equipped with a combined glass electrode.

4. Clifton (Nickel Electro LTD, UK) water bath.

5. Elektromag (Turkey) vortex stirrer.

2.2. Reagents and Buffers

1. Uric acid and ascorbic acid are purchased from Sigma Chemical Co., Steinheim, Germany.

2. Trolox (6-hydroxy-2,5,7,8-tetramethylchroman-2-carboxylic acid) from Aldrich Chem. Co., Steinheim, Germany.

3. Glutathione (reduced, GSH), α-tocopherol, albumin fraction V (from bovine serum, BSA), perchloric acid ($HClO_4$), dichloromethane, iron(II) chloride tetrahydrate, hydrogen peroxide (30%, by weight), sodium metabisulfite ($Na_2S_2O_5$), thiourea, sodium formate, mannitol, glucose, ammonium thiocyanate (NH_4SCN), sodium thiosulfate ($Na_2S_2O_3$), potassium hexacyanoferrate(II) ($K_4[Fe(CN)_6].3H_2O$), trisodium citrate pentahydrate, cerium(IV) sulfate, ethylacetate, dimethylsulfoxide (DMSO, 99.5%), methanol, sodium hydroxide, concentrated hydrochloric acid (HCl), and 96% ethanol from E. Merck, Darmstadt, Germany.

4. β-Carotene, L-lycine, disodium-EDTA (Na_2EDTA), 4-aminobenzoic acid, 2,4-dimethoxybenzoic acid, 3,5-dimethoxybenzoic acid, bilirubin, and ascorbate oxidase from Fluka Chemicals, Steinheim, Germany.

5. n-Hexane, KI, and 2-propanol from Riedel-deHaën, Steinheim, Germany.

6. Na_2HPO_4 and NaH_2PO_4 from J.T. Baker, Deventer, Holland.

7. CUPRAC assay reagents: Neocuproine (2,9-dimethyl-1,10-phenanthroline) is from Sigma Chemical Co., Steinheim, Germany; ammonium acetate (NH_4Ac), copper(II) chloride dihydrate are from E. Merck, Darmstadt, Germany.

8. ABTS assay reagents: ABTS (2,2'-azinobis(3-ethylbenzothiazoline-6-sulfonic acid) diammonium salt) radical reagent is purchased from Fluka, Steinheim, Germany (kept at +4°C); potassium persulfate, from E. Merck, Darmstadt, Germany.

9. TBARS assay reagents: Trichloroacetic acid (TCA) is purchased from Sigma Chem. Co., Steinheim, Germany; 2-thiobarbituric acid (TBA) from Aldrich Chem. Co., Steinheim, Germany; 2-deoxy-D-ribose from Fluka, Steinheim, Germany.

2.3. Supplies

1. Serum samples (i.e., liquid samples obtained as the centrifugate after blood clotting) from healthy adults are supplied freshly from Istanbul University, Cerrahpasa Faculty of Medicine, Central Laboratory, and citrated for a longer standing whenever necessary (*see* Note 2).

3. Methods

3.1. Preparation of Solutions

1. Preparation of CUPRAC assay solutions: $CuCl_2$ solution, 1.0×10^{-2} M, is prepared by dissolving 0.4262 g $CuCl_2 \cdot 2H_2O$ in water, and diluting to 250 mL. Ammonium acetate buffer at pH = 7.0, 1.0 M, is prepared by dissolving 19.27 g NH_4Ac in water and diluting to 250 mL. Neocuproine (Nc) solution, 7.5×10^{-3} M, is prepared daily by dissolving 0.039 g Nc in 96% ethanol, and diluting to 25 mL with ethanol. Trolox, 1.0×10^{-3} M, is prepared in 96% ethanol.

2. Preparation of ABTS assay solutions: The chromogenic radical reagent ABTS, at 7.0 mM concentration, is prepared by dissolving 0.1920 g of the compound in water, and diluting to 50 mL. To this solution is added 0.0331 g $K_2S_2O_8$ such that the final persulfate concentration in the mixture be 2.45 mM. The resulting ABTS radical cation solution is left to mature at room temperature in the dark for 12–16 h, and then used for TEAC assays.

3. Preparation of standard solutions of plasma antioxidants: The standard solutions of plasma antioxidants are prepared at 1.0×10^{-3} M concentration. α-Tocopherol and β-carotene are dissolved in dichloromethane (DCM), and the β-carotene solution is further diluted with the same solvent at 1:50 volume ratio. Ascorbic acid and glutathione (GSH) solutions are prepared in distilled water. Uric acid (0.0168 g) is dissolved in 20 mL of 0.01 M NaOH, the excess base is neutralized with the addition of 0.01 M HCl, and finally diluted to 100 mL with H_2O. Bilirubin (0.0146 g) is dissolved using 1 mL of 0.1 M NaOH, excess base is neutralized with 0.1 M HCl, and finally diluted to 25 mL with water.

4. Preparation of water soluble antioxidants as hydroxyl radical scavengers: Either 1.0×10^{-2} or 1.0×10^{-3} M aqueous solutions of thiourea, mannitol, Na-formate, ascorbic acid, DMSO, L-lysine, $K_4[Fe(CN)_6]$, trisodium citrate, glucose, sodium

metabisulfite, KI, NH_4SCN, and $Na_2S_2O_3$ are prepared in distilled water. Methanol, ethanol, and 2-propanol are diluted with H_2O prior to use.

5. Preparation of Fenton reaction solutions: The following buffer solutions are prepared: 0.2 M phosphate buffer at pH = 7.0, 10 mM probe buffers at pH = 7.0 are prepared by dissolving 0.137 g of 4-aminobenzoic acid, 0.182 g of 2,4-dimethoxybenzoic acid and 0.182 g of 3,5-dimethoxybenzoic acid in 0.1 M NaOH, adjusting the pH to pH = 7.0 with the addition of 0.1 M HCl, and diluting to 100 mL with H_2O. Fe(II) at 20 mM concentration is prepared by dissolving 0.1988 g $FeCl_2.4H_2O$ with 2 mL of 1 M HCl, some water, and diluting to 50 mL with water. Na_2-EDTA at 20 mM concentration is prepared by dissolving 0.372 g of the salt in water and diluting to 50 mL. Hydrogen peroxide at 10 mM concentration is prepared from a 0.5 M intermediary stock solution prepared from 30% H_2O_2, and then diluting this at a ratio of 1:50 with H_2O.

6. Preparation of TBARS assay solutions: Trichloroacetic acid (TCA) at 2.8% (by mass) is prepared in water, and thiobarbituric acid (TBA) at 1% in 50 mM aqueous NaOH. Deoxyribose at 10 mM is prepared in water.

3.2. Serum Extraction and Preparation for TAC Assay

1. Standard procedure for the separation of lipophilic and hydrophilic fractions: Serum samples are freshly collected and kept at +4°C in a refrigerator just prior to analysis (or stored at −70°C as necessary, are thawed slowly), mixed well on a vortex, and centrifuged if needed. Serum extraction is based on the procedure published by Aebischer et al. (29), and applied to plasma and serum samples as described by Prior et al. (30). One mL of such a (serum) sample is transferred to a centrifuge tube, 2 mL of 96% ethanol and 1 mL of distilled water are added, and mixed well. Four mL of n-hexane are added to the mixture, again mixed, and the final mixture is let to stand for a few min for the separation of phases. The solution is centrifuged at 5,000 rpm (1,500 g) for 5 min. The upper organic phase is separated and transferred to a dark tube. The hexane extraction procedure is repeated, i.e., 4 mL hexane are added to the remaining aqueous phase, mixed well, let to stand for a few min for the separation of phases, and centrifuged again at 5,000 rpm for 5 min. This second hexane extract is separated and transferred to the original dark tube so as to combine with the first extract. The organic solution comprising combined hexane extracts is dried down under N_2 flow, and the residue is taken up with 1 mL of dichloromethane (DCM) for the assay of lipophilic antioxidants. The above procedure is repeated eight times for the serum sample: the DCM phases are combined for the assay of lipophilic antioxidants.

The minute amounts of hexane remaining in the aqueous phase of each tube is removed by drying under nitrogen. Protein content of each tube is precipitated by adding 4 mL of 0.5 M $HClO_4$. The nature and relative quantity of this precipitant is as optimized by Prior et al. (30). The aqueous solution is centrifuged at 5,000 rpm (1,500 g) for 5 min. The upper clear phases of 8 tubes are combined for the assay of hydrophilic antioxidants. The combined acidic aqueous phase is neutralized with approximately 10.1 mL of 1.0 M NaOH prior to analysis. Thus, the serum is separated into two phases for the assay of lipophilic and hydrophilic antioxidants. For the application of standard addition technique (so as to observe whether the calibration curve of a given antioxidant compound in standard-added serum is parallel to the one obtained with the sole antioxidant), the lipophilic antioxidants: α-tocopherol and β-carotene are added one by one to the organic phase, and the hydrophilic antioxidants: bilirubin, uric acid, ascorbic acid, glutathione (GSH), and albumin are added one by one to the aqueous phase.

2. Preparation of serum for TAC assay without the separation of lipophilic and hydrophilic antioxidant fractions: A volume of 0.5 mL of serum is transferred to a centrifuge tube, 1 mL of 10% (w/v) TCA (in absolute ethanol) is added, and mixed well. The solution is centrifuged at 5,000 rpm (1,500 g) for 10 min. A 0.5-mL aliquot of the supernatant phase is withdrawn into a test tube, and neutralized with 2 mL NH_4Ac buffer. A suitable volume (0.2 mL) of the serum extract is taken for the CUPRAC analysis (*see* Note 3).

3.3. Standard Addition Method Applied to Organic Extract of Serum

To a test tube, 1 mL of copper(II) chloride solution, 1 mL of neocuproine solution, and 1 mL of NH_4Ac buffer solution are added in this order. A suitable aliquot (0.8 mL) of the combined organic extract (of serum) is added to this tube (such that the initial absorbance of this extract with respect to the CUPRAC spectrophotometric method would be around 0.2). To this mixture, 3.2 mL of DCM are added, shaken, and the organic phase is separated from the aqueous phase. Standard additions of α-tocopherol and β-carotene (*see* Fig. 15.1) in varying concentrations are made to the serum (organic) extract so as to construct the calibration curves of these lipophilic antioxidants in organic serum extract of initial absorbance around 0.2. Absorbance reading is made against a reagent blank at 450 nm. Since the vapor pressure of DCM at ambient temperature is high, the DCM used in the procedure is cooled to an initial temperature of +4°C to prevent evaporation losses. Elevated temperature incubation tests (as applied to hydrophilic antioxidants in the aqueous phase) are not carried out with the organic extract.

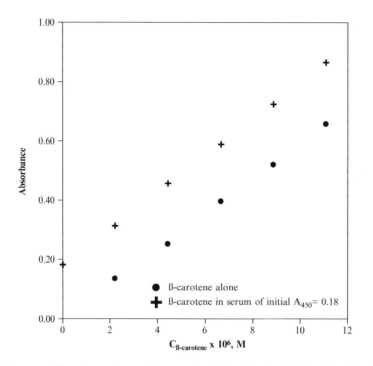

Fig. 15.1. Calibration curve of β-carotene alone and in serum (applied as "standard additions" technique).

3.4. Standard CUPRAC Antioxidant Capacity Assay Applied to Aqueous Serum Extract with Standard Additions

To a test tube, 1 mL of copper(II) chloride solution, 1 mL of neocuproine solution, and 1 mL of NH_4Ac buffer solution are added in this order. A suitable aliquot (1.5 mL) of the combined aqueous extract (of serum) is added to this tube (such that the initial absorbance of this extract with respect to the CUPRAC spectrophotometric method would be around 0.2). Standard additions of bilirubin, uric acid, ascorbic acid, and GSH (*see* Fig. 15.2) at varying concentrations are made to this extract so as to construct the calibration curves of these hydrophilic antioxidants in aqueous serum extract of initial absorbance around 0.2. If (x) mL of the standard antioxidant solution is taken, then (0.25-x) mL H_2O is added to make the final volume 4.75 mL (For dilution experiments of serum, 1.5 mL of the combined aqueous extract diluted with water at ratios varying between 1:1 and 1:10 is treated as the unknown sample, and 1.5 mL of this final diluted sample is subjected to CUPRAC analysis as stated above). Absorbance reading is made against a reagent blank at 450 nm. All hydrophilic antioxidants react fairly rapidly with the CUPRAC reagent (*see* Note 4) except uric acid and bilirubin which show a slight absorbance increase at room temperature. Therefore, absorbance readings are recorded 30 min after the mixing of analyte solution with reagents in the standard CUPRAC protocol (19). The results are evaluated by means of a calibration

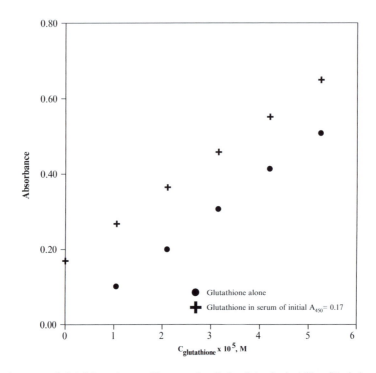

Fig. 15.2. Calibration curve of glutathione alone and in serum (applied as "standard additions" technique).

curve (line) for each antioxidant. Comparison with trolox as the reference compound is made using the room temperature molar absorptivity of trolox, i.e., $\epsilon = 1.67 \times 10^4 \, \text{L mol}^{-1} \text{cm}^{-1}$.

3.5. CUPRAC Assay Applied to Aqueous Serum Extract with Standard Addition and Incubation at Elevated Temperature

The standard addition method applied to aqueous extract of serum is followed with the single difference of extract volume taken for analysis (i.e., 0.7 mL of combined aqueous extract of serum is suitable so that it would yield an initial absorbance of 0.2 with respect to the CUPRAC method). After the addition of (x) mL of standard hydrophilic antioxidant (bilirubin, uric acid, GSH, and BSA) solutions, dilution is made with H_2O to 4.75 mL. The tubes are stoppered and incubated at 50°C in a water bath for 20 min. (Tests are also performed to follow the color development kinetics of hydrophilic antioxidants for longer incubation periods at this temperature). The incubation period is selected with respect to the kinetic behavior of bilirubin, which requires 20 min at 50°C for absorbance stabilization. Data for ascorbic acid are not collected at this stage, because ascorbic acid decomposes at elevated temperature incubation. The incubated tubes are let to cool to room temperature, and the 450 nm-absorbance is read as stated (19). The results are evaluated by means of a linear calibration curve for each antioxidant. Comparison with trolox as the reference compound is made using the 50°C – incubated molar absorptivity of trolox, i.e., $\varepsilon = 1.86 \times 10^4\,\mathrm{L\ mol^{-1}cm^{-1}}$.

3.6. Individual Determination of Ascorbic Acid Among Hydrophilic Serum Antioxidants

The original ascorbate oxidase enzyme solution of initial activity 328 U/mg is diluted with water to a concentration of 4 U/mL. To a separate test tube, 0.100 mL of 1.0×10^{-3} M ascorbic acid is added, and analyzed conventionally with the CUPRAC method to yield an absorbance of 0.30. To another tube containing the same amount (0.100 mL) of ascorbic acid solution, 0.200 mL of ascorbate oxidase solution is added, let to stand for 1 min, and analyzed with the CUPRAC method to observe at least 90% quenching of the absorbance due to ascorbic acid ($A_{450} = 0.03$).

A synthetic mixture of hydrophilic antioxidants is prepared to include 0.05 mL uric acid, 0.05 mL GSH, and 0.10 mL ascorbic acid standard solutions. The CUPRAC absorbance of this mixture is 0.57. To another tube containing the same antioxidants mixture (with identical amounts), 0.20 mL ascorbate oxidase solution is added, and subsequently analyzed with the CUPRAC method to yield an absorbance of 0.26, the absorbance difference corresponding to ascorbic acid content of the mixture. Thus, it is shown that ascorbic acid among hydrophilic antioxidants of serum can be individually quantified by the aid of two successive CUPRAC measurements of the antioxidant mixture with and without ascorbate oxidase, the ascorbic acid content being calculated from the difference.

3.7. ABTS Assay of Total Antioxidant Capacity of Serum Antioxidants

The preparation of ABTS/TEAC assay solutions is described in Subheading 15.3.1. The matured blue-green ABTS cation radical solution is diluted with 96% ethanol at a ratio of 1:10. The absorbance of the 1:10 diluted ABTS$^{\cdot+}$ radical cation solution is 1.28 ± 0.04 at 734 nm. To (x) mL of the sample solution (aqueous extract of serum) 1 mL of final ABTS$^{\cdot+}$ solution and (4-x) mL 96% EtOH are added, and the change of absorbance at the end of 6 min is recorded (usually the absorbance decrease at the 6th minute is used for calculations). As a convention, (x) is selected between 0.5 and 1.0 mL for the organic and aqueous extracts of serum, and the total volume is 5.0 mL.

3.8. Miniaturized CUPRAC Total Antioxidant Capacity Assay of Serum Extract Without Separation of Lipophilic and Hydrophilic Fractions

The preparation of the serum extract subject to this analysis is described in Subheading 15.3.2. Thus, 0.2 mL Cu(II) + 0.2 mL Nc + 0.2 mL buffer + 0.2 mL serum extract + 0.2 mL H_2O are mixed in a total volume of 1.0 mL. The absorbance of the mixture at 450 nm is measured against a reagent blank after 30 min of mixing the sample and reagents. This miniaturized system may cause slight turbidity problems in the ABTS assay, which can be overcome by filtering.

3.9. CUPRAC Assay for Water-Soluble Hydroxyl Radical Scavengers Using Substituted-Benzoate Probes

To a test tube, 3.0 mL of phosphate buffer (pH 7.0), 1.0 mL of 10 mM probe material (p-aminobenzoate, 2,4-dimethoxybenzo-ate, or 3,5-dimethoxybenzoate), 0.5 mL of 20 mM Na_2-EDTA, 0.5 mL of 20 mM $FeCl_2$ solution, (4-x) mL H_2O, (x) mL scavenger sample solution (x varying between 0.5 and 2.5 mL) are added at a concentration 1.0×10^{-2} or 1.0×10^{-3} M, and 1.0 mL of 10 mM H_2O_2 rapidly in this order, and the mixture in a total volume of 10 mL is incubated for 2 h in a water bath kept at 37°C. At the end of this period, the reaction is stopped with adding 0.5 mL of 2 M HCl, vortexed, 2 mL of this acidified solution is taken, 2 mL of ethylacetate (EtAc) added, and mixed again. To 1 mL of the EtAc extract, the modified CUPRAC assay (28) is applied in the following manner:

$$1.0 \text{ mL } Cu(II) + 1.0 \text{ mL Nc} + 1.0 \text{ mL } NH_4Ac$$
$$\text{buffer} + 1.0 \text{ mL EtAc extract} + 1.0 \text{ mL EtOH}$$
$$V_{total} = 5.0 \text{ mL}$$

The absorbance at 450 nm of the final solution in 5 mL total volume is recorded 30 min after mixing the sample and reagents against a reagent blank. None of the probes exhibited an initial CUPRAC absorbance.

The second-order rate constants of the scavengers are determined with competition kinetics (31, 32) by means of a linear plot of A_o/A as a function of $C_{scavenger}/C_{probe}$, where A_o and A are the CUPRAC absorbances of the system in the absence and presence of scavenger, respectively, and C is the molar concentration of relevant species (25, 33).

3.10. TBARS Assay for Hydroxyl Radical Scavengers Using a Deoxyribose Probe

The deoxyribose (TBARS) method is applied as the reference method of comparison for determining the rate constants (25). To a test tube, 3 mL of phosphate buffer (pH 7.0), 1.0 mL of 10 mM 2-deoxy-D-ribose (probe), 0.5 mL of 20 mM Na_2-EDTA, 0.5 mL of 20 mM $FeCl_2$ solution, (4-x) mL H_2O, (x) mL scavenger sample solution (x varying between 0.5 and 2.5 mL) are added at a concentration 1.0×10^{-2} or 1.0×10^{-3} M, and 1.0 mL of 10 mM H_2O_2 rapidly in this order, and the mixture in a total volume of 10 mL is incubated for 4 h in a water bath kept at 37°C. At the end of this period, the reaction is stopped with adding 5 mL of 2.8% TCA; 5 mL of 1% TBA is added, and the reaction mixture is kept in a 100°C-water bath for 10 min. The test tube containing the final mix is cooled under a flow of running tap water, and the absorbance at 520 nm is recorded.

The second-order rate constants of the ·OH scavengers are determined with competition kinetics (25, 33) by means of a linear plot of A_o/A as a function of $C_{scavenger}/C_{probe}$, where A_o and A are the TBARS absorbances of the system in the absence and presence of scavenger, respectively.

3.11. Results

1. Molar absorptivity, linear working range, and TEAC coefficients of serum antioxidants: The molar absorptivities and linear working ranges obtained from normal and incubated solutions of plasma antioxidants are listed in Table 15.1. Here, it is apparent that the highest molar absorptivities were obtained for bilirubin (5.3×10^4, in the aqueous phase) and β-carotene (5.6×10^4, in the organic phase). The TEAC coefficients of plasma antioxidants with respect to the CUPRAC method (i.e., the ratio of the molar absorptivity of antioxidant to that of trolox, measured under identical conditions) are listed and compared in Table 15.2 with those found by other widely-used methods currently employed, i.e., ORAC-peroxyl radicals (34), FRAP (15), and ABTS-persulfate (12) assays of total antioxidant capacity. The TEAC coefficients pertaining to the ABTS-persulfate method were simultaneously reported from the literature and experimentally found by us (see Table 15.2). Inspection of data in Table 15.2 reveals that the FRAP method cannot measure glutathione, as criticized for not being capable of measuring thiol-type antioxidants (3). In relation to cellular GSH and thiols metabolism, 2 molecules of GSH react with H_2O_2 or hydroperoxides through an enzymatic oxidation with glutathione peroxidase to form 1 molecule of glutathione disulfide (GSSG),

Table 15.1
The CUPRAC molar absorptivities and linear working ranges of plasma antioxidants

Antioxidant compound	ε (Lmol^{-1}cm^{-1})	Incubated ε (L mol^{-1}cm^{-1})	Linear range (M)
Ascorbic acid	$(1.59 \pm 0.03) \times 10^4$	(Decomposes)	5.6×10^{-6}–8.5×10^{-5}
Bilirubin	$(5.3 \pm 0.1) \times 10^4$	$(8.0 \pm 0.15) \times 10^4$	3.23×10^{-7}– 2.61×10^{-5}
Glutathione (GSH)	$(9.5 \pm 0.2) \times 10^3$	$(9.5 \pm 0.2) \times 10^3$	3.12×10^{-6}– 1.48×10^{-4}
Uric acid	$(1.60 \pm 0.03) \times 10^4$	$(2.8 \pm 0.05) \times 10^4$	7.07×10^{-7}– 8.64×10^{-5}
α-Tocopherol	$(1.83 \pm 0.03) \times 10^4$	–[a]	1.05×10^{-6}– 7.67×10^{-5}
β-Carotene	$(5.6 \pm 0.1) \times 10^4$	–[a]	3.37×10^{-7}– 2.49×10^{-5}

[a]Incubated absorptivity cannot be measured in organic solution. Bovine serum albumin (BSA) only reacts in incubated solution with an absorptivity of 9.24 mL mg^{-1}cm^{-1} reported as such since its molecular weight is too high (approximately given as 6.8×10^4 g mol^{-1}); the linear equation of its abs./concn. plot is: $A_{450} = 9.24 \times 10^{-3} C_{BSA} - 4.78 \times 10^{-3}$ ($r = 0.9996$) where C_{BSA} was in mg mL^{-1}

Table 15.2
Trolox equivalent antioxidant capacity (TEAC) coefficients of plasma antioxidants

Antioxidant compound	TEAC$_{CUPRAC}$	Inc. TEAC$_{CUPRAC}$	Measd. TEAC$_{ABTS}$	Lit. TEAC$_{ABTS}$	TEAC$_{ORAC}$	TEAC$_{FRAP}$
Ascorbic acid	0.96	–	1.03	1.05	0.52–1.12	0.95–1.05
Bilirubin	3.18	4.34	2.36	–	0.84	2.1–2.3
Glutathione (GSH)	0.57	0.57	1.51	1.28	0.68	Unmeasurable
Uric acid	0.96	1.54	1.11	1.01	0.92	1.0–1.2
α-Tocopherol	1.11	–	1.02	0.97	1.0	0.85–1.05
β-Carotene	3.35	–	2.80	2.57	0.64	Unmeasurable
Bovine serum albumin	–	0.033	–	–	–	0.05

Inc. TEAC$_{CUPRAC}$: TEAC measured in incubated solution (inc. at 50°C for 20 min); Measd. and Lit. TEAC$_{ABTS}$ values are experimentally measured and the literature – reported ABTS-persulfate values of TEAC coefficients, respectively; TEAC$_{ORAC}$ are extracted from the literature (ORAC-peroxyl radicals); TEAC$_{FRAP}$ values are calculated by dividing the literature FRAP values by 2, since original FRAP was reported as Fe(II) equivalents which is a 1-e reductant whereas conversion to trolox (2-e reductant) is required. The incubated TEAC$_{CUPRAC}$ values of ascorbic acid, α-tocopherol and β-carotene are not reported in Table 15.2 due to the reasons given in Table 15.1

where GSH acts as a 1 e-reductant (35). Likewise, for a structurally similar compound, cysteine, two cysteine residues in proteins may undergo a reversible oxidation to form a disulfide bond, which often plays an important structural role (2 RSH↔RSSR+2H$^+$+2e$^-$). In accord with these roles, the TEAC coefficients of GSH found by ORAC and Randox-TEAC (i.e., a commercialized version of the ABTS assay by Randox laboratories) assays are 0.59 and 0.66, respectively (3). Our (CUPRAC) TEAC coefficient of GSH is 0.57 (see Table 15.2), again consistent with its physiological role as a (1 e-reductant) antioxidant. However, metal-catalyzed reactions of H$_2$O$_2$ or peroxynitrite with a thiol may produce sulfinic (–SO$_2$H) or sulfonic (–SO$_3$H) acids through sulfenic acid (–SOH) intermediates, which is less likely in vivo (35). It is clear that the ABTS/persulfate assay treats GSH as a reductant capable of giving 2 or more electrons (The TEAC literature and experimental values of the latter assay for GSH were 1.28 and 1.51, respectively, as indicated in Table 15.2). We believe that our TEAC result of 0.57 is more reflective of the physiological role of GSH as an antioxidant. The exceptionally low TEAC values of the ORAC-peroxyl radical method for bilirubin and β-carotene in Table 15.2 is reminiscent of the fact that the fluorescent protein probe of the ORAC method, β-PE, as developed by Cao et al. (36), has interacted in a nonspecific manner –basically as hydrophobic interactions and H-bonding– with polyphenols, causing falsely low ORAC values for these polyphenols (37). Although it may be speculated that the high TEAC coefficient of bilirubin in the CUPRAC method might have arisen from the blue-green pigment, biliverdin, emerging as the oxidation product of bilirubin, we have shown with a separate $(Ce^{4+} + H_2SO_4)$ oxidation of bilirubin that the produced biliverdin does not exhibit a CUPRAC absorbance at 450 nm in neutral solution against a reagent blank.

The total antioxidant capacities of the serum samples using the CUPRAC, incubated CUPRAC, and ABTS-persulfate methods applied on the aqueous and organic fractions of serum are listed in Table 15.3. The results of hydrophilic and lipophilic antioxidants assays in Table 15.3 could be added to yield a sum as a measure of total antioxidant capacity of a sample (30) (*see* Note 5). Both CUPRAC and ABTS methods yield close results for the lipophilic antioxidants (organic fraction). The CUPRAC assay results of the aqueous extracts (for the hydrophilic antioxidants) as the outcome of room temperature measurements are significantly increased upon incubation. Since uric acid, bilirubin, and albumin constituting a great majority of total antioxidant capacity of serum (15, 38) and all show an absorbance increase upon elevated temperature incubation, the almost double-fold increase of CUPRAC capacities of aqueous extract of serum as a result of incubation is expectable.

Table 15.3
Lipophilic and hydrophilic total antioxidant capacities of serum samples using the CUPRAC (normal and incubated) and ABTS-persulfate assays ($N=5$)

Sample solutions	Method	Organic extract (mM TR)[a]	Aqueous extract (mM TR)[a]	Aqueous extract (inc) (mM TR)[a]
Serum 1	CUPRAC	0.08	0.27	0.54
	ABTS	0.08	0.84	–
Serum 2	CUPRAC	0.07	0.23	0.46
	ABTS	0.06	0.90	–
Serum 3	CUPRAC	0.08	0.19	0.39
	ABTS	0.08	0.73	–
Serum 4	CUPRAC	0.05	0.21	0.43
	ABTS	0.06	0.72	–
Serum 5	CUPRAC	0.06	0.25	0.51
	ABTS	0.06	0.78	–
Hydrophilic phase (comparison)	$A_{ABTS} = 0.472 + 1.4\, A_{CUPRAC}$ ($r=0.58$) at room temperature $A_{ABTS} = 0.483 + 0.67\, A_{CUPRAC}$ ($r=0.53$) in incubated solution			
inc: incubated at 50°C				

[a]All measurements show deviations approximately ±1.7% of the mean ($x = \bar{x} \pm \frac{t_{95}s}{\sqrt{N}}$)

The significantly higher results of the ABTS-persulfate assay compared to those of CUPRAC for the aqueous extract are probably due to the higher TEAC coefficients ascribed by ABTS to thiols of various origin (1.5 compared to 0.5, as seen in Table 15.2), and to the interaction of ABTS radical with the unidentified "dark" antioxidants possibly contributing at 1/3 ratio to the observed overall capacity (39).

2. Measurement of serum TAC without the separation of lipophilic and hydrophilic antioxidant fractions: After precipitation of serum proteins with 10% (w/v) TCA solution in ethanol, the centrifugate is analyzed with the miniaturized CUPRAC method described in Subheading 15.3.8 (*see* Note 3). With this modification, the TAC of both hydrophilic and lipophilic serum antioxidants is measured with an easier and more flexible method. The procedure is more rapid due to fewer steps taken, and the TAC results of healthy adult serum samples assayed by CUPRAC and ABTS methods correlate much better with each other ($r = 0.93$, *see* Fig. 15.3).

3. Conclusive remarks for the serum TAC assays: Only three methods are capable to simultaneously assay food and serum antioxidants; i.e., ORAC, ABTS, and CUPRAC respond to lipophilic as well as hydrophilic antioxidants. Of these three

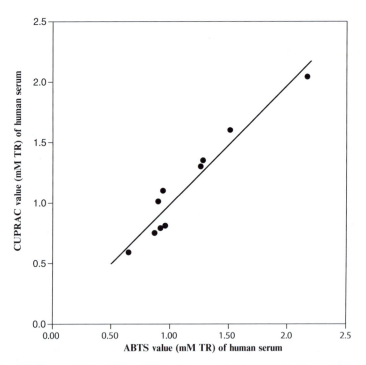

Fig. 15.3. The correlation of serum TAC results assayed with CUPRAC to those with ABTS in the miniaturized assay system where lipophilic and hydrophilic antioxidant fractions were not separated.

assays, the ORAC assay is lengthy and tedious and uses relatively expensive equipment (fluorescence detector) and reagents. The production of β-PE fluorescent probe (in the ORAC assay) varies from lot to lot, and this compound is not photostable. On the other hand, fluorescein as the alternative fluorescent probe of ORAC yields extremely high TEAC coefficients for antioxidants. The second assay, ABTS/TEAC, has a problem with thiol antioxidants yielding high values (e.g., TEAC for GSH was found as 1.5, although GSH acts as a 1-electron reductant). The radicalic reagent of ABTS is expensive and instable, and the blue-green ABTS radical cation produced by different modifications of the method gives entirely different results in TAC assays. Only CUPRAC is simple, low-cost, fairly rapid, and flexible. Its equipment and reagents are inexpensive (*see* Note 6). Beer's law is valid for a wide linear concentration range of antioxidants, and additivity of antioxidant capacities of serum antioxidants is conserved (*see* Note 5). The Randox-TEAC assay is adversely affected by the dilution of serum, while a plot of expected versus found CUPRAC values of serum upon dilution perfectly passes through the origin. The precision is rather high (*see* Note 7), and redox cycling of the reduced reagent (i.e., bis(neocuproine) copper(I) chelate) with H_2O_2 is not possible (*see* Note 8), compensating for a basic criticism directed to the FRAP assay. Applicability of CUPRAC at physiological pH, combined with rapid oxidizability of thiols that cannot be achieved with FRAP, is expected to make CUPRAC available to biological, biochemical, and medicinal scientists as a more frequently applied method for serum TAC measurement in the near future.

4. Measurement of water-soluble hydroxyl radical scavengers with the modified CUPRAC method: The substituted-benzoate probe absorbances almost reach saturation levels at the end of 1 h and basically remain unchanged for a total period of 2.5 h. The probe 3,5-dimethoxybenzoate yields the highest saturated absorbance (28).

The basic equations used in competition kinetics for calculating ·OH scavenging rate constants, as recommended by Halliwell et al. (25), can be summarized as follows:

$$\text{Probe (pr)} + \cdot\text{OH Radical (rad)} \rightarrow \text{Product}_1$$

$$\text{Scavenger (sc)} + \cdot\text{OH Radical (rad)} \rightarrow \text{Product}_2$$

The rates of formation of the products, i.e., $(d[\text{Product}_1]/dt)$, are proportional to the corresponding CUPRAC absorbances:

$$\Delta A = A_o - A = K\left(k_{sc}\,[\text{sc}][\text{rad}]\right) \qquad (15.1)$$

$$A = K\left(k_{pr}\,[\text{pr}][\text{rad}]\right) \qquad (15.2)$$

where K is the instrumental constant, A is the absorbance in the presence of hydroxyl radical scavengers (sc) at concentration [sc], and A_0 the absorbance in the absence of a scavenger; rate constants of reactions in the presence (k_{sc}) or absence of scavenger (k_{pr}), [pr] is the concentration of probe used in the experiment (CUPRAC absorbance arises from the reduction of the Cu(II)-neocuproine reagent to the Cu(I)-chelate (18, 19) by the hydroxylated probe). When equation (Eq. 15.1) is divided by equation (Eq. 15.2), we get;

$$\frac{A_0 - A}{A} = K\left(k_{sc}\,[\text{sc}][\text{rad}]\right)/K\left(k_{pr}\,[\text{pr}][\text{rad}]\right)$$

$$\Rightarrow \frac{A_0}{A} - 1 = k_{sc}\,[\text{sc}]/k_{pr}\,[\text{pr}]$$

$$\Rightarrow \frac{A_0}{A} = 1 + (k_{sc}\,[\text{sc}]/k_{pr}\,[\text{pr}]) \qquad (15.3)$$

Naturally, Eq. 15.3 is applicable to cases where the concentrations of the probe and scavenger are greatly in excess of that of the transient radical and remain effectively constant during the process. A rate constant for the reaction of the scavenger with hydroxyl radical can be deduced from the inhibition of color formation due to the hydroxylation of the probe. Competition kinetics involving a probe and a scavenger for ·OH reaction using various concentrations of the reactants should yield a straight line when A_0/A is plotted as a function of $C_{scavenger}/C_{probe}$, where the slope would yield the ratio of the associated rate constants (the means for $N=4$ or 5 data points regarding the slope differ by $\leq \pm$ 6–7%), and the intercept would be roughly equal to 1. The calculated rate constants of scavengers are proportional to the slopes of these lines, and the precision of data is associated with the linear correlation coefficients. We take thiourea as the reference compound, and using the second-order rate constant of thiourea for ·OH reaction: $k\,(\text{M}^{-1}\text{s}^{-1})=4.7\times10^9$ as given by Halliwell (40), we measure the hydroxyl radical scavenging rate constants of the probes as p-aminobenzoate: 6.85×10^9, 3,5-dimethoxybenzoate: 4.61×10^9, and 2,4-dimethoxybenzoate: $4.27\times10^9\,\text{M}^{-1}\text{s}^{-1}$. It is known from the literature (41) that the rate constant for benzoate is $k=5.9\times10^9\,\text{M}^{-1}\text{s}^{-1}$, generally in accord with these values. With the same reasoning, we find the k for 2-deoxy-D-ribose using the TBARS method as 2.97×10^9 against thiourea (the 2nd order rate constant of which is known), whereas its k is reported in the literature as $3.1\times10^9\,\text{M}^{-1}\text{s}^{-1}$ (27). With the use of Eq. 15.3, A_0/A plots as a function of $C_{scavenger}/C_{probe}$ are given in Fig. 15.4 for the 3,5-dimethoxybenzoates probe using the modified CUPRAC method. The linear equations for ·OH scavenger compounds are in accord with Eq. 15.3; their calculated second-order

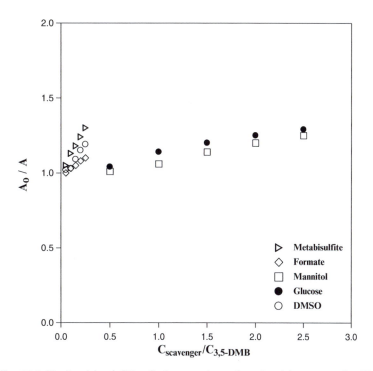

Fig. 15.4. Kinetic plots of ·OH radical scavenging action of certain compounds with respect to Eq. 15.3 using the modified CUPRAC method with a 3,5-dimethoxybenzoate (3,5-DMB) probe.

rate constants are given in Table 15.4 with the tested probes using the CUPRAC method, and with the deoxyribose probe using the TBARS method.

The rate constants calculated by the modified CUPRAC method using three substituted – benzoate probes (Table 15.4) are generally in accord with the literature values, and with those found by the TBARS method using a deoxyribose probe. Our k values for mannitol and Na-formate calculated according to the modified CUPRAC method with a p-aminobenzoate probe are in accord with those reported by Gutteridge with pulse radiolysis (42) and those by Halliwell (40), while formate showed an interference in the deoxyribose degradation method either in the presence or absence of EDTA (42). Consequently, the k for formate that we find with the TBARS method (Table 15.4) is much higher than those we calculate with modified CUPRAC (Table 15.4). Since ·OH scavengers may react with other intermediates of the reaction mixture, thiourea and formate –in the absence of EDTA– were reported by Gutteridge (42) to deviate from simple competition kinetics, a behavior not observed with CUPRAC using the three probes. Another drawback of the classical TBARS method is the possible formation of the interfering Prussian blue, i.e., $KFe[Fe(CN)_6]$, which may result from the reaction of excessive

Table 15.4
The ·OH scavenging activity of various antioxidant compounds using the modified CUPRAC method with *p*-aminobenzoate, 3,5-dimethoxybenzoate, 2,4-dimethoxybenzoate probes, and using the TBARS method with a 2-deoxy-D-ribose probe

·OH scavengers	Rate const. using PAB $(M^{-1}s^{-1})$	Rate const. using 2,4-DMB $(M^{-1}s^{-1})$	Rate const. using 3,5-DMB $(M^{-1}s^{-1})$	Rate const. using 2-DD $(M^{-1}s^{-1})$
$Na_2S_2O_5$ (10^{-3}M)	12.6×10^9	8.03×10^9	5.62×10^9	8.25×10^9
Na-formate (10^{-3}M)	2.81×10^9	3.42×10^9	1.89×10^9	10.23×10^9
Ascorbic Acid (10^{-2}M)	2.53×10^9	2.14×10^8	7.4×10^8	3.1×10^8
Mannitol (10^{-2}M)	1.46×10^9	8.11×10^8	6×10^8	8.1×10^8
Glucose (10^{-2}M)	1.73×10^9	4.3×10^8	5.1×10^8	1.27×10^9
KI (10^{-2}M)	6.64×10^9	16.6×10^9	12.8×10^9	1.43×10^9
$Na_2S_2O_3$ (10^{-2}M)	6.1×10^9	4.31×10^9	3.92×10^9	2.39×10^9
NH_4SCN (10^{-2}M)	–	2.69×10^9	4.84×10^9	4.8×10^9
Ethanol (1:10)	3.9×10^8	1.03×10^8	6.9×10^7	2.17×10^7
Methanol (1:10)	1.64×10^8	4.3×10^7	3.7×10^7	1.24×10^7
2-Propanol (1:10)	2×10^8	1.28×10^8	4.6×10^7	1.86×10^7
DMSO (10^{-3}M)	9.86×10^9	4.1×10^9	4.24×10^9	3.13×10^9
Na_3-citrate(10^{-3}M)	–	6.23×10^9	1.48×10^9	–
L-lysine (10^{-2}M)	1.64×10^9	7.8×10^8	2.49×10^8	9.3×10^8
$K_4Fe(CN)_6$ (10^{-2}M)	7.26×10^9	4.44×10^9	2.9×10^9	1.64×10^9

$Fe(CN)_6^{4-}$ with the Fe(III) product of the redox reactions, whereas in the modified CUPRAC method, Prussian blue would not be extracted into ethylacetate and therefore does not cause an interference with the final color due to Cu(I)-neocuproine. Consequently, the k value for hexacyanoferrate(II) found with TBARS is extracted from a nonlinear curve whereas data obtained with CUPRAC/substituted-benzoates conform well to the linear equation imposed by Eq. 15.3. As for other scavengers, the k values for formate (40) and KI (41) found with CUPRAC/2,4- and 3,5-dimethoxybenzoate probes and the *k* value for glucose (43) found with CUPRAC/*p*-aminobenzoate probe are more or less in accord with the literature values. Generally, in the modified CUPRAC method, iodide, metabisulfite, formate, thiourea, thiosulfate, hexacyanoferrate(II), citrate, and DMSO yield higher

·OH scavenging activities than simple alcohols (methanol, ethanol, and propanol), mannitol, glucose, and nonaromatic amino acids, in accord with the order given in various literature sources. The deviation of alcohol oxidation kinetics from linearity and of the intercept values from unity probably hint to a different ·OH reaction mechanism for MeOH, EtOH, and PrOH.

4. Notes

The advantages of the CUPRAC method over other similar assays and some details to be considered in performing the assay are summarized below:

1. The CUPRAC redox reaction producing colored species is carried out at nearly physiological pH (pH 7 of ammonium acetate buffer) as opposed to the acidic conditions (pH 3.6) of FRAP, or to the basic conditions (pH 10) of Folin-Ciocalteu (FCR) assay. At more acidic conditions than the physiological pH, the reducing capacity may be suppressed due to protonation on phenolics, whereas at more basic conditions, proton dissociation of phenolics (converted into phenolates) would enhance a sample's reducing capacity. Thus, the CUPRAC assay working at a pH close to that of physiological fluids gives a realistic estimate of antioxidants in a sample.

2. The CUPRAC method is based on the elevation of the redox potential of Cu(II, I) in the presence of a Cu(I)-stabilizing ligand, neocuproine. The preferential stabilization of Cu(I) with the formation of the stable $Cu(Nc)_2^+$ chelate raises this standard potential from 0.17 V to 0.6 V, thereby rendering the CUPRAC reagent capable to oxidize most serum antioxidants. However, if the serum samples are preserved with the addition of EDTA instead of citrate, Cu(II) is preferentially stabilized as the CuY^{2-} complex (where Y^{4-} is the tetracarboxylate anion of EDTA) and the Cu(II, I) redox potential decreases to a level at which serum antioxidants can no longer be oxidized. Thus, serum samples should be preserved with citrate instead of EDTA on longer standing.

3. When the lipophilic and hydrophilic antioxidant fractions of serum are not separated, the resulting TAC assay can be miniaturized by minimizing the treated volumes, thereby, preventing analyte losses. Currently, CUPRAC is one of the three TAC assays applicable to both lipophilic and hydrophilic antioxidants (the other two assays meeting this criterion are ABTS/TEAC and modified ORAC), thanks to the solubility of the CUPRAC chromophore, bis(neocuproine)copper(I) chelate, in both organic solvent and aqueous media. The CUPRAC method

can simultaneously measure hydrophilic as well as lipophilic antioxidants (e.g., β-carotene and α-tocopherol). BHA, BHT, α-tocopherol, and most other oil-soluble antioxidants may be easily assayed in MeOH, while β-carotene requires dichloromethane (DCM) for fully exhibiting its antioxidant potency (as the bis(neocuproine)copper(I) chelate is also soluble in DCM). As an advantage over the widely used Folin and FRAP reagents, CUPRAC can additionally measure lipophilic antioxidants (19, 44, 45).

4. The CUPRAC reagent is fast enough to oxidize thiol-type antioxidants (19, 44, 46), whereas according to the protocol developed by Benzie and Strain (15), the FRAP method may only measure with serious negative error certain thiol-type antioxidants like glutathione (i.e., the major low molecular-weight thiol compound of the living cell). The CUPRAC assay also responds much faster than FRAP to certain hydroxycinnamic acids. The possible reason for this with respect to electronic configurations is the kinetic inertness of high-spin d^5-Fe(III) having half-filled d-orbitals, while CUPRAC utilizing d^9-Cu(II) oxidant involves faster kinetics.

5. The total antioxidant capacity (TAC) values of antioxidants found with CUPRAC are perfectly additive, i.e., the TAC of a phenolic mixture is equal to the sum of TAC values of its constituent polyphenols. The parallellism of the linear calibration curves of pure antioxidants alone and in a given complex matrix (such as serum extract) demonstrates that there are no chemical interactions of interferent nature among the solution constituents, and that the antioxidant capacities of the tested antioxidants are additive, in conformity to the Beer's law (18, 19, 44). This additivity behavior is not valid for antioxidant activity assays performed on serum, because antioxidant activity is related to the rate constant of the reaction of antioxidants with the assay reagents, and rate constants or "lag times" are not additive.

6. The method is easily applicable to conventional laboratories using standard equipment like a colorimeter rather than more sophisticated but costly instrumental techniques of analysis. The method involves minimal sample preparation, the experimental procedure is flexible, and suitable for automation.

7. The within-run and between-run coefficients of variation (CV) of the CUPRAC method (0.7 and 1.5%) are much lower than those of most methods that find wide use in total antioxidant assays. The CV of CUPRAC is much lower than those of most kinetic-based assays (19), confirming high precision.

8. Since the Cu(I) ion produced as the product of the CUPRAC redox reaction is in chelated state (i.e., Cu(I)-Nc), it cannot

act as a pro-oxidant that may cause oxidative damage to lipids. The ferric ion-based assays were criticized for producing Fe^{2+}, which may act as a pro-oxidant to produce ·OH radicals as a result of its Fenton-type reaction with H_2O_2 (22). The stable Cu(I)-chelate was previously shown by us not to react with hydrogen peroxide, but the reverse reaction, i.e., oxidation of H_2O_2 with Cu(II)-Nc, is possible (47). Since a cascade of ROS-generating reactions oxidizing lipids is not possible with CUPRAC, there is no negative error of antioxidant determination that may arise from a hypothetical pro-oxidant role of Cu(I).

Acknowledgments

The authors would like to express their gratitude to Istanbul University Research Fund, Bilimsel Arastirma Projeleri Yurutucu Sekreterligi, for the funding of Project YOP-4/27052004 and to State Planning Organization of Turkey for the Advanced Research Project of Istanbul University (2005K120430). The authors would like to extend their gratitude to TUBITAK (Turkish Scientific and Technical Research Council) for the Research Projects 105T402, and 106T514.

References

1. Halliwell B, Gutteridge JMC (1989) Free radicals in biology and medicine. Oxford University Press, Oxford

2. Halliwell B, Aruoma OI (1991) DNA damage by oxygen-derived species: its mechanisms and measurements in mammalian systems. FEBS Lett 281:9–19

3. Cao G, Prior RL (1998) Comparison of different analytical methods for assessing total antioxidant capacity of human serum. Clin Chem 44:1309–1315

4. De Zwart LL, Meerman JH, Commandeur JN, Vermeulen NP (1999) Biomarkers of free radical damage applications in experimental animals and in humans. Free Radical Biol Med 26:202–226

5. Yu BP (1994) Cellular defenses against damage from reactive oxygen species. Physiol Rev 74:139–162

6. Ghiselli A, Serafini M, Natella F, Scaccini C (2000) Total antioxidant capacity as a tool to assess redox status: critical view and experimental data. Free Radical Biol Med 29:1106–1114

7. Packer JE, Slater TF, Willson RL (1979) Direct observation of a free radical interaction between vitamin E and vitamin C. Nature 278:737–738

8. Stocker R, Weidemann MJ, Hunt NH (1986) Possible mechanisms responsible for the increased ascorbic acid content of *Plasmodium vinckei*-infected mouse erythrocytes. Biochim Biophys Acta 881:391–397

9. Arnao MB (2000) Some methodological problems in the determination of antioxidant activity using chromogen radicals: a practical case. Trends Food Sci Technol 11:419–421

10. Miller NJ, Rice-Evans CA, Davies MJ, Gopinathan V, Milner A (1993) A novel method for measuring antioxidant capacity and its application to monitoring the antioxidant status in premature neonates. Clin Sci 84:407–412

11. Sanchez-Moreno C, Larrauri JA, Saura-Calixto F (1998) A procedure to measure the antiradical efficiency of polyphenols. J Sci Food Agric 76:270–276

12. Re R, Pellegrini N, Proteggente A, Pannala A, Yang M, Rice-Evans C (1999) Antioxidant activity applying an improved ABTS radical cation decolorization assay. Free Radical Biol Med 26:1231–1237

13. Wayner DD, Burton GW, Ingold KU, Locke S (1985) Quantitative measurement of the total peroxyl radical-trapping antioxidant capability of human blood plasma by controlled peroxidation. The important contribution made by plasma proteins. FEBS Lett 187:33–37

14. Rice-Evans CA, Miller NJ (1994) Total antioxidant status in plasma and body fluids. Methods Enzymol 234:279–293

15. Benzie IFF, Strain JJ (1996) The ferric reducing ability of plasma (FRAP) as a measure of "antioxidant power": the FRAP assay. Anal Biochem 239:70–76

16. Janaszewska A, Bartosz G (2002) Assay of total antioxidant capacity: comparison of four methods as applied to human blood plasma. Scand J Clin Lab Invest 62:231–236

17. Ou B, Huang D, Hampsch-Woodill M, Flanagan JA, Deemer EK (2002) Analysis of antioxidant activities of common vegetables employing oxygen radical absorbance capacity (ORAC) and ferric reducing antioxidant power (FRAP) assays: a comparative study. J Agric Food Chem 50:3122–3128

18. Apak R, Güçlü K, Özyürek M, Karademir SE (2004) A novel total antioxidant capacity index for dietary polyphenols, vitamins c and e, using their cupric ion reducing capability in the presence of neocuproine: CUPRAC method. J Agric Food Chem 52:7970–7981

19. Apak R, Güçlü K, Özyürek M, Karademir SE, Altun M (2005) Total antioxidant capacity assay of human serum using copper(II)-neocuproine as chromogenic oxidant: the CUPRAC method. Free Radic Res 39:949–961

20. Huang D, Ou B, Prior RL (2005) The chemistry behind antioxidant capacity assays. J Agric Food Chem 53:1841–1856

21. Halliwell B, Gutteridge JMC (1992) Biologically relevant metal ion-dependent hydroxyl radical generation. An update. FEBS Lett 307:108–112

22. Halliwell B, Gutteridge JMC (1984) Oxygen toxicity, oxygen radicals, transition metals and disease. Biochem J 219:1–14

23. Gutteridge JMC (1981) Thiobarbituric acid-reactivity following iron-dependent free-radical damage to amino acids and carbohydrates. FEBS Lett 128:343–346

24. Halliwell B, Gutteridge JMC (1981) Formation of a thiobarbituric-acid-reactive substance from deoxyribose in the presence of iron salts. The role of superoxide and hydroxyl radicals. FEBS Lett 128:347–352

25. Halliwell B, Gutteridge JM, Aruoma OI (1987) The deoxyribose method: a simple "test tube" assay for determination of rate constants for reactions of hydroxyl radicals. Anal Biochem 165:215–219

26. Buege JA, Aust SD (1978) Microssomal lipid peroxidation. Methods Enzymol 52:302–310

27. Cheng Z, Li Y, Chang W (2003) Kinetic deoxyribose degradation assay and its application in assessing the antioxidant activities of phenolic compounds in a Fenton-type reaction system. Anal Chim Acta 478:129–137

28. Bektaşoğlu B, Çelik SE, Özyürek M, Güçlü K, Apak R (2006) Novel hydroxyl radical scavenging antioxidant activity assay for water-soluble antioxidants using a modified CUPRAC method. Biochem Biophys Res Commun 345:1194–1200

29. Aebischer C, Schierle J, Schuep W (1999) Simultaneous determination of retinol, tocopherols, carotene, lycopene, and xanthophylls in plasma by means of reversed-phase high performance liquid chromatography. In: Packer L (ed) Methods in enzymology. Oxidants and antioxidants. Academic Press, New York, pp 348–362

30. Prior RL, Huang H, Gu L, Wu X, Bacchiocca M, Howard L et al (2003) Assays for hydrophilic and lipophilic antioxidant capacity (oxygen radical absorbance capacity (ORAC$_{FL}$) of plasma and other biological and food samples. J Agric Food Chem 51:3273–3279

31. McHatton RC, Espenson JH, Bakac A (1986) Carbon-carbon bond formation in the reaction of aliphatic radicals with alkylcobaloximes. J Am Chem Soc 108:5885–5890

32. Pronai L, Ichikawa Y, Ichimori K, Nakazawa H, Arimori S (1990) Hydroxyl radical-scavenging activity of slow-acting anti-rheumatic drugs. J Clin Biochem Nutr 9:17–23

33. Aruoma OI, Halliwell B, Hoey BM, Butler J (1988) The antioxidant action of taurine, hypotaurine and their metabolic precursors. Biochem J 256:251–255

34. Cao G, Sofic E, Prior RL (1997) Antioxidant and prooxidant behavior of flavonoids: structure-activity relationships. Free Radical Biol Med 22:749–760

35. Dickinson DA, Forman HJ (2002) Cellular glutathione and thiols metabolism. Biochem Pharmaco 64:1019–1026

36. Cao G, Verdon CP, Wu AHB, Wang H, Prior RL (1995) Automated oxygen radical

absorbance capacity assay using the COBAS FARA II. Clin Chem 41:1738–1744

37. Ou B, Hampsch-Woodill M, Prior RL (2001) Development and validation of an improved oxygen radical absorbance capacity assay using fluorescein as the fluorescent probe. J Agric Food Chem 49:4619–4626

38. Tubaro F, Ghiselli A, Rapuzzi P, Maiorino M, Ursini F (1998) Analysis of plasma antioxidant capacity by competition kinetics. Free Radical Biol Med 24:1228–1234

39. Tsai K, Hsu TG, Kong CW, Lin KC, Lu FJ (2000) Is the endogenous peroxyl-radical scavenging capacity of plasma protective in systemic inflammatory disorders in humans? Free Radical Biol Med 28:926–933

40. Halliwell B (1978) Superoxide-dependent formation of hydroxyl radicals in the presence of iron chelates. Is it a mechanism for hydroxyl radical production in biochemical systems? FEBS Lett 92:321–326

41. George VB (1988) Critical review of rate constants for reactions of hydrated electrons, hydrogen atoms and hydroxyl radicals (HO/ O⁻) in aqueous solution. J Phyl Chem Ref Data 17:513–886

42. Gutteridge JMC (1987) Ferrous-salt-promoted damage to deoxyribose and benzoate. The increased effectiveness of hydroxyl-radical scavengers in the presence of EDTA. Biochem J 243:709–714

43. Aruoma OI, Laughton MJ, Halliwell B (1989) Carnosine, homocarnosine and anserine: could they act as antioxidants *in vivo*? J Biochem 264:863–869

44. Apak R, Güçlü K, Demirata B, Özyürek M, Çelik SE, Bektaşoğlu B, Berker KI, Özyurt D (2007) Comparative evaluation of total antioxidant capacity assays applied to phenolic compounds, and the CUPRAC assay. Molecules 12:1496–1547

45. Çelik SE, Özyürek M, Güçlü K, Apak R (2007) CUPRAC Total antioxidant capacity assay of lipophilic antioxidants in combination with hydrophilic antioxidants using the macrocyclic oligosaccharide methyl-β-cyclodextrin as the solubility enhancer. React Func Polym 67:1548–1560

46. Tütem E, Apak R (1991) Simultaneous spectrophotometric determination of cystine and cysteine in amino acid mixtures using copper(II)-neocuproine reagent. Anal Chim Acta 255:121–125

47. Tütem E, Apak R, Baykut F (1991) Spectrophotometric determination of trace amounts of copper(I) and reducing agents with neocuproine in the presence of copper(II). Analyst 116:89–94

Chapter 16

Analysis of Antioxidant Activities in Vegetable Oils and Fat Soluble Vitamins and Biofactors by the PAO-SO Method

Kazuo Sakai, Satoko Kino, Masao Takeuchi, Tairin Ochi, Giuseppe Da Cruz, and Isao Tomita

Abstract

Accumulated evidences indicate that reactive oxygen species (ROS) are involved in the pathophysiology of aging process. Antioxidants are believed to play an important role in the defense system to counteract ROS in the body. While excess hydrophilic antioxidants can be excreted easily in urine, lipophilic antioxidants can penetrate into blood lipoproteins and cell membranes, and may maintain long and high bioavailability. These lipophilic antioxidants are thus expected to contribute greatly to the prevention of age-related diseases.

Oils extracted from plant seeds are known to contain various lipophilic antioxidants such as vitamin E (α-tocopherol) and carotenoids. They are known to not only decrease serum low-density-lipoprotein (LDL) level, but also prevent oxidation of LDL. In addition to vitamin E (α-tocopherol) and carotenoids, other lipophilic antioxidants such as γ-oryzanol and sesaminol (from sesamolin) are in rice bran and sesame, respectively. They are sometimes called "vitamin-like food factors" or "biofactors."

Although there are several methods for measuring the total antioxidant activities for various plant extracts, most of these methods are designed for hydrophilic antioxidants, and not for lipophilic antioxidants present in various plant seed oils.

In this report, we present an assay method for the total potency of antioxidants that are soluble in oil (PAO-SO) utilizing bathocuproine (BC) as a chromogen. BC-based antioxidant activity assay shows good linearity ($r^2 = 0.9986$), good reproducibility (CV < 10%), and good recovery (86–91%) when dl-α-tocopherol, for example, is added to sesame oil. Total antioxidant activity of rape-seed oil, olive oil, and sesame oil could also be successfully measured.

Key words: Total antioxidant activity assay, Bathocuproine, Reduction of cupricion, Antioxidants and biofactors in plant seed oils, Potency of antioxidants soluble in oil (PAO-SO)

1. Introduction

Accumulated evidences indicate that reactive oxygen species (ROS) are involved in the pathophysiology of aging process and various age-related diseases such as diabetes, cancer, Alzheimer's

D. Armstrong (ed.), *Advanced Protocols in Oxidative Stress II,* Methods in Molecular Biology, vol. 594
DOI 10.1007/978-1-60761-411-1_16, © Humana Press, a part of Springer Science + Business Media, LLC 2010

disease, and Parkinson's disease. Antioxidants are believed to play an important role in the defense system to counteract ROS, and they can be classified into the following two groups: (1) Hydrophilic antioxidants such as glutathione, ascorbic acid, and various polyphenolics; and (2) Lipophilic antioxidants such as dl-α-tocopherol, oryzanols, and carotenoids. While excess hydrophilic antioxidants can be excreted in urine, lipophilic antioxidants can penetrate into blood lipoproteins and cell membranes, and may show high bioavailability (1). These lipophilic antioxidants are thus expected to contribute greatly to disease prevention and to anti-aging processes. For example, lycopene, which is one of the lipophilic antioxidants from tomato, is reported to target the prostate tissues, decrease tissue 8-hydroxy-2'-deoxyguanosine (8-OHdG: a biomarker for oxidative damage of DNA), and also decrease the serum concentration of prostate-specific-antigen (PSA: a biomarker for prostate cancer) (2). It has been reported that the bioavailability of lycopene is improved when tomato paste, a major source of lycopene, is mixed with oil (3). Oils extracted from plant seeds such as olive oil contain various lipophilic antioxidants, and have been reported to decrease serum low-density-lipoprotein (LDL) level and also prevent oxidation of LDL (4). In addition to dl-α-tocopherol, lipophilic antioxidants such as γ-oryzanol and sesaminol (from sesamolin) are contained in rice bran and sesame, respectively.

Several methods for measuring antioxidant activities have been reported: Trolox equivalent antioxidant capacity assay (TEAC), total radical absorption potentials (TRAP), ferric reducing antioxidant power (FRAP), and oxygen radical absorption capacity assay (ORAC) (5). Most of these methods, however, are mainly designed for the determination of hydrophilic antioxidants. Antioxidant assay for lipophilic substances, especially for plant seed oils such as olive oil and sesame oil, has not been established.

In this report, we present an assay method for the total potency of antioxidants soluble in oil (PAO-SO), utilizing reduction of cupric ion. Apak et al. (6) have first developed copper reducing antioxidant capacity (CUPRAC) assay utilizing neocuproine as the chromogenic chelator for Cu^+ and have been successfully applied for polyphenols in fruit juices and tea samples (7). We have newly developed copper reducing antioxidant activity assay using bathocuproine (BC: Fig.16.1) (8,9). BC is a Cu^+-specific chromogenic chelating agent. When antioxidants or oil itself is mixed with Cu^{++} in the presence of BC, copper ion will be reduced to form Cu^+-BC complex, which shows maximum absorbance at around 492 nm (Fig.16.2). The overall reactions proceed fast, and assay can be completed in 3 min (Fig. 16.3). BC-based antioxidant activity assay is sensitive to

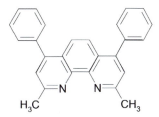

Fig. 16.1. Chemical structure of bathocuproine (BC).

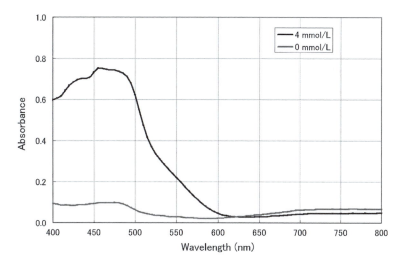

Fig. 16.2. Spectra of the reaction mixture of BC, Cu^{2+} and dl-α-tocopherol (0 and 4 mmol/L).

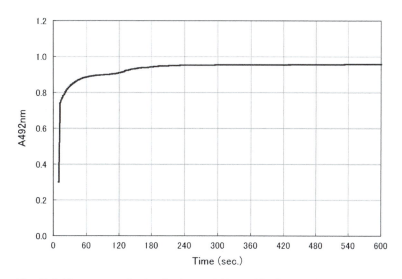

Fig. 16.3. Time course of color development (4 mmol/L dl-α-tocopherol). 10 μL of standard (4 mmol/L), 390 μL of dilution buffer and 100 μL of Cu^{2+} solution were mixed, and absorbance at 492 nm were measured by photo spectrometer (Shimadzu model UV-160, Kyoto, Japan). Temperature was controlled at $22 \pm 2°C$.

Table 16.1
Antioxidant activities determined by BC-based antioxidant assay (PAO-SO)

dl-α-Tocopherol	3,606 μmol/L
β-carotene	402 μmol/L
Ascorbic acid	2,627 μmol/L
(+)-Catechin	5,991 μmol/L
(−)-Epicatechin	7,594 μmol/L
Oryzanol	777 μmol/g
Billilubin	110 μmol/L
Glutathione	1,852 μmol/L
Quercetin	1,836 μmol/L
Gallic acid	7,197 μmol/L
Probucol	1,999 μmol/L
2(3)-t-Butyl-4-hydroxyanisole (BHA)	3,779 μmol/L

Other antioxidant activities may be determined by PAO-SO

various antioxidants (Table 16.1) in food extracts and lipophilic antioxidants in plant seed oils. It can also be applied to the human serum. The method is very simple, fast, and reliable to detect total antioxidant activity.

2. Materials

2.1. Equipment

1. Micro plate reader set up for 492 nm detection with reference wavelength 620 nm (Model MPR-A4i Tosoh corporation, Tokyo, Japan).
2. Vortex mixer.
3. Pipettes and tips.
4. Multichannel pipettes and tips.
5. 96 wells – micro titer plate.

2.2. Reagents

1. Dilution buffer (BC solution in toluene and ethanol).
2. Cu^{2+} solution (<0.05% $CuSO_4$ solution containing ethanol).
3. Standard solution (4.0 mmol/L *dl*-α-tocopherol in ethanol).
4. Ethanol >99.8% (Nacalai Tesque, Inc., Kyoto, Japan).

2.3. Samples	1. *dl*-α-Tocopherol 97% (Alfa Aesar, MA, USA). 2. Rape-seed oil and olive oil (Nisshin OilliO Group Ltd, Tokyo, Japan). 3. Sesame oil (Kadoya Sesame Mills Inc., Tokyo, Japan).

3. Methods

3.1. Preparation of Standards	1. A 400 mmol/L standard stock solution can be prepared by dissolving 200 μL of *dl*-α-tocopherol with 1,000 μL of ethanol. 2. Standard solutions are prepared as 0, 1, 2, and 4 mmol/L by diluting standard stock solution with ethanol.
3.2. Predilution of Samples	1. Plant seed oils should be prediluted by dilution buffer to prevent turbidity or phase separation when mixed with dilution buffer and Cu^{2+} solution. Dilution fold should be determined depending on the nature of each sample. For example, 1:1 dilution may be suitable for some sesame oils, which are sold as food in Japanese market. 2. Antioxidant that is dissolved in ethanol should be diluted by ethanol, if needed.
3.3. Assay Protocol	1. Allow all reagents and samples to be kept at room temperature before use. 2. Mix 10 μL of diluted oil samples or standards to 390 μL of dilution buffer, and mix for 5 s. 3. Place 200 μL of diluted samples or standards to micro plate wells. 4. Read absorbance at 492 nm with reference at 620 nm(Read1). 5. Add 50 μL of Cu^{2+} solution to all wells, mix, and incubate for 10 min. 6. Read absorbance at 492 nm with reference at 620 nm(Read 2). 7. Prepare standard curve by plotting [(read 2) – (read 1)] as Y axis and [standard conc.] as X axis. A linear regression algorithm ($Y = a \times X + b$) may be recommended (Fig. 16.4). 8. Calculate *dl*-α-tocopherol equivalent conc. using standard curve. 9. Convert *dl*-α-tocopherol equivalent to antioxidant activity (μmol/L) by multiplying with 1938.
3.4. Results *3.4.1. Linearity Test*	1. Prepare *dl*-α-tocopherol solution at the concentration of 0, 0.25, 0.5, 1, 1.5, 2, 2.5, 3, 3.5, and 4 mmol/L in ethanol. As shown in Fig. 16.5, a good linearity up to 4 mmol/L was observed ($r^2 = 0.9986$).

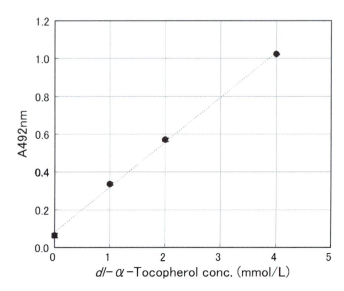

Fig. 16.4. Standard plot of BC-based antioxidant assay.

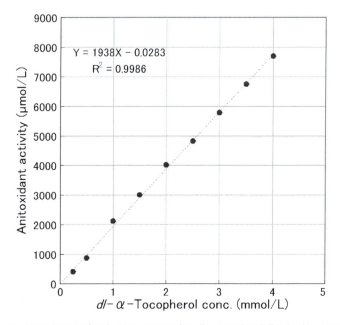

Fig. 16.5. Linearity test of antioxidant assay using dl-α-tocopherol dissolved in ethanol.

3.4.2. Recovery Test

1. Recovery test against sesame oil was assessed using dl-α-tocopherol. Sesame oil was 2:3 prediluted by dilution buffer, and used as reference sample (40% sesame oil solution). dl-α-Tocopherol was added to 40% sesame oil solution at final concentration 0, 0.8, 1.6, and 2.4 mmol/L, and total antioxidant activity was assessed.

2. Results are shown in Fig. 16.6 and Table 16.2. Antioxidant activity of *dl*-α-tocopherol can be recovered 86–91% when *dl*-α-tocopherol is mixed with sesame oil.

3.4.3. Reproducibility

1. To test the reproducibility of this assay, *dl*-α-tocopherol solution at the concentrations of 0, 2, and 4 mmol/L, and 2:3 prediluted rape-seed oil, olive oil and sesame oil samples were measured repeatedly.

2. Results are summarized in Table 16.3. Good reproducibility with $SD < 220\,\mu mol/L$ and $CV < 10\%$ was observed. Lower detection limit of this assay is 100 μmol/L (approx. $3 \times SD$ of 0 mmol/L standard).

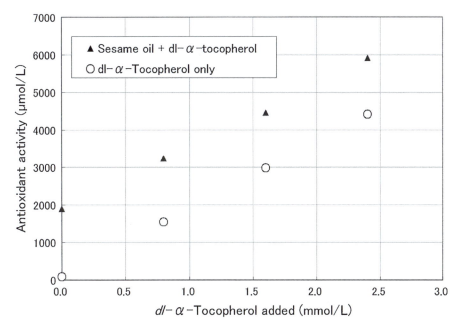

Fig. 16.6. Recovery test of *dl*-α-tocopherol added to sesame oil.

Table 16.2
Recovery of *dl*-α-tocopherol added to sesame oil

dl-α-Tocopherol conc. (mmol/L)	Antioxidant activity measured (μmol/L)		Recovered $(C = B_n - B_0)$	Recovered ratio (C/A, %)
	Buffer only (A)	Seasame oil (B)		
0	85	1,896	0	–
0.8	1,545	3,235	1,339	87
1.6	2,986	4,452	2,555	86
2.4	4,406	5,909	4,013	91

Table 16.3
Reproducibility test of standards and plant seed oils

Antioxidant activity measured (μmol/L, $N=12$)

													Mean	SD	CV (%)
Standard (0 mmol/L)	269	196	139	220	204	228	204	187	163	155	171	163	191	36	–
Standard (2 mmol/L)	3,911	3,943	4,024	4,016	3,935	4,106	4,065	3,870	3,870	4,089	3,984	3,968	3,982	80	2.0
Standard (4 mmol/L)	7,634	7,432	7,561	7,586	7,464	7,399	7,488	7,423	7,456	7,505	7,602	7,286	7,486	99	1.3
Rape-seed oil	2,780	2,699	2,659	2,517	2,659	2,902	2,963	2,578	2,294	2,841	2,618	2,841	2,969	186	6.9
Olive oil	2,415	2,091	2,375	2,517	2,192	2,638	2,476	2,253	2,679	2,375	2,152	2,720	2,407	209	8.7
Seasame oil	4,910	5,092	4,768	4,849	4,748	4,930	4,930	4,991	4,748	4,383	5,255	5,052	4,888	219	4.5

3.4.4. Detection of
Antioxidant Activity
in Plant Seed Oils

1. In order to measure the optimum condition for each oil, samples (rape-seed oil, olive oil, and sesame oil) were prediluted for 12.5, 25.0, 37.5, 50.0, 62.5, and 75.0% using dilution buffer. They are mixed for 10 s vigorously, and applied to BC-based antioxidant assay.

2. As shown in Fig. 16.7, good linearity was observed for all three samples tested ($r^2 = 0.9537$, 0.9853, and 0.9925 for rape-seed oil, olive oil, and sesame oil, respectively). Antioxidant activity

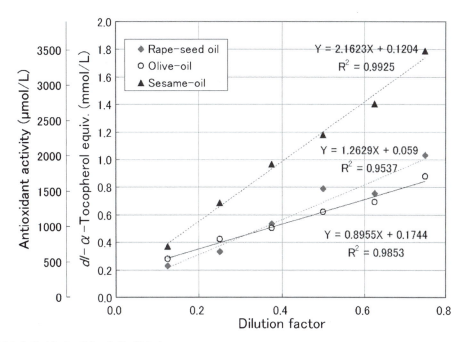

Fig. 16.7. Antioxidant activity of oils diluted.

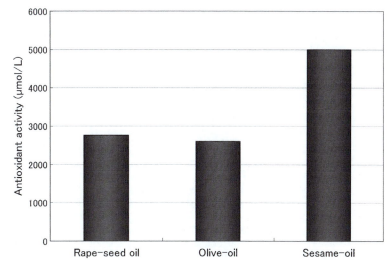

Fig. 16.8. Total antioxidant activity of various plant seed oils determined by BC-based antioxidant assay (PAO-SO).

of all three samples are estimated using 37.5% prediluted data, because slight turbidity was observed when some diluted samples (dilution factor at 50, 62.5, and 75%) were mixed with Cu^{2+} solution. Finally, total antioxidant activity of rape-seed oil, olive oil, and sesame oil is estimated to be 2,760, 2,598, and 4,999 µmol/L, respectively (Fig. 16.8).

4. Note

1. Total antioxidant capacity assay utilizing reduction of cupric ion and detected by bathocuproine (BC) was first described by Giuseppe Da Cruz in 1998 and 1999 (Italian patent No. 01309421 (8) and US patent No. 6613577 (9)), respectively, and have been used for some clinical researches (10–12). BC-based total antioxidant capacity assay kits are available from Japan Institute for the Control of Aging (JaICA) as the test kit for "Potency of Anti-Oxidant Soluble in Oil: PAO-SO."

References

1. Huang D, Ou B, Hampsch-Woodill M, Flanagan JA, Deemer EK (2002) Development and validation of oxygen radical absorbance capacity assay for lipophilic antioxidants using randomly methylated beta-cyclodextrin as the solubility enhancer. J Agric Food Chem 50:1815–1821

2. Chen L, Stacewicz-Sapuntzakis M, Duncan C, Sharifi R, Ghosh L, van Breemen R, Ashton D, Bowen PE (2001) Oxidative DNA damage in prostate cancer patients consuming tomato sauce-based entrees as a whole-food intervention. J Natl Cancer Inst 93:1872–1879

3. Gartner C, Stahl W, Sies H (1997) Lycopene is more bioavailable from tomato paste than from fresh tomatoes. Am J Clin Nutr 66: 116–122

4. Vissers MN, Zock PL, Katan MB (2004) Bioavailability and antioxidant effects of olive oil phenols in humans: A review. Eur J Clin Nutr 58:955–965

5. Huang D, Ou B, Prior RL (2005) The chemistry behind antioxidant capacity assays. J Agric Food Chem 53:1841–1856

6. Apak R, Guclu K, Ozyurek M, Karademir SE (2004) Novel total antioxidant capacity index for dietary polyphenols and vitamins C and E, using their cupric ion reducing capability in the presence of neocuproine: CUPRAC method. J Agric Food Chem 52:7970–7981

7. Apak R, Guclu K, Ozyurek M, Bektasoglu B, Bener M (2008) Cupric ion reducing antioxidant capacity assay for food antioxidants: vitamins, polyphenolics, and flavonoids in food extracts. In: Armstrong D (ed) Advanced protocols in oxidative stress I. Humana, NY, pp 163–193

8. Giuseppe DC (1998) Impiego della batocuproina (acido disolfonico) per la valutazione del potere antiossidante (PAO) nei liquidi e soluzioni. Italian patent 01309421

9. Giuseppe DC (1999) Use of bathocuproine for the evaluation of the antioxidant power in liquids and solutions. US patent 6,613,577

10. Martino M, Chiarelli F, Moriondo M, Torello M, Azzari C, Galli L (2001) Restored antioxidant capacity parallels the immunologic and virologic improvement in children with perinatal human immunodeficiency virus infection receiving highly active antiretroviral therapy. Clin Immunol 100:82–86

11. Preget P, Bollo E, Cannizzo FT, Biolatti B, Contato E, Biolatti PG (2005) Antioxidant capacity as a reliable marker of stress in dairy calves transported by road. Vet Rec 156: 53–54

12. Falaschini A, Marangoni G, Rizzi S, Trombetta MF (2005) Effects of the daily administration of a rehydrating supplement to trotter horses. J Equine Sci 16:1–9

Chapter 17

Measuring Antioxidant Capacity Using the ORAC and TOSC Assays

Andrew R. Garrett, Byron K. Murray, Richard A. Robison, and Kim L. O'Neill

Abstract

Recent epidemiological studies have shown that there may be a link between oxidative stress and the development of several types of chronic diseases. Studies have also shown that diets rich in fruits and vegetables may decrease the incidence of cancer and other chronic diseases. The antioxidant activity of the phytochemicals these foods contain may be partially responsible for the decreased incidence of these diseases in people who regularly consume them. While there are several assays currently used to assess the antioxidant activity of phytochemicals and other antioxidant compounds, two are reviewed here in detail. The first is the oxygen radical absorbance capacity (ORAC) assay, which measures the decrease in fluorescence decay caused by antioxidants, and the second is the total oxyradical scavenging capacity (TOSC) assay, which measures the decrease in ethylene gas production caused by the inhibition of the thermal hydrolysis of ABAP (2,2′-Azobis(2-methyl-(propionamidine) dihydrochloride) by KMBA (α-keto-γ-(methylthio)butyric acid sodium salt) in the presence of antioxidant compounds. These two assays are discussed here, with an in depth review of their methodology and correlation.

Key words: ORAC, Oxygen radical absorbance capacity assay, TOSC, Total oxyradical scavenging capacity assay, Antioxidant, Phytochemical, Oxidative stress, Fluorescein, Trolox equivalents

1. Introduction

It has been theorized that the development and progression of several chronic degenerative diseases may be caused or augmented by oxidative stress (1-11). Much of this oxidative stress is generated as a by-product of fundamental metabolic processes that occur constantly. It is known that reactive oxygen species (ROS) react readily with biological substances, and can destroy or damage such molecules as nucleic acids, proteins, and lipids (12). This damage, in turn, can lead to a variety of degenerative diseases, including cancer, heart disease, Alzheimer's Disease,

D. Armstrong (ed.), *Advanced Protocols in Oxidative Stress II,* Methods in Molecular Biology, vol. 594
DOI 10.1007/978-1-60761-411-1_17, © Humana Press, a part of Springer Science+Business Media, LLC 2010

and Parkinson's Disease (1). While many of these biological systems are somewhat equipped to cope with this oxidative stress, oxidative damage may still occur.

Several different assays have been specifically designed to accurately describe the antioxidant activity of various phytochemicals, vitamins, and other compounds. Some assays measure the inhibition of a test reaction by various antioxidant compounds (22, 23). These include the TEAC (Trolox Equivalent Antioxidant Capacity), Radox TEAC, FRAP (Ferric Reducing/Antioxidant Power), HORAC (Hydroxyl Radical Averting Capacity), FOX (Ferrous Oxide-Xylenol Orange), and TRAP (Total Radical Antioxidant Potential) assays. Other assays involve electron spin resonance (ESR) spectroscopy and use techniques such as time-resolved pulse radiolysis and spin trapping. Many others are also used (8, 14-16, 18, 19, 22-31).

However, these existing assays are not without limitation (1, 8, 24, 25). First, not all of these assays give the same trends for antioxidant activity. For example, poor correlation has been observed between the FRAP and TEAC, the ORAC and the FRAP, and the ORAC and TEAC assays (1, 8). Second, some of the data taken in vitro are difficult to extrapolate to an in vivo model because some of these assays are not run at physiological pH. Third, some of these assays do not measure the bioavailability of antioxidant metabolites in vivo. Thus, multiple assays should be investigated and validated continuously to resolve questions about the true antioxidant capacity of a given sample. The use of multiple assays will undoubtedly lead to the most accurate depiction of antioxidant capacity.

Two important assays, the TOSC and the ORAC, will be examined here, including an in depth look at methodology and correlation.

1.1. Oxygen Radical Absorbance Capacity Assay

The ORAC assay is a widely accepted tool for measuring the antioxidant activity of various vitamins, phytochemicals, and other organic and inorganic compounds (18). The ORAC assay measures the oxidative degradation of a fluorescent molecule (usually fluorescein sodium salt or beta-phycoerythrin) after being mixed with an oxygen radical initiator (14, 16). Fluorescence intensity decreases as oxidative degradation increases, with fluorescence typically recorded over 100 min from the addition of the oxygen radical initiator. When an antioxidant compound is mixed with the fluorescent molecule and the oxygen radical initiator, however, it can protect the fluorescent molecule from degradation, and the fluorescence intensity is thus maintained over time. Thus, the longer the fluorescent molecule maintains its intensity, the less it is being degraded by the oxygen radicals present, or the more it is being protected by the antioxidant compound that is present.

A phycoerythrin fluorescence-based assay for ROS was originally developed (20) and was later modified (21) to create the ORAC assay. Since then, the ORAC assay has become the current food industry standard for assessing antioxidant capacity of food additives, whole foods, juices, and raw vitamins (15, 16, 19).

The ORAC assay has several advantages, including reliability, high sensitivity, and the ability to measure the antioxidant activity of chain-breaking antioxidants (8, 18, 19). Also, the ORAC assay can accurately measure both the inhibition time and the inhibition degree of antioxidants to provide an accurate measurement of antioxidant activity (18).

Some problems with the early ORAC assays, however, were their sensitivity to temperature variations (a problem solved by preheating well plates) and their inability to measure a complete range of biologically relevant radicals, since they can only measure peroxyl radicals (18, 19, 27). In addition, it is important that the pH of the fluorescein be maintained above 7, as it is pH sensitive. Despite these small obstacles, the ORAC remains among the most preferred assays for measuring antioxidant capacity, and has even been suggested to be the most useful for ranking antioxidants. It is recommended that the ORAC be used to quantify peroxyl radical scavenging capacity (14).

1.2. Total Oxyradical Scavenging Capacity Assay

The Total Oxyradical Scavenging Capacity (TOSC) assay is another effective assay used to assess antioxidant activity. In contrast to the ORAC, the TOSC measures the decrease of ethylene production caused by antioxidants (32). Moreover, its advantages include its ability to measure the antioxidant capacity of biological tissues and to measure both lipid and water soluble antioxidants, similar to the ORAC. Unlike the ORAC, however, the TOSC has the capability of distinguishing between faster acting and slower acting antioxidants, as it is a kinetically based assay (28, 29).

2. Materials

2.1. Equipment

1. HP5890A gas chromatograph along with the Hp ChemStation GC 5890 online program

2. Fluorescence Microplate Reader used with a COBAS FARA II spectroflourometric analyzer (Roche Diagnostic Systems), or BMG FLUOstar Optima fluorescence microplate reader (*see* Note 1).

3. Clear-bottom well plates: The plate size should be selected based on the instrument capabilities

2.2. Reagents

2.2.1. ORAC Assay

1. AAPH: For the ORAC assay, AAPH (2,2'-azobis(2-amidino-propane) dihydrochloride) is used as the free-radical initiator because it forms free radicals at a constant rate (26). Antioxidants quench the oxygen radicals generated by AAPH, thereby inhibiting fluorescence decay (14, 27).

2. Flourescein Sodium Salt: A modified ORAC that used fluorescein instead of β-phycoerythrin was developed after the original ORAC assay was validated (15). Fluorescein was preferred because of its several advantages: it is less expensive than β-phycoerythrin, does not interact with other compounds, and does not photobleach (15).

3. Trolox: Trolox (±)-6-Hydroxy-2,5,7,8-tetramethylchromane-2-carboxylic acid, 97% is a widely recognized water soluble Vitamin E derivative used as a baseline against which relative antioxidant activity is measured. Thus, results for test samples (including chemical compounds and foods) are reported and published as μmol Trolox equivalents/mg or TE.

4. Sample: Samples that have been run in the ORAC assay include a variety of antioxidant compunds, such as retinols, phenols, carotenoids, tocopherols, vitamins, phytochemicals, and whole fruits, vegetables, seeds, spices, grains, and legumes (1, 18).

2.2.2. TOSC Assay

1. ABAP: In the TOSC assay, the thermal homolysis of ABAP (2,2'-Azobis(2-methyl-propionamidine), dihydrochloride) generates peroxyl radicals.

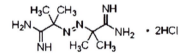

2. KMBA (α-keto-γ-(methylthio)butyric acid sodium salt): KMBA is oxidized by ABAP to produce ethylene gas, which is then measured by gas chromatography. When antioxidants are present, however, they quench the peroxyl radicals formed by ABAP, and, consequently, ethylene gas production is inhibited (31, 32).

$$CH_3SCH_2CH_2-\overset{\overset{\displaystyle O}{\|}}{C}-\overset{\overset{\displaystyle O}{\|}}{C}-ONa$$

3. Trolox: Similar to the ORAC, Trolox is used as a standard for determining antioxidant activity in the TOSC assay. As with the ORAC assay, results from the TOSC are also reported as μmol Trolox equivalents/mg. KMBA, ABAP, and Trolox can all be obtained from Sigma-Aldrich (St. Louis, MO) and other commercial chemical manufacturers and distributers.

4. Sample: Many types of samples can be run in the TOSC assay, including vitamins, cellular enzymes, organic and inorganic

compounds, fruit and vegetable juices, and other antioxidant compounds (32, 33). Recent studies have included TOSC analyses of Acai fruits, Hydrothermal Vent Mussels, and pyrene from marine invertebrates (34-36). TOSC samples must be water soluble or DMSO (dimethyl sulfoxide) soluble.

3. Methods

3.1. ORAC Methods

The ORAC assay is an important assay because of its practicality and ease of use. It can be confusing to some researchers, however, because many investigators have incorporated their own variations on the assay (15, 17, 18, 30, 32). Steps below are taken from Tomer, et al. (32) (*see* Note 2).

1. AAPH is prepared at a concentration of 79.65 mmol/L by adding 216 mg to 10 mL phosphate buffer (75 mM, pH 7.4). Caution must be used when handling AAPH, as it can cause apoptosis, and damage to the liver and kidney, as well as to capillaries and lymphocytes (26).

2. Fluorescein is then prepared by dissolving 22.5 mg into 50 mL PBS to make a first stock solution. A second stock solution is prepared by adding 50 μL of the first fluorescein stock solution to 10 mL of phosphate buffer. Finally, 320 μL of the second fluorescein stock solution is added to 20 mL of phosphate buffer.

3. Experimental samples and control samples (including Trolox) are also prepared at this point. Special care should be taken to ensure that each sample is prepared and stored properly. Samples may be either water soluble or DMSO soluble.

4. 400 μL of fluorescein and 40 μL of sample are first pipetted into each well. Phosphate buffer is used as a blank, and a Trolox dilution series (50, 25, 12.5, and 6.25 μM) is used as a standard.

5. When these aliquots are added to each well, fluorescence readings are taken for time 0. During cycle 4, the reaction is initiated by injecting 150 μL of AAPH into the respective wells (3.5 min and 90 s time intervals can also be used). Fluorescence readings may be taken until the flourescence counts have decreased to negative control amounts. Previous studies have been performed using various cycle times and varying time intervals (15, 17, 32).

6. Fluoresence readings are taken every 3.5 min, up to 35 cycles. Flourescence readings for each cycle are saved, as they will all be used to calculate areas under the curves and Trolox Equivalents.

3.2. TOSC Assay

The time required for a complete run is between 2½ and 3 h (1½ h actual run time in gas chromatograph). Additional time should be expected for comprehensive data analysis.

1. Reagent Preparation: A 200 mM concentration of ABAP is prepared by adding 434 mg ABAP and ddH$_2$O up to 8 mL total volume. A 20 mM concentration of KMBA is prepared by adding 2.7 mg KMBA and ddH$_2$O up to 8 mL total volume. Following their preparation, the respective vials of ABAP and KMBA should be maintained on ice. A 100 mM potassium phosphate buffer at pH 7.4 is also prepared (32). Experimental and control samples to be tested should also be prepared at the desired concentrations at this point.

2. Vial Preparation: In 10 mL glass vials, 100 μL PBS (Phosphate Buffered Saline) is added to the bottom of each. Next, 690 μL ddH$_2$O is added to the sample vials. Vials are then vortexed. After vortexing, 100 μL KMBA are added to all vials. Next, vials are capped and vortexed a second time. After each sample is vortexed the second time, a 10 μL sample is then added to each sample vial, and the vials are again capped.

3. Incubation & ABAP Injection: The septa-sealed vials should next be incubated for 5 min in a 37°C water bath to equilibrate. After the 5 min incubation period, 100 μL ABAP is added into the vials to initiate the reaction (ABAP should be vortexed before each injection).

4. Gas Chromatograph Preparation: GC temperatures are set as follows: 160°C for the injector port, 60°C for the oven, and 220°C for the flame ionization detector (FID). It is recommended that the injection port septa be changed before each run. The hydrogen/air balance at the detector exhaust should be set to 8–12%. Mobile phase helium can be used to push the other gases though the gas chromatograph at a flow rate of 30 mL/min. Once lit, the baseline signal value is normally around 10. Slight variations may be due to the amount of air/hydrogen intake. Once the GC injection syringe barrel is cleaned and the machine prepped, injections and measurements may begin.

5. Gas Chromatography: Twelve minutes after the first ABAP injection, a 1 mL aliquot of gas from each vial should be injected into the GC at staggered intervals. At time 10 min, data should be saved, as area under the curve for these data will later be calculated. At time 12 min, the 8 injections are repeated at the same time intervals 7 times, for a total of 8 passes that make up a 96 min run. Data saved from every cycle will be used to calculate area under the curve.

3.3. Results

3.3.1. Trolox Equivalents

The most important aspect of both of these assays is calculating Trolox equivalents, or TE. As mentioned, Trolox equivalents represent μmol Trolox equivalents/mg sample.

The area under the flourescence kinetic curve (AUC) for each sample is calculated as

$$AUC = (0.5 + f_1/f_1 + f_2/f_1 + f_3/f_1 + \ldots + f_n/f_1) \times CT$$

where f_1 = initial fluorescence reading at cycle 1, f_n = flourescence reading at cycle n, and CT = cycle time in min (15, 32).

In order to calculate Trolox equivalents, a Trolox standard curve must first be obtained. This is done by performing the ORAC or TOSC assay first on varying concentrations of Trolox (most preferably a serial dilution, which will make calculating a standard curve simpler and more accurate), and calculating the individual AUC values for each concentration using the method described above. Once the AUC values for the Trolox concenrations have been calculated, the AUC for the blank well must also be calculated and then subtracted from the total AUC for each Trolox concentration. Subtracting the blank AUC from the Trolox (and later from each sample) AUC gives the net AUC, which accounts for the background readings given by the flouresence reader. This concept is shown in Fig. 17.1. The area *between* the two curves represents the net AUC:

Net AUC = Sample AUC – Control AUC

A Trolox standard curve is then created from the net AUC values for each concentration of Trolox used. The slope of the standard curve is then calculated and used to convert net area under the sample curve to Trolox Equivalents per liter (TE/L). TE/L are then converted to Trolox Equivalents per miligram sample for reporting.

3.3.2. ORAC/TOSC Correlation

To date, only a few studies have been performed that assess the correlations among the differing antioxidant assays (8, 32). Of these, Cao and Prior (8) have found weak but significant

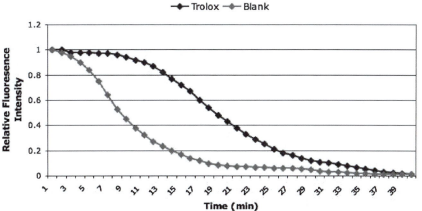

Fig. 17.1. Subtracting the blank AUC from the Trolox or sample AUC gives the net AUC.

correlation between the ORAC assay and the FRAP assay, but no correlation between the ORAC and the TEAC assays.

One study has also been performed to assess the correlations between the ORAC and TOSC assays (32). While precise correlation was not observed, the trends observed in each case were the same: Green Tea Polyphenols and MegaNatural Gold Grape seed extract yielded high TOSC values and high ORAC values; Lemon Fruit 12:1 and cirus bioflavonoids yielded low TOSC values and low ORAC values; and Grape skin extract, GSKE-40 grape seed extract, quercitin, pycogenol, and pine bark theraplant all yielded moderate TOSC values and moderate ORAC values.

Figure 17.2 contains some of the data from the study (32). This figure demonstrates the relative activity of the 11 phytochemicals measured by the TOSC assay (light bars) and by the ORAC assay (dark bars). Minimum and maximum values for each assay were scaled to 0% and 100%, respectively, to faclilitate inter-assay comparisons. The numbers represent the phytochemicals as follows: (1) lemon fruit, (2) citrus bioflavonoids, (3) pomegranate, (4) grape seed extract, (5) pine bark theraplant, (6) quercitin, (7) pycnogenol, (8) grape seed extract, (9) rutin, (10) alpha lipoic acid, and (11) green tea polyphenols. Values reflect the mean and standard deviation.

Moderate correlation between average TOSC and average ORAC values was observed among the 11 phytochemicals ($R^2 = 0.60$). TOSC and ORAC values are both in μmol Trolox equivalents/mg and are the average of several replicates using both the TOSC and ORAC assays (Fig. 17.3).

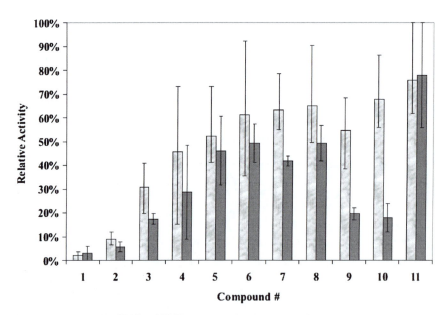

Fig. 17.2. Correlation between the ORAC and TOSC assays across 11 compounds.

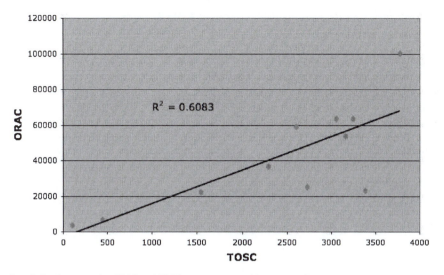

Fig. 17.3. Correlation between the ORAC and TOSC assays acorss 11 compounds.

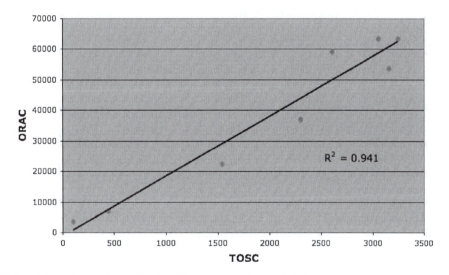

Fig. 17.4. Correlation between the ORAC and TOSC assays (7 outliers excluded).

When the outliers rutin, alpha lipoic acid, and green tea polyphenols were removed from the dataset, high correlation between the remaining TOSC and ORAC assay values was observed ($R^2 = 0.94$). Again, TOSC and ORAC values are represented in µmol Trolox equivalents/mg and are the average of several replicates using both the TOSC and ORAC assays (Fig. 17.4).

Included are tables that show the measured average TOSC and ORAC values obtained over several replicates (Fig. 17.5).

Comparing the two assays revealed that more variability in antioxidant activity could be seen for the TOSC assay than for

Phytochemical	TOSC[a]	ORAC
Lemon Fruit	103	3885
Citrus Bioflavonoids	446	7410
Pomegranate	1543	22530
Grape Seed Extract	2294	37125
Pine Bark Theraplant	2608	59483
Quercitin	3059	63750
Pycnogenol	3154	53970
Grape Skin Extract	3246	63668
Rutin	2727	25440
Alpha Lipoic acid	3381	23235
Green tea polyphenols	3377	100658

Fig. 17.5. The measured average TOSC and ORAC values for 11 compounds.

the ORAC assay. Such variability may have been due to slight procedural errors, including error in the injection process into the gas chromatograph. Establishing specific numbers for the antioxidant activity of certain phytochemical compounds also proved difficult because the conditions under which the fruit or vegetable samples were cultivated and stored varied greatly from sample to sample. Despite small variations in sampling, the antioxidant activity ranking of these phytochemical compounds was similar among replicates (32).

Absolute values of the antioxidant counts for the ORAC and TOSC assays varied greatly in this study because different radical sources are used for the two assays (AAPH is used for the ORAC assay, while ABAP is used for the TOSC assay), and also because samples measured in the ORAC and TOSC assays were prepared at different concentrations because of the enhanced sensitivity of the ORAC assay.

While the measurement of any antioxidant compound in vitro is a valuable first step for identifying good antioxidants for further use in in vivo testing, it should not be assumed that the antioxidant activity of a compound in vitro will necessarily reflect the antioxidant activity in vivo. As mentioned, studies have shown that consumption of fruits and vegetables results in elevated plasma ORAC values in humans (13), and synergy is often observed between multiple antioxidant compounds in vivo (7, 13).

The discrepancies (32) seen between the ORAC and TOSC assays suggest that in some cases, the results from both assays may be useful in determining an antioxidant's true activity. As other antioxidant assays are also used and developed, discrepancies between these assays pertaining to the relative antioxidant activity of various samples will likely be greatly diminished. As more research is performed, the realm of predictive power will increase, and the ORAC and TOSC assays will continue to be useful tools for assessing antioxidant activities.

4. Notes

1. Microplate readers may vary somewhat in their available excitation and emission filters. Cao and Prior used an excitation wavelength of 540 nm and an emission wavelength of 565 nm (8); Tomer used an excitation wavelength of 485 nm and an emission wavelength of 520 nm (32). Ou used an excitation wavelength of 493 nm and an emission wavelength of 515 nm (15). The wavelengths available on the instruments must be set to be able to read fluorescein emissions in order for ORAC data to be collected and analyzed properly.

2. There have been many different methods used in published research articles for the ORAC assay. Some alternative methods include, for example, the following: Ou used an AAPH concentration of 12.8 mM and a fluorescein concentration of 48 nM. When putting aliquots into each well, Ou used 365 μL Fluorescein, and 20 μL sample (15). Huang (18) used a final AAPH concentration of 153 mM by dissolving 0.414 g AAPH into 10 mL of 75 mM phosphate buffer (pH 7.4). A flourescein stock solution of 4×10^{-3} mM was prepared in 75 mM phosphate buffer (pH 7.4), which was stored and wrapped in foil at 5°C. This stock solution was diluted 1:1,000 with 75 mM phosphate buffer (ph 7.4) immediately prior to use.

References

1. Oxygen Radical Absorbance Capacity (ORAC) of Selected Foods (2007); Nutrient Data Laboratory, Agricultural Research Service, US Department of Agriculture, United States Department of Agriculture.

2. Ames BN, Gold LS, Willet WC (1995) The Causes and Prevention of cancer. Proc Natl Acad Sci USA 92:5258–5265

3. Ames BN, Shigenaga MK, Hagen TM (1993) Oxidants, antioxidants, and the degenerative diseases of aging. Proc Natl Acad Sci USA 90:7915–7922

4. Christen Y (2000) Oxidative stress and Alzheimer's disease. Am J Clin Nutr 71:621S–629S

5. Diaz MN, Frei B, Keaney JF Jr (1997) Antioxidants and atherosclerotic heart disease. N Eng J Med 337:408–416

6. Lang AE, Lozano AM (1998) Parkinson's disease. First of two parts. N Eng J Med 339:111–114

7. Packer L, Colman C (1999) The anti-oxidant miracle. Wiley, New York

8. Cao G, Prior RL (1998) Comparison of different analytical methods for assessing total antioxidant capacity of human serum. Clin Chem 44(6): 1309–1315

9. Halliwell B, Gutteridge JMC, Cross CE (1992) Free Radicals, antioxidants, and human disease: where are we now? J Clin Lab Med 119:598–620

10. Tomer DP (2003) Measuring parameters that are important in cancer prevention and treatment: assessing antioxidant activity and angiogenesis. Brigham Young University, Master's Thesis

11. McLeman LD (2004) Evidence of synergistic intracellular antioxidant networking and antioxidant regulation of cellular lipid peroxidation in HL-60 acute promyelogenous leukemia. Master's Thesis. Brigham Young University. 1-37.

12. O'Neill K, Murray B (2002) Power plants. Woodland Publishing, Utah

13. Cao G, Booth SL, Sadowski JA, Prior RL (1998) Increases in human plasma antioxidant capacity after consumption of controlled diets high in fruit and vegetables. Am J Clin Nutr 68:1081–1087

14. Huang D, Ou B, Prior RL (2005) The chemistry behind antioxidant capacity assays. J Agric Food Chem 53:1841–1856

15. Ou B, Hampsch-Woodill M, Prior RL (2001) Development and validation of an improved oxygen radical absorbance capacity assay using fluorescein as the fluorescent probe. J Agric Food Chem 49(10):4619–4626

16. Cao G, Alessio H, Cutler R (1993) Oxygen-radical absorbance capacity assay for antioxidants. Free Radic Biol Med 14(3):303–311

17. Cao G, Sofic E, Prior RL (1996) Antioxidant capacity of tea and common vegetables. J Agric Food Chem 44:3426–3431

18. Huang D, Ou B, Hampsch-Woodill M, Flanagan J, Prior RL (2002) High-throughput assay of oxygen radical absorbance capacity (orac) using a multichannel liquid handling system coupled with a microplate fluorescence reader in 96-well format. J Agric Food Chem 50:4437–4444

19. Ou B, Huang D, Hampsch-Woodill M, Flanagan JA, Deemer EK (2002) Analysis of antioxidant of common vegetables employing oxygen radical absorbance capacity (ORAC) and ferric reducing antioxidant power (FRAP) assays: a comparative study. J Agric Food Chem 50:3122–3128

20. Glazer AN (1990) Phycoerythrin fluorescence based assay for reactive oxygen species. Methods Enzymol 186:161–168

21. Cao G, Prior RL (1999) Measurement of oxygen radical absorbance capacity in biological samples. Methods Enzymol 299:50–62

22. Pellegrini N, Del Rio D, Colombi B, Bianchi M, Brighenti F (2003) Application of the 2, 2'-azinobis(3-ethylbenzothiazoline-6-sulfonic acid) radical cation assay to a flow injection system for the evaluation of antioxidant activity of some pure compounds and beverages. J Agric Food Chem 51:260–264

23. Pulido R, Bravo L, Suara-Calixto F (2000) Antioxidant activity of dietary polyphenols as determined by a modified ferric reducing/antioxidant power assay. J Agric Food Chem 48:3396–3402

24. Aruoma OI (2003) Methodological considerations for characterizing potential antioxidant actions of bioactive components in plant foods. Mutat Res 523–524:9–20

25. Gorinstein S, Martin-Belloso O, Katrich E, Lojek A, Ciz M, Gligelmo-Miguel N, Haruenkit R, Park YS, Jung ST, Trakhtenberg S (2003) Comparison of the contents of the main biochemical compounds and the antioxidant activity of some Spanish olive oils as determined by four different radical scavenging tests. J Nutr Biochem 14:154–159

26. Yokozawa T, Cho EJ, Hara Y, Kitani K (2000) Antioxidative activity of green tea treated with radical initiator 2, 2'-azobis(2-amidinopropane) dihydrochloride. J Agric Food Chem. 48:5068–5073

27. Prior RL, Hoang H, Gu L, Wu X, Bacchiocca M, Howard L, Hampsch-Woodill M, Huang D, Ou B, Jacob R (2003) Assays for hydrophilic and lipophilic antioxidant capacity (oxygen radical absorbance capacity (ORAC(FL))) of plasma and other biological and food samples. J Agric Food Chem 51:3273–3279

28. Dugas AJ Jr, Castaneda-Acosta J, Bonin GC, Price KL, Fischer NH, Winston GW (2000) Evaluation of the total peroxyl radical-scavenging capacity of flavonoids: structure-activity relationships. J Nat Prod 63:327–331

29. Regoli F, Winston GW (1999) Quantification of total oxidant scavenging capacity of antioxidants for peroxynitrite, peroxyl radicals, and hydroxyl radicals. Toxicol Appl Pharmacol 156:96–105

30. Adapted from http://www.bmglabtech.com/application-notes/fluorescence-intensity/orac-148.cfm. Consulted 7/9/08.

31. Somogyi A, Rosta K, Pusztai P, Tulassay Z, Nagy G (2007) Antioxidant Measurements. Physiol Meas 28:R41–R55

32. Tomer DP, McLeman LD, Ohmine S, Scherer PM, Murray BK, O'Neill KL (2007) Comparison of the total oxyradical scavenging capacity and oxygen radical absorbance capacity assays. J Med Food 10(2):337–344

33. Lichtenthaler R, Marx F (2005) Total oxidant scavenging capacities of common european fruit and vegetable juices. J Agric Food Chem 53:103–110

34. Lichtenthaler R, Rodrigues RB, Maia JGS, Papagiannopoulos M, Fabricius H, Marx F (2005) Total oxidant scavenging capacities of Euterpe oleracea Mart. (Acai) fruits. Int J Food Sci Nutr 56(1):53–64

35. Company R, Serafim A, Cosson R, Camus L, Shillito B, Fiala-Medioni A, Bebianno MJ (2006) The effect of cadmium on antioxidant responses and the susceptibility to oxidative stress in the hydrothermal vent mussel Bathymodiolus azoricus. Marine Biol 148:817–825

36. Giessing AMB, Mayer LM (2004) Oxidative coupling during gut passage in marine deposit-feeding invertebrates. Limnol Oceanogr 49(3):716–726

Chapter 18

Assessing the Neuroprotective Effect of Antioxidant Food Factors by Application of Lipid-Derived Dopamine Modification Adducts

Xuebo Liu, Naruomi Yamada, and Toshihiko Osawa

Summary

Advances in understanding the neurodegenerative pathologies are creating new opportunities for the development of neuroprotective therapies, such as antioxidant food factors, lifestyle modification and drugs. However, the biomarker by which the effect of the agent on neurodegeneration is determined is limited. We here address hexanoyl dopamine (HED), one of novel dopamine adducts derived from brain polyunsaturated acid, referring to its in vitro formation, potent toxicity to SH-SY5Y cells, and application to assess the neuroprotective effect of antioxidative food factors. Dopamine is a neurotransmitter, and its deficiency is a characterized feature in Parkinson's disease (PD); thus, HED provides a new insight into the understanding of dopamine biology and pathophysiology of PD and a novel biomarker for the assessment of neuroprotective therapies. We have established an analytical system for the detection of HED and its toxicity to the neurobistoma cell line, SH-SY5Y cells. Here, we discuss the characteristics of the system and its applications to investigate the neuroprotective effect of several antioxidants that originate from food.

Key words: HED, Parkinson's disease, Biomarker, Food factors, Neuroprotective effect

1. Introduction

Increasing evidence suggests that oxidative stress play a crucial role in the majority of neurodegenerative diseases. Parkinson's disease (PD) is a neurodegenerative disorder characterized by a dramatic loss of dopaminergic neurons in the substantia nigra, and the subsequent deficiency of dopamine in the brain areas (1). Until now, very little is known about why and how the PD neurodegenerative process begins and progresses; however, recent

D. Armstrong (ed.), *Advanced Protocols in Oxidative Stress II*, Methods in Molecular Biology, vol. 594
DOI 10.1007/978-1-60761-411-1_18, © Humana Press, a part of Springer Science+Business Media, LLC 2010

studies indicate that there are high levels of basal oxidative stress in the substantia nigra pars compacta (SNc) in the normal brain, and this is increased in PD (2).

Oxidative stress in the brain easily leads to the lipid peroxidation reaction due to a high concentration of polyunsaturated fatty acids (PUFA), such as docosahexaenoic acid (DHA, $C22:6/\omega-3$) and arachidonic acid (AA, $C18:4/\omega-6$), which are present in the brain (3). We have recently found that lipid hydroperoxides, the primary peroxidative products, can universally react with primary amino groups to form N-acyl-type (amide-linkage) adducts (4–9). We have previously described the in vitro and in vivo formation of DHA-derived adducts, N^{ε}-(succinyl) lysine and N^{ε}-(propanoyl) lysine, by LC-MS/MS or immunochemical analysis. In addition, during the reaction of oxidized arachidonic acid (AA) with the lysine residue, the formation of N^{ε}-(hexanoyl) lysine and N^{ε}-(glutaryl) lysine were also detected. The N-acyl-type adducts are specific to the peroxidation of polyunsaturated fatty acids; therefore, their formations are the useful markers for the lipid peroxidation, protein modification, and related dysfunction that occur in these fatty acids enriched tissues.

Dopamine is the endogenous neurotransmitter produced by nigral neurons. Dopamine loss can trigger not only prominent secondary morphological changes, such as density reduction of the dendritic spines, but also changes in the density and sensitivity of dopamine receptors (1); therefore, it is a sign of PD development. The reasons for dopamine loss are attributed to dopamine's molecular instability. Some possible causes of dopamine loss are abnormalities of dopaminergic neurons (10), dopamine degradation by monoamine oxidase A (MAO-A) (11) or auto-oxidation (12), and the reaction with amino acid cysteine (13). Dopamine is a member of catecholamine family. The catechol structure contributes to high oxidative activation of dopamine. Additionally, the NH_2-teminals in dopamine's structure may represent another reactive spot; however, little experimental evidence proves this. Based on our previously described reaction between lipid hydroperoxides and NH_2 residues, we examined the reaction of dopamine with reactive LOOH species derived from lipid peroxidation.

We particularly report here on hexanoyl dopamine (HED), a dopamine modified adduct derived from arachidonic acid, referring to its formation, effect on SH-SY5Y cells and applications to investigate the neuroprotective effect of antioxidant foods, such as tocopherol (14, 15), curcumin (16, 17), and sesamin analogs (18–21) and astaxanthin (22, 23). We have demonstrated that HED was present in rat brain (data not shown) and toxic to the neuroblastoma SH-SY5Y cells, thereby representing a novel biomarker for the assessment of neuroprotective therapies against PD.

2. Materials

2.1. Equipment

1. Reverse Phase HPLC (Nomura Chemical Inc., Japan).
   Column: Develosil ODS-HG-5 column (20×250 mm).

2. HPLC-Tandem Mass Spectrometry (MS/MS).
   Spectrometer: API 2000 triple quadrupole mass spectrometer (Applied Biosystems) through a TurboIonSpray source.

   Chromatography: Develosil ODS-HG-3 column (2.0×250 mm) with an Agilent 1100 HPLC system.

2.2. Reagents

Arachidonic acid and lipoxidase were obtained from the Sigma-Aldrich Co. (St. Louis, mo, USA). Dopamine·HCl was purchased from Nacalai Tesque, Inc. (Kyoto, Japan). Hexanoic anhydride was obtained from Wako Pure Chemical Industries, Ltd. (Osaka, Japan). The antibodies against poly(ADP-ribose) polymerase (PARP) was purchased from the Cell Signaling Technology, Inc. (Beverly, MA). Active caspase-3 rabbit monoclonal antibody was purchased from the Epitomics, Inc. (California).

2.3. Standards

Hexanoyl dopamine (HED) was prepared in our laboratory. Briefly, HED was chemically synthesized by incubating dopamine (0.5 mM) with hexanoic anhydride (0.5 mM) in 5 ml of 100 mM sodium phosphate buffer (pH 7.4)-saturated sodium acetate (1:1, v/v) for 60 min at room temperature. The synthesized HED were purified by reverse-phase HPLC using a Develosil ODS-HG-5 column (20×250 mm) in an isocratic system of 15% or 50% acetonitrile containing 0.1% trifluoroacetic acid at the flow rate of 6 ml/min. The elution profiles were monitored by absorbance at 280 nm. The amino residues in the dopamine adducts were identified by the ninhydrin reaction. The mass, structure, and formula of the synthesized molecule were identified by HPLC–MS, NMR, and ESI-TOF-MS analyses, respectively.

2.4. Cell Cultures

SH-SY5Y human dopaminergic neuroblastoma cells were kindly donated by Dr. Maruyama (National Institute for Longevity Science, Japan). SH-SY5Y cells and NIH-3 T3 cells were grown in Cosmedium-001 (Cosmo-Bio, Tokyo, Japan) containing 5% FBS and Dulbecco's modified Eagle's medium (DMEM) containing 10% FBS , respectively, and maintained at 37°C in an atmosphere of 5% CO_2 in air.

2.5. Food Factors

Tocopherol analogs (α-tocopherol, β-tocopherol, γ-tocopherol, and δ-tocopherol), tocotrienol (α-tocotrienol, β-tocotrienol, γ-tocotrienol, and δ-tocotrienol) curcumin anaglogs (crucumin and tetrahydrocrucumin), and sesamin analogs (sesamin, sesamolin, sesaminol, sesaminol-6-catechol, and sesaminal triglucoside) and astaxanthin were used in this study. All of them were from our laboratory.

3. Methods

3.1. HED Formation

1. HPLC–MS/MS Conditions
   The chromatographic separation was performed by a gradient elution as follows: 0–10 min, linear gradient from 0.1% formic acid to 50% aqueous acetonitrile containing 0.1% formic acid; 10–15 min, hold; 15–20 min, linear gradient to 0.1% formic acid; flow rate = 0.2 ml/min. The instrument response was optimized by infusion experiments with the standard compounds using a syringe pump at the flow rate of 5 μl/min. HED formation was determined using electrospray ionization MS/MS in the multiple reaction monitoring mode.

2. *In Vitro* formation
   The level of AA hydroperoxide (15-HPETE) was determined using a lipid hydroperoxide assay kit (Cayman, Michigan). Dopamine (2 mM) was incubated with 10 mM AA-hydroperoxides in phosphate buffer (pH 7.4) at 37°C for different times. The reaction was terminated by immediate freezing at –80°C. The detection was carried out by HPLC–MS/MS. 15-HPETE was prepared as previously described in ref. 24.

3.2. Cellular Toxicity

1. Assessment of Cell Viability
   Cell viability was evaluated by an MTT assay. SH-SY5Y cells in 96-well plates were incubated with drugs for different times, followed by further incubation with 500 μg/ml MTT at 37°C for 2 h. Cell viability in some experiments was also measured using PI and Hoechst 33258 staining.

2. ROS Measurement
   Endogenous ROS level was detected by flow cytometry using H_2DCF-DA (Molecular Probes). Briefly, the drug-treated cells were incubated with H_2DCF-DA for 30 min, and the fluorescence of dichlorofluorescein (DCF) was measured using an EPICS Elite Flow Cytometer.

3. Identification of Apoptosis Induction
   Western blot analysis— The cells were washed twice with phosphate-buffered saline, pH 7.0, and lysed with lysis buffer (50 mM Tris–HCl, pH 7.5, 150 mM NaCl, 1% Triton X-100, 0.5% sodium deoxycholate, 0.1% SDS, 100 μg/ml phenylmethylsulfonyl fluoride). After protein quantification, equal amounts of the protein (total protein, 20–50 μg) were boiled

with Laemmli sample buffer for 5 min at 100°C. The samples were run on 10% SDS–polyacrylamide gels, transferred to a nitrocellulose membrane, incubated with 5% skim milk in TTBS (Tris–buffered saline containing 10% Tween 20) for blocking, washed, and treated with the primary antibodies. After washing with TTBS, the blots were further incubated for 1 h at room temperature with the IgG antibody coupled to horseradish peroxidase in TTBS. Blots were then washed 3 times in TTBS before visualization. An ECL kit was used for detection.

3.3. Assay for HED In Vitro *Formation* Inhibition by Food Factors

10 μl of 1 mM 15-HPETE (in ethanol) and 20 μl of 0.2 mM dopamine (in H_2O, *see* Note 1) were added into 60 μl of PB (pH 7.4) with 10 μl of agent (in DMSO) or DMSO. The reaction was performed at 37°C for 24 h, and stopped immediately by putting into –80°C deep freezer. The HED detection was carried out following the protocols described in Subheading 18.3.1, step 2.

3.4. Statistics

All data were analyzed using Bonferroni/Dunn's multiple comparison procedure.

3.5. Results

Figure 18.1 shows the proposed chemical formation scheme and HPLC–MS/MS analysis of HED, including structure (a), mass (b), and formula (c). Figure 18.2 shows the HED formation. Figure 18.2a shows the in vitro HED detection by LC-MS/MS analysis; Fig. 18.2b shows the dose-dependent formation of HED in the reaction of 15-HPETE with dopamine. We also detected HED formation in vivo (*see* Note 2). Figure 18.3 shows the effect of HED on SH-SY5Y cellular toxicity. Figure 18.3a shows the effect of HED on cell viability by MTT assay; Figure 18.3b shows the effect of HED on ROS generation in the cells; Figure 18.3c shows the effect of HED on apoptosis induction using apoptosis hallmarks, including PARP cleavage and accumulation of active caspase-3. In addition, monoamine transporters were needed for HED cytotoxicity (*see* Note 3). Figure 18.4 shows the chemical structure of antioxidant food factors used in this study. Figure 18.5 shows the effect of food factors on the in vitro HED formation. Sesaminol-6-catechol (SMLC) showed the most significant inhibitive effect on the in vitro HED formation (*see* Note 4)

4. Notes

1. To assess protective effect of antioxidant food factors on HED formation, the method shown in Subheading 18.3.3 should be referred. But, it is critical that dopamine solution in H_2O

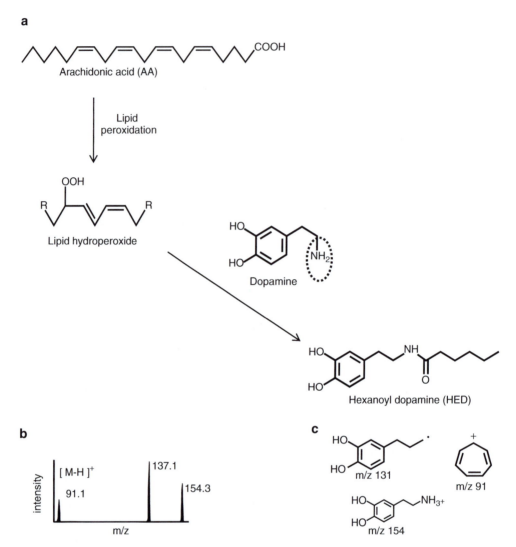

Fig. 18.1. Proposed chemical formation scheme and HPLC–MS/MS analysis of HED formation. (**a**) Proposed reaction scheme of HED formation. (**b**) The [MH]⁺ ion m/z 252 of HED, was subjected to CID, and the daughter ions were scanned. (**c**) The proposed structures of individual ions are shown.

should be prepared on the day because dopamine is to be easily oxidized.

2. We have also demonstrated that HED was present in rat brain (data not shown), in addition to the toxicity to the neuroblastoma SH-SY5Y cells, thereby representing a novel biomarker for the assessment of neuroprotective therapies against PD.

3. HED cytotoxicity has been demonstrated to be selective to neuronal cells with monoamine transports, such as dopamine transporter (DAT), norepinephrine transporter (NET) and 5-HT transporter (5-HTT).

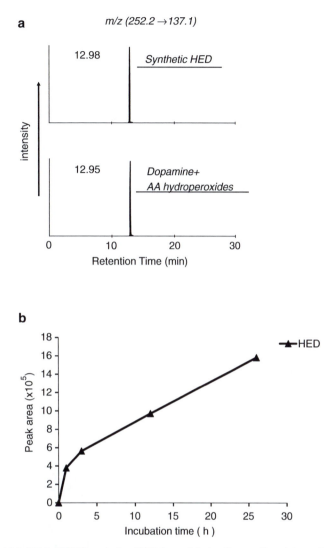

Fig. 18.2. HPLC–MS/MS analysis of HED formed during the reaction of dopamine with AA hydroperoxides. Dopamine (2 mM) was incubated with AA hydroperoxides (10 mM) in 0.1 M phosphate buffer (pH 7.4) at 37°C. Ion monitoring of HED transition was m/z 252. (**a**) authentic dopamine adduct and reaction mixture of AA hydroperoxides with dopamine. (**b**) HED formation in a time-dependent manner.

4. In Fig. 18.5, sesaminol-6-catechol (SMLC) showed the most significant inhibitive effect on the *in vitro* HED formation. We also recently found that SMLC markedly prevented HED-induced ROS generation and cell death in SH-SY5Y cells (data not shown). In addition, SMLC was detected in the brain of rats fed with SMLC-contained foods, suggesting that it could cross brain-blood barrier (BBB) and exhibit the neuroprotective effect *in vivo*.

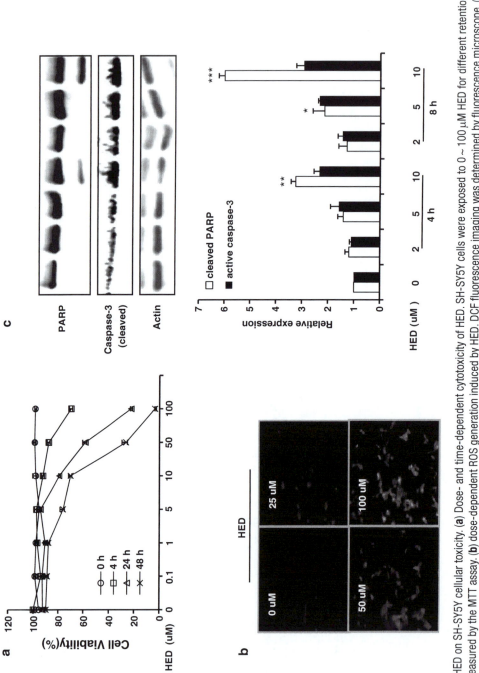

Fig. 18.3. Effect of HED on SH-SY5Y cellular toxicity. (a) Dose- and time-dependent cytotoxicity of HED. SH-SY5Y cells were exposed to $0 \sim 100 \,\mu M$ HED for different retention times. Cell viability was measured by the MTT assay. (b) dose-dependent ROS generation induced by HED. DCF fluorescence imaging was determined by fluorescence microscope. (c) PARP cleavage and active caspase-3 expression in SH-SY5Y cells exposed to $0 \sim 10 \,\mu M$ HED for 4 h and 8 h. The cleavage of PARP and expression of active caspase-3 were tested by western blotting and statistically analyzed.

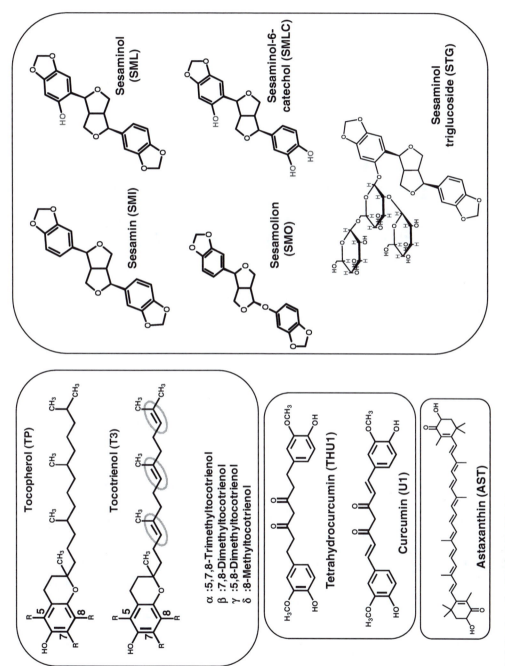

Fig. 18.4. Chemical structure of antioxidant food factors.

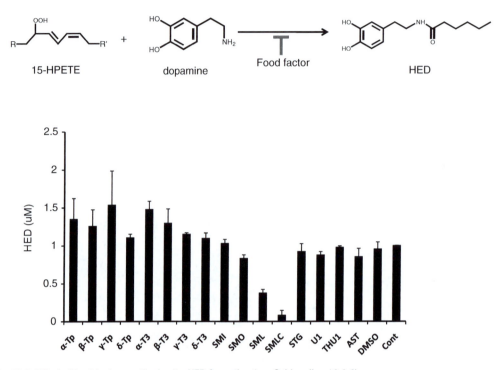

Fig. 18.5. Effect of food factors on the *in vitro* HED formation (*see* Subheading 18.3.3).

References

1. Galvan A, Wichmann T (2008) Pathophysiology of parkinsonism. *Clin Neurophysiol* **119**(7): 1459–1474

2. Jenner P (2003) Oxidative stress in Parkinson's disease. *Ann Neurol* **53**(Suppl 3):S26–S38

3. Porter NA, Caldwell SE, Mills KA (1995) Mechanisms of free radical oxidation of unsaturated lipids. Lipids 30(4):277–290

4. Kato Y, Makino Y, Osawa T (1997) Characterization of a specific polyclonal antibody against 13-hydroperoxyoctadecadienoic acid-modified protein: formation of lipid hydroperoxide-modified apoB-100 in oxidized LDL. *J Lipid Res* **38**(7):1334–1346

5. Kato Y, Osawa T (1998) Detection of oxidized phospholipid–protein adducts using anti-15-hydroperoxyeicosatetraenoic acid-modified protein antibody: contribution of esterified fatty acid-protein adduct to oxidative modification of LDL. *Arch Biochem Biophys* **351**(1): 106–114

6. Kato Y, Mori Y, Makino Y, Morimitsu Y, Hiroi S, Ishikawa T, Osawa T (1999) Formation of Nepsilon-(hexanonyl)lysine in protein exposed to lipid hydroperoxide. A plausible marker for lipid hydroperoxide-derived protein modification. *J Biol Chem* 274(29):20406–20414

7. Kawai Y, Kato Y, Fujii H, Mkino Y, Mori Y, Naito N, Osawa T (2003) Immunochemical detection of a novel lysine adduct using an antibody to linoleic acid hydroperoxide-modified protein. *J Lipid Res* 44(6):1124–1131

8. Kawai Y, Fujii H, Kato Y, Kodama M, Naito N, Uchida K, Osawa T (2004) Esterified lipid hydroperoxide-derived modification of protein: formation of a carboxyalkylamide-type lysine adduct in human atherosclerotic lesions. *Biochem Biophys Res Commun* 313(2):271–276

9. Kawai Y, Fujii H, Okada M, Tsuchie Y, Uchida K, Osawa T (2006) Formation of Nepsilon-(succinyl)lysine in vivo: a novel marker for docosahexaenoic acid-derived protein modification. *J Lipid Res* 47(7):1386–1398

10. Bove J, Prou D, Perier C, Przedborski S (2005) Toxin-induced models of Parkinson's disease. *NeuroRx* 2(3):484–494

11. Gotz ME, Kunig G, Riederer P, Youdim MB (1994) Oxidative stress: free radical production in neural degeneration. Pharmacol Ther 63(1):37–122

12. Hald A, Lotharius J (2005) Oxidative stress and inflammation in Parkinson's disease: is there a causal link? *Exp Neurol* **193**(2): 279–290

13. LaVoie MJ, Hastings TG (1999) Peroxynitrite- and nitrite-induced oxidation of dopamine: implications for nitric oxide in dopaminergic cell loss. *J Neurochem* **73**(6):2546–2554

14. Yamagishi M, Osakab N, Takizawa T, Osawa T (2001) Cacao liquor polyphenols reduce oxidative stress without maintaining alpha-tocopherol levels in rats fed a vitamin E-deficient diet. *Lipids* **36**(1):67–71

15. Atkinson J, Epand RF, Epand RM (2008) Tocopherols and tocotrienols in membranes: a critical review. *Free Radic Biol Med* **44**(5): 739–764

16. Osawa T, Kato Y (2005) Protective role of antioxidative food factors in oxidative stress caused by hyperglycemia. *Ann N Y Acad Sci* **1043**:440–451

17. Kitani K, Osawa T, Yokozawa T (2007) The effects of tetrahydrocurcumin and green tea polyphenol on the survival of male C57BL/6 mice. *Biogerontology* **8**(5):567–573

18. Yokota T, Matsuzaki Y, Koyama M, Hitomi T, Kawanaka M, Enoki-Konishi M, Okuyama Y, Takayasu J, Nishino H, Nishikawa A, Osawa T, Sakai T (2007) Sesamin, a lignan of sesame, down-regulates cyclin D1 protein expression in human tumor cells. *Cancer Sci* **98**(9): 1447–1453

19. Sheng HQ, Hirose Y, Hata K, Zheng Q, Kuno T, Asano A, Yamada Y, Hara A, Osawa T, Mori H (2007) Modifying effect of dietary sesaminol glucosides on the formation of azoxymethane-induced premalignant lesions of rat colon. *Cancer Lett* **246**:63–68

20. Miyake Y, Fukumoto S, Okada M, Sakaida K, Nakamura Y, Osawa T (2005) Antioxidative catechol lignans converted from sesaminol triglucoside by culturing with *Aspergillus*. *J Agric Food Chem* **53**(1):22–27

21. Kang M-H, Naito M, Sakai K, Uchida K, Osawa T (2000) Mode of action of sesame lignans in protecting low-density lipoprotein against oxidative damage in vitro. *Life Sci* **66**:161–171

22. Liu X, Osawa T (2007) Cis astaxanthin and especially 9-cis astaxanthin exhibits a higher antioxidant activity *in vitro* compared to the all-trans isomer. *Biochem Biophys Res Commun* **357**:187–193

23. Aoi W, Naito Y, Takanami Y, Ishii T, Kawai Y, Akagiri S, Kato Y, Osawa T, Yoshikawa T (2008) Astaxanthin improves muscle lipid metabolism in exercise via inhibitory effect of oxidative CPT I modification. *Biochem Biophys Res Commun* **366**:892–897

24. Liu XB, Shibata T, Hisaka S, Osawa T (2008) DHA hydroperoxides as a potential inducer of neuronal cell death: A mitochondrial dysfunction-mediated pathway. *J Clin Biochem Nutr* **43**(1):26–33

Chapter 19

LIBS-Based Detection of Antioxidant Elements: A New Strategy

Geeta Watal, Bechan Sharma, Prashant Kumar Rai, Dolly Jaiswal, Devendra K. Rai, Nilesh K. Rai, and A.K. Rai

Abstract

The present study deals with the scientific evaluation of antioxidant potential of aqueous extract of *Trichosanthes dioica* fruits on diabetes-induced oxidative stress of diabetic rats. The most effective dose of mg/kg bw of fruit aqueous extract was given orally to diabetic rats for 30 days. Different oxidative stress parameters were analyzed in various tissues of control and treated diabetic rats. The observed elevated level of lipid peroxidation (LPO) comes down significantly ($p < 0.05$) and decreased activities of antioxidant enzymes such as catalase (CAT), superoxide dismutase (SOD), glutathione peroxidase (GPx), and glutathione-S-transferase (GST) got increased ($p < 0.05$) significantly of diabetic rats on extract treatment. Laser-Induced Breakdown Spectroscopy (LIBS) has been used as an analytical tool to detect major and minor elements like Mg, Fe, Na, K, Zn, Ca, H, O, C, and N present in the extract. The higher concentration of Ca^{2+}, Mg^{2+}, and Fe^{2+}, as reflected by their intensities are responsible for antioxidant potential of *T. dioica*.

Key words: *Trichosanthes dioica*, Diabetes, LIBS , Indian herbs

1. Introduction

Laser-induced breakdown spectroscopy (LIBS) is a modern analytical technique used for qualitative and quantitative analysis of trace elements present in any material. The principle of LIBS is based on the spectral analysis of radiation emanating from micro plasma generated by focusing a high power pulsed laser beam on surface of the sample. The characteristic emission from plasma is recorded as spectrum, which provides a fingerprint of constituents of target material. Thus, the spectrum contains qualitative and quantitative information, which can be correlated with sample identify or can be used to determine the amount of its constituents.

D. Armstrong (ed.), *Advanced Protocols in Oxidative Stress II*, Methods in Molecular Biology, vol. 594
DOI 10.1007/978-1-60761-411-1_19, © Humana Press, a part of Springer Science + Business Media, LLC 2010

It is a versatile, sensitive, real-time, and in-situ elemental analysis technique having microanalyses capability for any kind of material in any phase solid, liquid, or gas (1, 2). This technique is unique in the sense that it requires no sample preparation, as sample treatment is often prone to induced errors due to contamination and losses.

LIBS can scientifically unfold the mystery enclosed in medicinal plants based on their inorganic constituents. Apart from having various organic molecules, plants are a rich source of inorganic mineral elements implicated in various health concerns. The crude material of these medicinal herbs is used in Ayurvedic medicines (3, 4). LIBS is a powerful and convenient method to detect the elements present in plant materials responsible for their medicinal value (*see* Note 1). It is potentially a useful method for rapidly analyzing multiple samples on site at low cost. A number of medicinal plants with antidiabetic activity have been screened and indexed (3–5) for their glycemic elements.

LIBS is a value-added efficient technique for preparing a high quality and most up-to-date elemental database of medicinal plants serving as a discovery tool to correlate a wide diversity in medicinal properties of herbs based on their elemental composition. The world's largest and most comprehensive collection of medicinal plants and their elements-related information can be prepared as a state-of-the-art information service with the help of this advanced analytical technique and can be explored further in the pharmaceutical industry.

The present study describes the LIBS-based detection of major and minor elements present in *Trichosanthes dioica* (pointed gourd) fruits responsible for their antioxidant potential, based on screening of the best set of oxidative elements involved. The assessment of antioxidant efficacy of *T. dioica* fruits has also been assessed in terms of their impact on various oxidative stress parameters in various key tissues such as brain, liver kidney, pancreas, and spleen of normal as well as streptozotocin (STZ)-induced diabetic rats in order to validate the antioxidant potential of elements present in *T. dioica* fruits. In brief, LIBS helps in assessing the impact of elemental composition changes on different medicinal properties of plants such as antioxidant, antidiabetic, etc.

2. Materials

2.1. Equipment

1. Q- switched Nd: YAG laser pulsed laser (Continuum Surellite III-10).
2. Multichannel spectrometer (Ocean Optics LIBS2000+) equipped with CCD.

3. Software for analysis of LIB spectra, such as OOI LIBS software.

4. ANOVA Software for statistical analysis, such as SPSS computer software, version 7.5.

5. Centrifuge.

6. Lyophilizer.

7. ELICO VU-VIS Double beam spectrophotometer.

8. Vortex mixture.

9. Parafilm (American National Can, Greenwich CT).

10. Micro pipit.

11. Soxhlet.

2.2. Reagents

1. Phosphate buffer (100 mM, pH 7.4) containing 150 mM KCl.

2. Isotonic ice-cold NaCl (0.9%, w/v) solution.

3. Pyrogallol.

4. H_2O_2 (10 mM).

5. Malondialdehyde (MDA)/Thiobarbituric acid reactive substances (TBARS).

6. Bovine serum albumin (BSA).

7. Tris Succinate (50 mM, pH 8.2).

3. Methods

3.1. Material Processing

Fresh fruits of *T. dioica* were cut into small pieces and shade dried. The dried pieces were mechanically crushed and extracted with distilled water using Soxhlet at 80–100°C up to 36 h. The extract was filtered and concentrated in rotatory evaporator at 35 ± 5°C under reduced pressure, to obtain semisolid material, which was then lyophilized to get the material in the form of powder (yield 14.9% w/w).

3.2. Sample Preparation

The sample was prepared by dissolving1 g of the extract powder of *T. dioica* fruits in 10 ml of distilled water at room temperature.

3.3. Assessment of Antioxidant Potential

Diabetes was induced by a single intraperitonial injection of freshly prepared Streptozotocin (purchased from Sigma Aldrich Chem. Co. USA.) 55 mg/kg bw in 0.1 m citrate buffer (pH 4.5) to a group of rats fasted overnight. After 3 days of STZ administration, depending on their glucose levels (BGL), the animals were marked as a severely diabetic group. A thirty-day long treatment with the dose of 1000 mg/kg bw of *T. dioica* extract powder was given to normal and STZ-induced diabetic rats. After thirty days of treatment, the rats were fasted overnight and sacrificed by cervical dislocation causing minimal pain. The different key organs were surgically removed, rinsed with isotonic ice-cold NaCl (0.9%, w/v) solution, blotted dry and weighed. A 10% (w/v) homogenate of each tissue was prepared in phosphate buffer (100 mM, pH 7.4) containing 150 mM KCl and centrifuged at $9,000 \times g$ for 30 min. The pellet was discarded and the cell-free supernatant was used for estimation of different oxidative stress parameters such as lipid peroxidation (LPO) and the activities of free radical scavenging enzymes such as catalase (CAT), glutathione peroxidase (GPx), superoxide dismutase (SOD), and glutathione-S-transferase (GST) using standard protocols (6–10). Data obtained was statistically evaluated using one-way ANOVA, followed by a post hoc Newman–Keuls Multiple Comparison Test. The values are considered significant at ($p < 0.05$).

3.4. Detection of Trace Elements

A schematic diagram of the experimental setup for recording the LIBS spectra is shown in Fig. 19.1. The LIBS spectra of *T. dioica* fruits extract powder, dissolved in distilled water, was recorded for identifying the presence of best set of elements responsible for its antioxidant potential. The 4-channel spectrometer equipped with CCD (Ocean optics LIBS 2000+) consisting 4-grating was used to get the dispersed light from the plasma. A pulsed laser

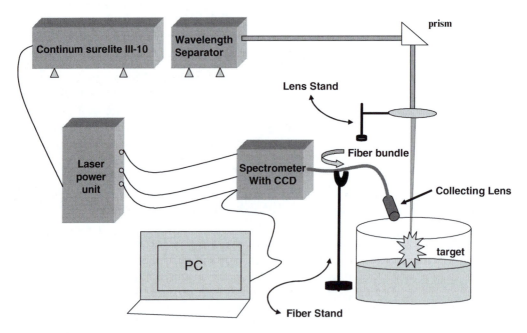

Fig. 19.1. LIBS experimental setup.

beam from a Q- switched Nd: YAG laser (Continuum Surellite III-10) was focused on the sample using a converging lens (Quartz) of 30 cm focal length, the temperature of the locally heated region rose rapidly and resulted in plasma formation on sample surface. The emitted light from micro-plasma was collected using an optical fiber tip placed in the vertical plane at 45° with respect to the laser beam and finally fed into an entrance slit of the multichannel spectrometer (Ocean Optics LIBS2000+) equipped with CCD and 4 gratings. The spectra presented in each Figs. 19.2 and 19.3 are the average of 100 scans (100 shots). The initial three gratings had the resolution of 0.1 nm covering the wavelength range from 200 to 310 nm, 310 to 400 nm and 400 to 510 nm, respectively, while the fourth grating, called as broad band grating, covered the wavelength range from 200 to 1100 nm and had a resolution of 0.75 nm. All the four gratings

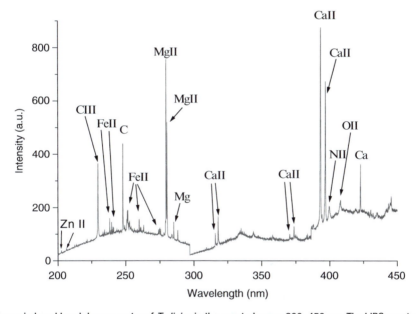

Fig. 19.2. Laser-induced breakdown spectra of *T. dioica* in the spectral range 200–450 nm. The LIBS spectra of the fruit extract of *T. Dioica* samples were recorded at 2 Hz repetition rate of and 175-mJ-laser energy using the experimental arrangement shown in Fig. 19.1. Nd:YAG (Continuum Surelite III-10) laser delivers maximum pulsed laser energy of 425 mJ at 532 nm, with pulse width 3–4 ns and repetition rate 10 Hz. The laser beam is focused by a plano-convex quartz lens of focal length 30 cm onto the sample surface, the temperature of the locally heated region rises rapidly and results in plasma formation. The light emission from microplasma is collected using optical fiber placed in the horizontal plane containing the laser beam (and at 45° with respect to laser beam) and fed to 4-channel spectrometer (Ocean Optics LIBS2000+) equipped with CCD. The spectrometer had initial three gratings with the resolution of 0.1 nm covering the wavelength range from 200 to 310 nm, 310–400 nm and 400–510 nm, respectively, while the fourth grating covered the wavelength range from 200 to 1100 nm and had a resolution of 0.75 nm. All the four gratings were used simultaneously to record the LIBS spectra. Each spectra presented in Figs. 19.2 and 19.3 is the average of 100 number of scans (100 shots). Figures 19.2 and 19.3 clearly revealed the presence of Mg, Fe, Na, K, Zn, Ca, H, O, C, and N elements in the spectral range λ 200 nm-900 nm.

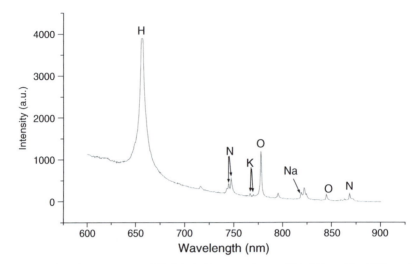

Fig. 19.3. Laser-induced breakdown spectra of *T. dioica* in the spectral range 200–900 nm. The LIBS spectra of the fruit extract of *T. dioica* samples were recorded at 2 Hz repetition rate of and 175-mJ-laser energy using the experimental arrangement shown in Fig. 19.1. Nd:YAG (Continuum Surelite III-10) laser delivers maximum pulsed laser energy of 425 mJ at 532 nm, with pulse width 3–4 ns and repetition rate 10 Hz. The laser beam is focused by a plano-convex quartz lens of focal length 30 cm onto the sample surface, the temperature of the locally heated region rises rapidly and results in plasma formation. The light emission from microplasma is collected using optical fiber placed in the horizontal plane containing the laser beam (and at 45° with respect to laser beam) and fed to 4-channel spectrometer (Ocean Optics LIBS2000+) equipped with CCD. The spectrometer have initial three gratings had the resolution of 0.1 nm covering the wavelength range from 200 to 310 nm, 310–400 nm and 400–510 nm, respectively, while the fourth grating, covered the wavelength range from 200 to 1100 nm and had a resolution of 0.75 nm. All the four gratings were used simultaneously to record the LIBS spectra. Each spectra presented in Figs. 19.2 and 19.3 is the average of 100 number of scans (100 shots). Figures 19.2 and 19.3 clearly revealed the presence of Mg, Fe, Na, K, Zn, Ca, H, O, C, and N elements in the spectral range λ 200 nm-900 nm.

were used simultaneously to record the LIBS spectra. LIBS spectra were recorded for aqueous extract of *T. dioica* at 2 Hz repetition rate of and 175-mJ-laser energy.

3.5. Results

3.5.1. Impact on Oxidative Stress Parameters

A significant increase was observed in the activities of free radical scavenging enzymes like SOD, CAT, GPx, and GST on long-term treatment with *T dioica* (Table 19.1). However, the diabetes-induced LPO recovered its normal values indicating thereby a high antidiabetic activity of *T. dioica* fruits.

3.5.2. Analysis of Antioxidative Elements

The spectra of *T. dioica* fruit extract, shown in Figs. 19.2 and 19.3 were taken at optimized experimental conditions (*see* Notes 4–6). It clearly revealed the presence of Mg, Fe, Na, K, Zn, Ca, H, O, C, and N elements in the spectral range λ 200–900 nm. According to the Boltzmann distribution law, intensity is directly related to concentration (11) therefore, the intensity of observed spectral lines corresponding to major and minor elements present in the extract speaks about their concentrations and helped in defining their role in diabetes-induced oxidative stress management.

Table 19.1

Effect of most effective dose of *Trichosanthes dioica* fruit aqueous extract on BGL and lipid profile of severely diabetic rats (mean ± SD)

Treatment groups	Treatment	Oxidative stress parameters in different tissues of rats				
		Liver	Kidney	Brain	Pancreas	Spleen
		Activity of superoxide dismutase (SOD), IU/mg protein				
NC	DW	12.3 ± 2.5	11.4 ± 1.8	9.5 ± 2.6	4.3 ± 2.8	3.2 ± 1.9
NT	TE	12.9 ± 3.1	10.9 ± 2.7	10.4 ± 2.5	4.5 ± 1.1	3.4 ± 0.9
DC	DW	4.2 ± 0.7	4.7 ± 0.9	3.5 ± 1.2	1.7 ± 1.0	2.2 ± 0.8
DT	TE	6.8 ± 1.5	5.9 ± 2.6*	4.7 ± 1.5**	2.5 ± 0.6	2.4 ± 0.5
		Activity of catalase (CAT), IU/mg protein				
NC	DW	7.32 ± 2.2	6.41 ± 1.7	5.29 ± 1.9	3.42 ± 1.1	2.56 ± 0.6
NT	TE	8.2 ± 2.6	7.12 ± 2.3	5.32 ± 1.4	3.67 ± 1.1	2.49 ± 0.7
DC	DW	4.2 ± 1.8	4.7 ± 1.3	3.5 ± 1.0	1.7 ± 0.5	2.2 ± 0.8
DT	TE	6.84 ± 2.5	5.98 ± 2.2*	4.72 ± 1.2*	2.51 ± 0.4*	2.43 ± 1.2**
		Activity of glutathione S transferase (GST), IU/mg protein				
NC	DW	1.52 ± 0.4	1.30 ± 0.8	0.91 ± 0.9	0.72 ± 0.4	0.59 ± 0.2
NT	TE	1.42 ± 0.3	1.51 ± 0.4	1.25 ± 0.2	0.68 ± 0.1	0.62 ± 0.3
DC	DW	0.92 ± 0.5	0.41 ± 0.2	0.35 ± 0.2	0.50 ± 0.3	0.24 ± 0.1
DT	TE	1.40 ± 0.8	0.63 ± 0.6*	0.56 ± 0.3*	0.62 ± 0.2*	0.31 ± 0.2***
		Activity of glutathione peroxides (GPx), IU/mg protein				
NC	DW	9.31 ± 2.2	7.96 ± 1.5	6.53 ± 1.7	4.27 ± 1.9	3.52 ± 1.1
NT	TE	9.9 ± 2.4	7.62 ± 2.1	6.84 ± 1.9	4.31 ± 1.3	3.47 ± 1.1
DC	DW	5.4 ± 2.8	5.2 ± 2.2	3.6 ± 2.6	2.3 ± 0.9	2.1 ± 0.6
DT	TE	8.84 ± 2.5	6.72 ± 1.9*	6.01 ± 1.7*	3.67 ± 1.6***	2.95 ± 0.8**
		Lipid peroxidation (LPO) in terms of nM of malondialdehyde (MDA) released/ mg protein				
NC	DW	1.43 ± 1.1	1.35 ± 0.4	0.89 ± 0.3	1.17 ± 0.6	1.20 ± 0.3
NT	TE	1.46 ± 0.5	1.24 ± 0.7	0.81 ± 0.6	1.02 ± 0.5	1.10 ± 0.4
DC	DW	1.98 ± 0.4	1.85 ± 0.8	1.62 ± 0.7	1.42 ± 0.9	1.68 ± 0.2
DT	TE	1.32 ± 0.2	1.14 ± 0.4*	1.12 ± 0.2*	1.13 ± 0.7**	1.21 ± 0.5***

Effects of *T. dioica* fruit extract treatment on the oxidative stress parameters in different tissues of rats. The treatment groups were normal control (NC), normal treated (NT), Diabetic control (DC), and diabetic treated (DT). The treatment regimen and doses have been discussed in materials and method section. DW and TE denote distilled water and *T. dioica* fruit extract respectively. The oxidative stress parameters were estimated in terms of activities of SOD, CAT, GST, GPX, and lipid peroxidation were determined. One international unit (IU) of enzyme activity has been defined as fifty percent inhibition of pyrogallol auto oxidation per min. One international unit (IU) of catalase activity was defined as micromoles of H_2O_2 decomposed per minute. One international unit (IU) of GST activity was defined as expressed as nanomoles of GSH–CDNB conjugate formed per minute whereas, glutathione peroxidase activity was expressed as μg of GSH consumed/min/mg protein. Specific activity of enzymes is expressed as activity (IU)/mg protein. Lipid peroxidation (LPO) has been expressed in terms of nM of malondialdehyde (MDA) produced/mg protein. Values are expressed as mean ± SD; $n = 6$, where n = number of determinations. ***Indicates values significantly different from DC at p 0.001, **means values are significantly different from DC at p 0.01 and *indicates value significantly different from DC at p 0.05.

Table 19.2
Intensity ratio of different elements with respect to Carbon (247.8 nm)

Elements	Ratio
Zn (202.5 nm)/C (247.8 nm)	0.02511
Zn (206.2 nm)/C (247.8 nm)	0.01868
C III (229.62 nm)/C (247.8 nm)	1.38261
C (247.8 nm)/C (247.8 nm)	1
Fe II (234.3 nm)/C (247.8 nm)	0.04655
Fe II (238.2 nm)/C (247.8 nm)	0.10883
Fe II (239.5 nm)/C (247.8 nm)	0.08793
Fe II (240.4 nm)/C (247.8 nm)	0.03884
Fe II (249.3 nm)/C (247.8 nm)	0.02752
Fe II (258.5 nm)/C (247.8 nm)	0.0392
Fe II (259.8 nm)/C (247.8 nm)	0.18889
Fe II (260.7 nm)/C (247.8 nm)	0.05208
Fe II (261.1 nm)/C (247.8 nm)	0.29323
Fe II (273.9 nm)/C (247.8 nm)	0.04585
Fe II (274.9 nm)/C (247.8 nm)	0.04928
Fe II (275.5 nm)/C (247.8 nm)	0.05908
Mg II (279.5 nm)/C (247.8 nm)	1.48591
Mg II (280.2 nm)/C (247.8 nm)	1.13031
Ca II (315.8 nm)/C (247.8 nm)	0.10521
Ca II (317.9 nm)/C (247.8 nm)	0.34345
Ca II (393.3 nm)/C (247.8 nm)	3.38716
Ca II (396.8 nm)/C (247.8 nm)	2.02546
Ca II (422.6 nm)/C (247.8 nm)	0.55465
Mg (285.2 nm)/C (247.8 nm)	0.16581
K (766.4 nm)/C (247.8 nm)	0.02045
K (769.9 nm)/C (247.8 nm)	0.01339

The ratio of intensities of detected elements (Mg, Fe, Na, K, Zn, Ca, C, H, O, and N) to the intensity of reference lines C and O, which were the essential constituents of plant materials, was estimated to evaluate their proportional concentration. Since

Table 19.3
Intensity ratio of different elements with respect to Oxygen (777.2 nm)

Elements	Ratio
O (777.2 nm)/O (777.2 nm)	1
O (844.6 nm)/O (777.2 nm)	0.07293
H (656.2 nm)/O (777.2 nm)	7.05396
Na (818.3 nm)/O (777.2 nm)	0.0419
Na (589.5 nm)/O (777.2 nm)	0.00796
N (744.2 nm)/O (777.2 nm)	0.05028
N (746.8 nm)/O (777.2 nm)	0.45277
N (868.3 nm)/O (777.2 nm)	0.06613

gratings of different resolutions were used, the whole spectra was divided in to two parts: the first, covered the wavelength range from 200 to 510 nm with 0.1 nm resolution and the second lied in the wavelength range 510–1100 nm with 0.75 nm resolution. To find the intensity ratios of spectral lines, the C line (247.88 nm) as reference line for the spectral range of λ 200–510 nm and O line (844.10 nm) as the reference line for spectral wavelength range of 500–1100 nm had been selected. Thus, the intensity ratios of Zn/C, Fe/C, Mg/C, Ca/C, K/O, Na/O, H/O, and N/O were calculated and are given in Tables 19.2 and 19.3. The higher concentration of Ca^{2+}, Mg^{2+}, and Fe^{2+} as reflected by their intensities are responsible for antioxidant potential of *T. dioica*. The salient features of the protocol are illustrated in Notes.

4. Notes

1. It has been observed that the elemental composition of medicinal plants changes with change in their biological activities (5), therefore, changes in spectral characteristics of herbal samples can be correlated with their medicinal properties.

2. In liquid phase, repetition rate also becomes an important parameter due to shock wave generation and splashing phenomenon.

3. In order to enhance the signal to background and signal to noise ratio and also to get reproducibility, average spectra should be preferred instead of single shot spectra.

4. LIBS spectra is sensitive to experimental protocols such as laser power, lens to sample distance, and position of emission collection optics with respect to plasma plume, which have to be optimized before recording.

5. The most suitable experimental protocols found for this study were laser energy – 175 mJ, lens to sample distance of 30 cm and tip of the fiber bundle at 45° with respect of laser beam.

6. Depending upon the sample the laser energy and laser frequency can be changed from 1 to 425 mJ and 0.001 to10 Hz, respectively.

7. Association of oxidative stress with deficiency of Ca^{2+}, Fe^{2+}, and Mg^{2+} has been validated by the present report, which in its turn confirms the antioxidant property of *T. dioica* fruits (12–15) (*see* Note 7).

Acknowledgements

The authors are grateful to DRDO, National Medicinal Plants board (NMPB) New Delhi, India, for providing the financial assistance. PKR and DKR are thankful to Indian Council of Medical Research (ICMR) and Council of Scientific and Industrial Research (CSIR) respectively, for the award of SRF to them. The drawing of the experimental setup by Mr. Vivek Kumar Singh, Department of Physics, University of Allahabad is acknowledged.

References

1. Pasquini C, Cortez J, Silva LMC, Gonzaga FB (2007) Laser induced breakdown spectroscopy. Braz Chem Soc 18:463–512

2. Rai AK, Rai VN, Rai DK, Thakur SN, Yueh FU, Singh JP (2007) Laser induced breakdown spectroscopy of solid and molten materials. Elesvier Science, B.V. Chemistry and Chemical Engineering, The Netherlands, pp 255–285

3. Rai PK, Jaiswal D, Diwaker S, Watal G (2008) Antihyperglycemic profile of *Trichosanthes dioica* seeds in experimental models. Pharm Biol 46(5):1–6

4. Rai PK, Rai NK, Rai AK, Watal G (2007) Role of LIBS in elemental analysis of *Psidium guajava* responsible for glycemic potential. Instrum Sci Tech 35:507–522

5. Kar A, Choudhary BK, Bandyopadhyay NG (2003) Comparative evaluation of hypoglycemic activity of some Indian medicinal plants in alloxan diabetic rats. J Ethnopharmacol 84:105–108

6. Beers RF, Sizer IW (1952) A spectrophotometric method for measuring the breakdown of hydrogen peroxide by catalase. J Biol Chem 195:133–140

7. Niehaus WG, Samuelsson B (1968) Formation of malondialdehyde from phospholipids arachidonate during microsomal lipid peroxidation. Eur J Biochem 6:126–130

8. Ellman GL (1959) A simple, sensitive and reliable method for determining free sulfhydryl content in peptides protein. Arch Biochem Biophys 82:70–77

9. Warholm M, Guthenberg C, Vonbahr C, Mannervik B (1985) Glutathione transferase from human liver. Meth Enzymol 113:499–504

10. Lowry OH, Rosebrough NJ, Farr AL, Randall RJ (1951) Protein measurement with Folin's phenol reagent. J Biol Chem 193:265–275

11. Sabsabi M, Cielo P (1995) Quantitative analysis of aluminum alloys by laser-induced breakdown spectroscopy and plasma characterization. Appl Spect 49:499–507

12. Giugliano M, Bove M, Grattaro M (2000) Insulin release at the molecular level: metabolic-electrophysiological modeling of the pancreatic beta cells. IEEE Transact Biomed Eng 47:611–623

13. Chiang WL, Hsiesh YS, Yang SF, Lu TA, Chu SC (2007) Differential expression of glutathion-S-transferase isoenzymes in various types of anemia in Taiwan. Clin Chem Acta 375:110–114

14. Touyz RM, Pu Q, He G, Chen X, Yao G, Neves MF, Viel E (2002) Effects of low dietary magnesium intake on development of hypertension in stroke-prone spontaneously hypertensive rats: role of reactive oxygen species. J Hypertens 20:2221–2232

15. Bussiere FI, Gueux E, Rock E, Girardeau JP, Tridon A, Mazur A, Rayssiguier Y (2002) Increased phagocytosis and production of reactive oxygen species by neutrophils during magnesium deficiency in rats and inhibition by high magnesium concentration. Brit J Nutr 87:107–113

Chapter 20

A Method for Evaluation of Antioxidant Activity Based on Inhibition of Free Radical-Induced Erythrocyte Hemolysis

Jun Takebayashi, Jianbin Chen, and Akihiro Tai

Abstract

There are many in vitro methods for evaluating antioxidant activity. In this chapter, we describe an operationally simple cell-based assay, oxidative hemolysis inhibition assay (OxHLIA). OxHLIA is based on inhibition of free radical-induced membrane damage in erythrocytes by antioxidants. The advantage of this method is that it uses peroxyl radicals as pro-oxidants and erythrocytes as oxidizable targets so that the results obtained reflect biologically relevant radical-scavenging activity and microlocalization of antioxidants. We also present here a comparison of OxHLIA with other common methods (DPPH, ABTS$^{\bullet+}$, and ORAC assays).

Key words: Antioxidant, Radical scavenger, Oxidative hemolysis, Peroxyl radicals, Cell-based antioxidant assay, Free radical-induced membrane damage, Erythrocytes

1. Introduction

Many in vitro methods have been developed for assessing antioxidant activities of pure compounds and food extracts (reviewed in refs. (1–4)). Of these, 1,1-diphenyl-2-picrylhydrazyl (DPPH) assay (5), 2,2′-azinobis(3-ethylbenzothiazoline-6-sulfonic acid) radical cation (ABTS$^{\bullet+}$) assay (6), and oxygen radical absorbance capacity (ORAC) assay (7) seem to have been widely used in recent years. DPPH assay and ABTS$^{\bullet+}$ assay (also known as Trolox equivalent antioxidant capacity (TEAC) assay) are based on direct scavenging of relatively stable radicals by antioxidants. The specific colors of DPPH radical and ABTS$^{\bullet+}$ disappear when these radicals are quenched, which enables monitoring of the radical-scavenging reaction spectrophotometrically. These methods are

D. Armstrong (ed.), *Advanced Protocols in Oxidative Stress II,* Methods in Molecular Biology, vol. 594
DOI 10.1007/978-1-60761-411-1_20, © Humana Press, a part of Springer Science+Business Media, LLC 2010

convenient, so they are very useful for the first screening test. However, since DPPH radical and ABTS$^{•+}$ do not exist in vivo, the results are not necessarily relevant to biological antioxidant activities. ORAC assay is based on inhibition of the free radical-induced oxidation of fluorescein. It utilizes a biologically relevant radical source, 2,2′-azobis(2-amidinopropane) dihydrochloride (AAPH)-derived peroxyl radicals. Thus, it is a more physiologically relevant assay than DPPH assay and ABTS$^{•+}$ assay, but the oxidizable target for peroxyl radicals, fluorescein, is not a mimic of a certain biomolecule. Therefore, it is better to ascertain biological efficacy of the sample concerned by in vivo or cell-based bioassays in early stages of investigation. In this chapter, we describe an operationally simple cell-based antioxidant assay that does not require special equipment, oxidative hemolysis inhibition assay (OxHLIA).

This antioxidant assay system based on inhibition of free radical-induced hemolysis of erythrocytes was originally developed by Niki et al. (8). The oxidation of erythrocyte membranes by AAPH-derived peroxyl radicals induces oxidation of lipid and proteins and eventually causes hemolysis, and this hemolysis can be inhibited by antioxidants (9). OxHLIA is a good experimental model for free radical-induced biomembrane damage and its inhibition by antioxidants. In many studies, erythrocytes were freshly prepared from blood samples for each assay, but we use preserved sheep erythrocytes to enhance the convenience of the method (*see* Note 1).

2. Materials

2.1. Equipment

1. Centrifuge (hematocrit centrifuge, table-top centrifuge, and microcentifuge).
2. Water bath shaker (such as Taitec Personal-11, Saitama, Japan).
3. Visible region absorbance spectrophotometer or microplate reader that can measure absorbance at exactly 524 nm (*see* Note 2).

2.2. Reagents

1. AAPH is obtained from Wako Pure Chemical (Osaka, Japan).
2. Milli-Q water (18 MΩ) obtained by a MilliQ system (Millipore, Bedford, MA) is used as water.
3. All other reagents including NaCl, Na_2HPO_4, and NaH_2PO_4 are special grade or better.

2.3. Buffer and Solution

1. Phosphate-buffered saline (PBS): 150 mM NaCl, 8.1 mM Na_2HPO_4, and 1.9 mM NaH_2PO_4.
2. AAPH solution (400 mM in PBS) (*see* Note 3).

2.4. Erythrocytes

Preserved sheep erythrocytes can be purchased from several animal vendors such as Nippon Bio-Supp (Tokyo, Japan) (*see* Note 4). To reduce the influence of individual differences in sheep (10), the use of pooled erythrocytes from several sheep (more than four) is recommended.

3. Methods

3.1. Preparation of Erythrocyte Suspension

1. Wash the erythrocytes three times with PBS (2,000 × *g* for 10 min).

2. Resuspend the washed erythrocytes with PBS at 10 times the final concentration. The concentration of the erythrocytes is determined by preliminary experiments to achieve good linearity in lysis phase of the hemolysis curve of the blank (in our experience, a final concentration of 0.5–1% (v/v) is optimal) (*see* Note 5).

3.2. OxHLIA

1. Prepare blank or sample solutions (1.25 times the final concentration, 12.0 mL, in PBS) in a flat-bottomed test tube (3.0 cm in diameter, 12.0 cm in depth) with a nonairtight cap (*see* Note 6). A small amount of organic solvents such as ethanol and dimethyl sulfoxide (DMSO) can be added if desired (*see* Note 7).

2. Add 1.5 mL of the erythrocyte suspension (*see* Subheading 20.3.1.2) and mix, and then incubate in a water bath at 37 °C with shaking for 10 min (*see* Note 8).

3. Add 1.5 mL of ice-cold AAPH solution (*see* Subheading 20.2.3, step 2) and mix, and then incubate in a water bath at 37 °C with shaking (*see* Note 8).

4. Withdraw an aliquot (~1 mL) of the reaction mixture into a 1.5 mL microcentrifuge tube at 15-min intervals and centrifuge (10,000 × *g* for 2 min).

5. Measure the absorbance of the supernatant at 524 nm (*see* Note 9) using a spectrophotometer. In the case of using a microplate reader, transfer exactly 200 μL of the supernatant to a flat-bottomed 96-well plate (*see* Note 10) and read the absorbance at 524 nm (*see* Note 2).

6. Calculate the degree of hemolysis using the following equation:

$$\text{Hemolysis}(\%) = \frac{\text{Absorbance of supernatant}}{\text{Absorbance of complete hemolysis}} \times 100$$

Absorbance of complete hemolysis is obtained from the supernatant by adding 900 μL of water to 100 μL of erythrocyte suspension (*see* Subheading 20.3.1.2).

3.3. Results

3.3.1. Typical Results and Data Presentation

Typical hemolysis curves with or without Trolox, a commonly used reference antioxidant, are shown in Fig. 20.1. As the concentration of Trolox is increased, the hemolysis is inhibited more strongly, and the hemolysis curves are sequentially shifted rightward. The results are expressed as delayed time of hemolysis (ΔT), which is calculated by

$$\Delta T \, (\min) = HT_{50} \, (\text{sample}) - HT_{50} \, (\text{blank}),$$

where HT_{50} is the 50% hemolysis time (min) graphically obtained from the hemolysis curve (*see* Fig. 20.2). The typical relationship between the concentration of Trolox and ΔT value is shown in

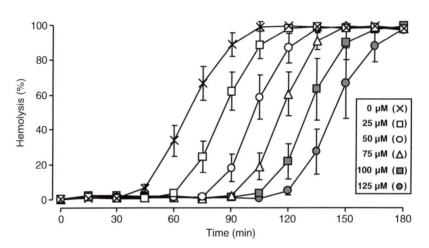

Fig. 20.1. Typical hemolysis curves of the blank and Trolox at various concentrations. Sheep erythrocytes suspended at a concentration of 0.7% (v/v) in PBS were incubated with 40 mM of AAPH in the presence of Trolox (0, 25, 50, 75, 100, and 125 µM) at 37 °C with shaking. The value at 0 time was obtained without AAPH. Each value is the mean ± S.E. of three separate experiments.

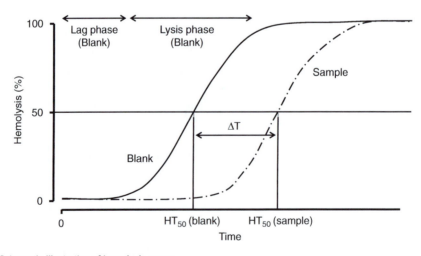

Fig. 20.2. Schematic illustration of hemolysis curve.

Fig. 20.3. Good linearity was observed within the range of 25–400 μM ($R^2 = 0.998$). The ΔT values of several common antioxidants at the concentration of 50 μM are shown in Table 20.1 (*see* Note 11).

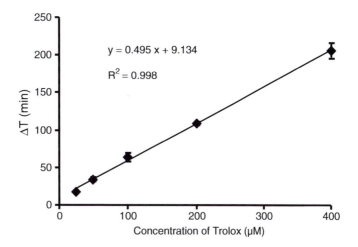

$$y = 0.495\ x + 9.134$$

$$R^2 = 0.998$$

Fig. 20.3. Typical relationship between concentration of Trolox and ΔT value. Sheep erythrocytes suspended at a concentration of 0.7% (v/v) in PBS were incubated with 40 mM of AAPH in the presence of Trolox (0, 25, 50, 100, 200, and 400 μM) at 37 °C with shaking. The average HT$_{50}$ (blank) in this experiment was 68 ± 6 min (*n* = 4). Each value is the mean ± S.D. of three or four separate experiments

Table 20.1
ΔT values for several common antioxidants

	Compounds	ΔT (min)	
		PBS	1% ethanol/99% PBS
Reference antioxidants	Trolox	28 ± 3	30 ± 4
Antioxidant vitamins	Ascorbic acid	15 ± 4	–
	α-Tocopherol	–	71 ± 11
Plasma antioxidants	Uric acid	28 ± 2	–
	Glutathione	11 ± 1	–
Polyphenols	Hesperetin	125 ± 8	–
	Catechin	89 ± 4	–
Pharmaceuticals	Edaravone[a]	96 ± 6	–

Sheep erythrocytes suspended at a final concentration of 0.7% (v/v) in PBS or in 1% ethanol/99% PBS were incubated with 40 mM AAPH in the presence of 50 μM antioxidants at 37 °C with shaking. The average HT$_{50}$ (blank) in PBS and in 1% ethanol/99% PBS was 73 ± 2 min (*n* = 4) and 88 ± 8 min (*n* = 3), respectively (*see* Note 7). Mean ± S.D. values of three or four separate experiments are shown

[a]Edaravone (MCI-186, 3-methyl-1-phenyl-2-pyrazolin-5-one) is a free-radical scavenger, which has already been clinically used to reduce the neuronal damage following ischemic stroke in Japan (14)

3.3.2. Comparison of OxHLIA with Other Commonly Used Antioxidant Assays and Data Interpretation

The antioxidants shown in Table 20.1 were assessed by the DPPH assay (Fig. 20.4a), ABTS$^{\bullet+}$ assay (Fig. 20.4b), and ORAC assay (Fig. 20.4c). Some important implications for data interpretation are as follows:

1. Antioxidants that cannot rapidly scavenge DPPH radical or ABTS$^{\bullet+}$ are not always weak antioxidants in OxHLIA (*see* hesperetin in DPPH assay (Fig. 20.4a)). In addition, antioxidants that show similar characteristics in the DPPH assay or ABTS$^{\bullet+}$ assay do not always show the same activities in OxHLIA (*see* Trolox, ascorbic acid, α-tocopherol, uric acid, and edaravone).

2. The ORAC assay and OxHLIA use the same radical source, AAPH-derived peroxyl radicals, and the results obtained by using these assays are therefore correlated to some extent. However, in the case of α-tocopherol and Trolox, the orders of activities shown by these assays are opposite. Trolox is a water-soluble vitamin E analog that retains the chemical structure of α-tocopherol essential for radical-scavenging reactions. The ORAC assay and OxHLIA utilize the same "hydrophilic" peroxyl radicals but different oxidizable targets, i.e., "hydrophilic" fluorescein or "lipophilic" biomembrane of erythrocytes. OxHLIA reflects the microlocalization of each antioxidant, and its results agree with the idea that lipophilic antioxidants are superior to hydrophilic antioxidants in protection against free radical-induced membrane damage.

3. ΔT in OxHLIA is a relative value, not an absolute value. For example, we found that ascorbic acid is a "relatively" weak antioxidant using intact erythrocytes but that it becomes a "relatively" strong antioxidant using preoxidized erythrocytes (10). This is an extreme case, but experimental conditions, especially states of erythrocytes, affect the results to some extent and may even alter the order of activity. Thus, it is important to assess the antioxidant activity in comparison with multiple reference antioxidants.

4. Samples containing surfactants may facilitate hemolysis, as if they act as pro-oxidants. In some cases, evaluation of antioxidants with detergent properties requires care.

5. The OxHLIA is a good assay for estimating "biologically relevant" antioxidant activity. However, it is not versatile, because free radical-induced membrane damage is important but is only one kind of oxidative damage in vivo. Thus, as many investigators have pointed out (1–4, 11, 12), it is essential to evaluate antioxidant activities using a combination of several methods (*see* Note 12). In addition, absorption, distribution, metabolism, and excretion (ADME) of antioxidants greatly affect antioxidant activity in vivo. Therefore, antioxidant efficacy continues to be assessed by in vivo assay systems.

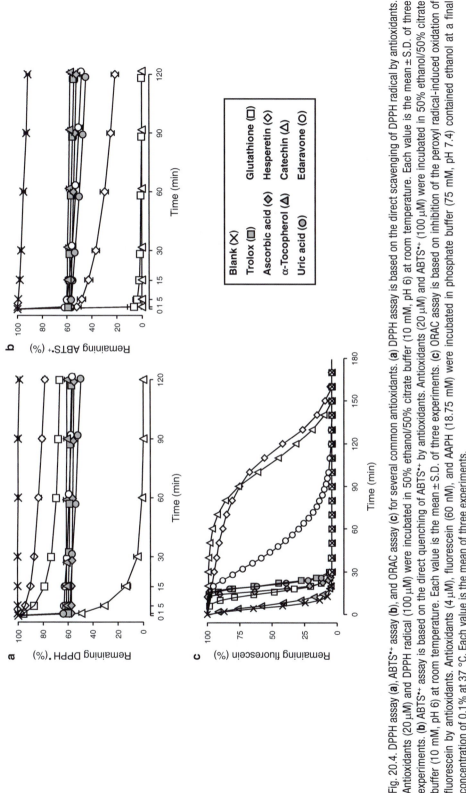

Fig. 20.4. DPPH assay (**a**), ABTS•+ assay (**b**), and ORAC assay (**c**) for several common antioxidants. (**a**) DPPH assay is based on the direct scavenging of DPPH radical by antioxidants. Antioxidants (20 μM) and DPPH radical (100 μM) were incubated in 50% ethanol/50% citrate buffer (10 mM, pH 6) at room temperature. Each value is the mean ± S.D. of three experiments. (**b**) ABTS•+ assay is based on the direct quenching of ABTS•+ by antioxidants. Antioxidants (20 μM) and ABTS•+ (100 μM) were incubated in 50% ethanol/50% citrate buffer (10 mM, pH 6) at room temperature. Each value is the mean ± S.D. of three experiments. (**c**) ORAC assay is based on inhibition of the peroxyl radical-induced oxidation of fluorescein by antioxidants. Antioxidants (4 μM), fluorescein (60 nM), and AAPH (18.75 mM) were incubated in phosphate buffer (75 mM, pH 7.4) contained ethanol at a final concentration of 0.1% at 37 °C. Each value is the mean of three experiments.

4. Notes

1. Erythrocytes from humans or from animals other than sheep, such as rats, rabbits, cattle, and horses, are also used for OxHLIA. Because preserved sheep erythrocytes of relatively consistent quality are commercially available at low cost throughout the year, we use erythrocytes from sheep.

2. In OxHLIA, the absorbance must be measured at exactly 524 nm (*see* Note 9). Therefore, a microplate reader equipped with a tunable monochromator (such as Powerscan HT, DS Pharma Biomedical, Osaka, Japan) may be required to read the absorbance at 524 nm.

3. Since AAPH continuously generates peroxyl radicals by thermal decomposition (13), AAPH solution must be fleshly prepared in PBS prior to use and kept on ice.

4. Preserved sheep erythrocytes can be used for a few weeks if stored aseptically at 4 °C.

5. Decrease in erythrocyte concentration steepens the slope of lysis curve but does not alter duration of the lag phase. Susceptibility of erythrocytes usually differs between lots to some extent. Thus, experimental conditions must be optimized at the onset of one series of experiments and should not be changed thereafter.

6. Since AAPH generates peroxy radicals in the presence of oxygen in the atmosphere (13), an airtight cap should not be used. In addition, variations in test tubes, especially differences in diameter, affect the results. We found that a flat-bottomed test tube (3.0 cm in diameter and 12.0 cm in depth, originally used for plant tissue culture) can improve miscibility and thus reproducibility compared to a round-bottomed test tube (1.5 cm in diameter and 15.0 cm in depth).

7. We have confirmed that ethanol (final concentration of less than 1%) and DMSO (final concentration of less than 5%) can be used. However, these solvents delayed hemolysis (*see* the legend to Table 20.1), so appropriate solvent blanks must be taken. Furthermore, changes in the concentration of organic solvents may affect the activities of lipophilic antioxidants, because localization of lipid-soluble antioxidants to erythrocyte membranes might be altered.

8. Subheading 20.3.2.2 is the pre-incubation step, and the time when AAPH solution is added is the zero time (Subheading 20.3.2.3). The incubation time after adding AAPH solution depends on the antioxidant ability of tested samples (*see* Note 11). It is important not to precipitate erythrocytes throughout the assay. Occasional mixes by hand may be required.

9. In general, degree of hemolysis is determined from the absorbance of the supernatant after centrifugation at 540–544 nm, which is the maximum absorption wavelength of oxyhemoglobin. However, in OxHLIA, AAPH-derived peroxyl radicals oxidize oxyhemoglobin to methemoglobin, which changes the absorption spectrum of hemoglobin (10). To determine the exact degree of hemolysis, we measured the absorbance at 524 nm, which is not affected by changes in the absorption spectrum (10).

10. It is important to dispense exactly $200\,\mu L$ of the supernatants for accurate measurement, since the path length in the 96-well plate is dependent on the dispense volume. The supernatants are particularly viscous after hemolysis, so we strongly recommend dispensing the supernatants using microsyringes (such as a $250\,\mu L$ Hamilton syringe, model 725SNR, Hamilton, Reno, NV) rather than micropipettes.

11. When HT_{50} is roughly expected, this value can be determined only by measuring the degree of hemolysis in the lysis phase (in one example, 45, 60, 75, 90, 105, 120, 135 and 150 min for blank and weak antioxidants, and 165, 180, 195, 210, 225, 240, 255 and 270 min for strong antioxidants).

12. Because there are a variety of oxidative stresses, the kind of "effective" antioxidant depends on the situation. Therefore, there seems to be no assay that is sufficient alone for estimating overall antioxidant activity, even in vivo assays. A combination of several methods based on different principles should therefore be used.

Acknowledgments

This work was partly supported by a Grant-in-Aid for Young Scientists (B) (No. 17780103) from the Ministry of Education, Culture, Sports, Science, and Technology of Japan.

References

1. Aruoma OI (2003) Methodological considerations for characterizing potential antioxidant actions of bioactive components in plant foods. Mutat Res 523–524:9–20

2. Prior RL, Wu X, Schaich K (2005) Standardized methods for the determination of antioxidant capacity and phenolics in foods and dietary supplements. J Agric Food Chem 53:4290–4302

3. Pérez-Jiménez J, Arranz S, Tabernero M, Díaz-Rubio ME, Serrano J, Goñi I, Saura-Calixto F (2008) Updated methodology to determine antioxidant capacity in plant foods, oils and beverages: extraction, measurement and expression of results. Food Res Intern 41:274–285

4. Magalhães LM, Segundo MA, Reis S, Lima JL (2008) Methodological aspects about in vitro evaluation of antioxidant properties. Anal Chim Acta 613:1–19

5. Brand-Williams W, Cuvelier ME, Berset C (1995) Use of a free radical method to evaluate

antioxidant activity. Lebenson Wiss Technol 28:25–30

6. Re R, Pellegrini N, Proteggente A, Pannala A, Yang M, Rice-Evans C (1999) Antioxidant activity applying an improved ABTS radical cation decolorization assay. Free Radic Biol Med 26:1231–1237

7. Prior RL, Hoang H, Gu L, Wu X, Bacchiocca M, Howard L, Hampsch-Woodill M, Huang D, Ou B, Jacob R (2003) Assays for hydrophilic and lipophilic antioxidant capacity (oxygen radical absorbance capacity ($ORAC_{FL}$)) of plasma and other biological and food samples. J Agric Food Chem 51:3273–3279

8. Niki E, Komuro E, Takahashi M, Urano S, Ito E, Terao K (1988) Oxidative hemolysis of erythrocytes and its inhibition by free radical scavengers. J Biol Chem 263:19809–19814

9. Sato Y, Kamo S, Takahashi T, Suzuki Y (1995) Mechanism of free radical-induced hemolysis of human erythrocytes: hemolysis by water-soluble radical initiator. Biochemistry 34:8940–8949

10. Takebayashi J, Kaji H, Ichiyama K, Makino K, Gohda E, Yamamoto I, Tai A (2007) Inhibition of free radical-induced erythrocyte hemolysis by 2-O-substituted ascorbic acid derivatives. Free Radic Biol Med 43:1156–1164

11. Niki E, Noguchi N (2000) Evaluation of antioxidant capacity. What capacity is being measured by which method? IUBMB Life 2000:323–329

12. Frankel EN, Meyer AS (2000) The problems of using one-dimensional methods to evaluate multifunctional food and biological antioxidants. J Sci Food Agric 80:1925–1941

13. Niki E (1990) Free radical initiators as source of water- or lipid-soluble peroxyl radicals. Methods Enzymol 186:100–108

14. Yoshida H, Yanai H, Namiki Y, Fukatsu-Sasaki K, Furutani N, Tada N (2006) Neuroprotective effects of edaravone: a novel free radical scavenger in cerebrovascular injury. CNS Drug Rev 12:9–20

Chapter 21

Design and Synthesis of Antioxidant α-Lipoic Acid Hybrids

Maria Koufaki and Anastasia Detsi

Abstract

The design and synthesis of hybrid molecules encompassing two pharmacophores in one molecular scaffold is a well-established approach to the synthesis of more potent drugs with dual activity. In this chapter, we will present the most important synthetic methodologies we have applied for the preparation of hybrid compounds containing the "universal antioxidant" α-lipoic acid. Experimental details for the synthesis and purification techniques of specific examples of molecules will be given. The synthesized molecules combine antioxidant activity with a variety of other biological activities such as protection against reperfusion arrhythmias, neuroprotective, and anti-inflammatory activity.

Key words: α-Lipoic acid, Hybrid molecules, Antioxidants, Synthetic methodology

1. Introduction

α-Lipoic acid (LA) is a naturally occurring nonenzymatic antioxidant, referred to as the "universal antioxidant." It exhibits its antioxidant activity not only by direct radical trapping and/or metal chelation, but also by regenerating other antioxidants (such as vitamins E and C and glutathione). Once taken up by cells, LA is reduced to dihydrolipoic acid (DHLA) in various tissues. DHLA retains the powerful antioxidant activity, although it has also pro-oxidant properties. LA/DHLA redox couple is a unique example of molecules acting as antioxidants in both their oxidized and reduced form (1–3).

The cellular redox imbalance caused by excessive production of Reactive Oxygen Species (ROS), which cannot be quenched by the normal defensive mechanisms of the organism, is called oxidative stress. This imbalance is an important feature of a plethora of pathophysiological conditions and is responsible for

D. Armstrong (ed.), *Advanced Protocols in Oxidative Stress II,* Methods in Molecular Biology, vol. 594
DOI 10.1007/978-1-60761-411-1_21, © Humana Press, a part of Springer Science+Business Media, LLC 2010

multicellular and organ damages. The beneficial effects of lipoic acid have been reported in numerous studies implicating oxidative stress, such as ischemia-reperfusion injury, neurodegenaration, inflammation, diabetes, and radiation injury (4–8).

The design and synthesis of hybrid molecules encompassing two pharmacophores in one molecular scaffold is a well-established approach to the synthesis of more potent drugs with dual activity (9). Using this approach, various research groups have recently designed and synthesized hybrid compounds by covalently bonding lipoic acid to other bioactive molecules. These efforts resulted in new molecules with antioxidant activity hyphenated with a wide variety of other activities such as protection against reperfusion arrhythmias (10, 11), nitric oxide synthase inhibition (12), erythrocyte protection from hemolysis (13), antiproliferative activity (14), acetylcholinesterase inhibition (15), EGFR inhibition (16), radioprotection (17), neuroprotective activity (18), anti-inflammatory activity (19), and inhibition of butyrylcholinesterase (20).

In this report, we will focus on our contribution to the growing field of the design and synthesis of lipoic acid hybrids by presenting part of the methodology we have developed for the synthesis of novel molecules with a dual mode of action. Our approach involves appropriate functionalization of lipoic acid (activated as lipoyl chloride or N-hydroxysuccinimidyl-ester) and subsequent coupling with the second pharmacophore. In this context, we have prepared three categories of hybrid molecules: (A) trolox–lipoic acid and (B) quinolinone–lipoic acid conjugates, in which the two molecules are connected through amine spacers, and (C) 1,2-dithiolane-3-alkyl analogs containing protected or free catechol moieties connected through heteroaromatic rings such as triazole, 1,2,4-oxadiazole, 1,3,4-oxadiazole, tetrazole or thiazole in order to explore the influence of the bioisosteric replacement of the amide group on the neuroprotective activity of the lipoic acid/dopamine conjugate (Fig. 21.1).

Compounds of category (A) are potent antioxidants in microsomal lipid peroxidation and totally suppress reperfusion arrhythmias during reoxygenation (10, 11). Quinolinone–lipoic acid hybrids (category B) combine antioxidant and anti-inflammatory activity (19) and, finally, the molecules of category (C) are strong neuroprotective agents (18).

2. Materials

2.1. Equipment

The described reactions and purification protocols have been performed using standard equipment available in synthetic organic chemistry laboratories.

Category A: Trolox-α-lipoic acid hybrids

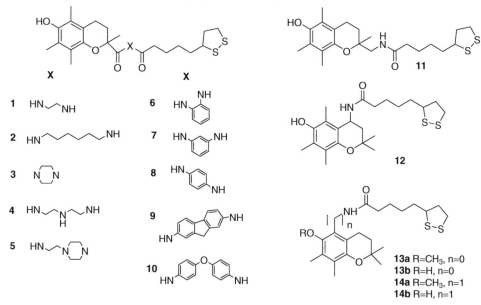

X

1 HN⌒NH   6 HN⌒NH (NH)
2 HN⌒⌒⌒NH   7 HN⌒⌒NH
3 N⌒N (piperazine) 8 HN⌒⌒NH
4 HN⌒N(H)⌒NH   9
5 HN⌒N⌒N   10

Category B: Quinolinone-α-lipoic acid hybrids

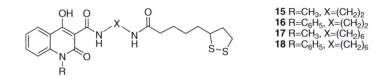

15 R=CH₃, X=(CH₂)₂
16 R=C₆H₅, X=(CH₂)₂
17 R=CH₃, X=(CH₂)₆
18 R=C₆H₅, X=(CH₂)₆

Category C: 1,2-dithiolane-catechol hybrids

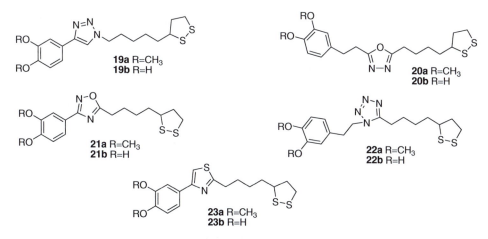

19a R=CH₃
19b R=H
20a R=CH₃
20b R=H
21a R=CH₃
21b R=H
22a R=CH₃
22b R=H
23a R=CH₃
23b R=H

Fig. 21.1. Categories of lipoic acid hybrid molecules described in this chapter.

2.2. Reagents

1. All the reagents used for the synthesis of the new molecules were commercially available and of analytical grade. The reagents were purchased from Sigma-Aldrich, Fluka or ACROS Organics.

2. The racemic form of *a*-lipoic acid (*a*-dl-lipoic acid) was used in all reactions and was purchased from FLUKA.

3. All organic solvents were of analytical grade and were obtained from LABSCAN or Fisher Scientific. Solvents were dried prior to use by literature reported methods.

4. Starting materials were prepared according to procedures developed in our laboratories (*see* corresponding references).

5. Silica gel plates Macherey–Nagel Sil G-25 UV$_{254}$ were used for thin layer chromatography. Chromatographic purification was performed with silica gel (200–400 mesh) obtained from Macherey–Nagel.

3. Methods

3.1. Synthesis of α-Lipoyl Chloride

Lipoyl chloride was prepared by treatment of sodium lipoate with oxalyl chloride in benzene, following a literature procedure (21) (*see* Note 1). Sodium lipoate was synthesized by adding lipoic acid (1 eq) to a mixture of isopropanol, NaOH (1.5 eq) and drops of H$_2$O and stirring the mixture at 50 C for 1 h. Evaporation of the solvent and trituration with acetone afforded the pure sodium lipoate as a yellow solid (yield 95%).

3.2. Synthesis of N-(Lipoyloxy) succinimide

To a solution of lipoic acid (0.5 g, 2 mmol) and *N*-hydroxysuccin-imide (0.23 g, 2 mmol) in 4 mL dichloromethane, cooled at 0 C (ice-water bath), was added dropwise a solution of dicyclohexyl-carbodiimide (0.5 g, 2.5 mmol) in 3 mL dichloromethane, over a period of 30 min. The resulting mixture was kept in the refrigera-tor overnight in order to ensure full precipitation of dicyclohexy-lurea (DCU). DCU was filtered and the filtrate was evaporated in vacuo. The residue was triturated with diethyl ether and a yellow-ish solid was formed, which was filtered and washed with ether to give the pure *N*-(lipoyloxy)succinimide in very good yield (98%) (*see* Note 2).

3.3. Synthesis of 5-(1,2-Dithiolan-3-yl) pentan-1-ol

To a solution of lipoic acid (0.200 g, 0.97 mmol) in 4 mL dry CH$_2$Cl$_2$, 1 M catecholborane in tetrahydrofuran (4.85 mmol) was added dropwise. The mixture was refluxed (70–80 C) for about 8 h. Cold water was then added dropwise and the organic solvent was evaporated in vacuo. The resulting mixture was extracted with CH$_2$Cl$_2$ and H$_2$O followed by 5-6 aliquots of 1 M NaOH to remove the catechol. The organic portion was washed with saturated aqueous NaCl, dried over sodium sulfate and the solvent was evaporated in vacuo. Yield: 0.090 g (48 %), yellow viscous oil.

3.4. Synthesis of 5-(1,2-Dithiolan-3-yl) pentyl Methanesulfonate

To a solution of 5-(1,2-dithiolan-3-yl)pentan-1-ol (0.040 g, 0.21 mmol) in 3 mL dry CH$_2$Cl$_2$ and 0.5 mL pyridine, CH$_3$SO$_2$Cl (0.03 mL, 0.42 mmol) was added at 0 C. After the addition, the mixture was stirred at ambient temperature for 3 h. The resulting mixture was extracted with ethyl acetate and water. The organic portion was washed with saturated aqueous NaCl, dried over sodium sulfate and the solvent was evaporated in vacuo. Yield: 0.040 g (70 %), yellow viscous oil.

3.5. Synthesis of 3-(5-Azidopentyl)-1,2-dithiolane

To a solution of 5-(1,2-dithiolan-3-yl)pentyl methanesulfonate (0.040 g, 0.15 mmol) in 2 mL dry DMF, sodium azide (1.5 mmol) was added and the mixture was heated at 60 C overnight. After the completion of the reaction, the mixture was extracted with ethyl acetate and water. The organic portion was washed with saturated aqueous NaCl, dried over sodium sulfate, filtered and the solvent was evaporated in vacuo (*see* Note 3). Yield: 0.030 g (88%), yellow viscous oil.

3.6. Synthesis of Lipothioamide

To a solution of lipoamide (0.100 g, 0.49 mmol) in 3 mL dry THF was added Lawesson's reagent (0.197 g, 0.49 mmol). The resulting mixture was refluxed for 2 h and after evaporating the solvent, the residue was purified by flash-column chromatography (CH$_2$Cl$_2$/CH$_3$OH 95:5). Yield: 0.100 g (92 %), white waxy solid.

3.7. Synthesis of Compounds of Categories A and B

3.7.1. Coupling of Lipoic Acid Chloride to Trolox Aminoamides (10, 11)

To a mixture of THF (3 mL) and H$_2$O (2 mL) were sequentially added the appropriate aminoamide (0.3 mmol), lipoic acid chloride (135 mg, 0.6 mmol), and NaHCO$_3$ (100 mg, 1.2 mmol) (*see* Note 4); the mixture was stirred at ambient temperature for 2 h. The reaction mixture was diluted with H$_2$O and extracted with EtOAc. The organic extracts were combined, washed with saturated aqueous NaCl, and dried (Na$_2$SO$_4$), and the solvent was evaporated in vacuo. The products were purified using flash column chromatography. The following compounds were obtained according to this procedure N-(3,4-Dihydro-6-hydroxy-2,5,7,8-tetramethyl-2H-1-benzopyran-2-carbonyl)-N'-(1,2-dithiolane-3-pentanoyl)-piperazine (3); N-(2-{4[5-(1,2-Dithiolan-3-yl)pentanoyl] piperazino-ethyl)}-6-hydroxy-2,5,7,8-tetramethyl-2-chromanecarboxamide (5); N-(3,4-Dihydro-6-hydroxy-2,5,7,8-tetramethyl-2H-1-benzopyran-2-carbonyl)-N'-(1,2-dithiolane-3-pentanoyl)-1,2-phenylenediamine (6); N-(3,4-Dihydro-6-hydroxy-2,5,7,8-tetramethyl-2H-1-benzopyran-2-carbonyl)-N'-(1,2-dithiolane-3-pentanoyl)-1,3-phenylenediamine (7); N-(3,4-Dihydro-6-hydroxy-2,5,7,8-tetramethyl-2H-1-benzopyran-2-carbonyl)-N'-(1,2-dithiolane-3-pentanoyl)-1,4-phenylenediamine (8); N-(3,4-Dihydro-6-hydroxy-2,5,7,8-tetramethyl-2H-1-benzopyran-2-carbonyl)-N'-(1,2-dithiolane-3-pentanoyl)-2,7-diaminofluorene (9); N-(3,4-Dihydro-6-hydroxy-2,5,7,8-tetramethyl

-2*H*-1-benzopyran-2-carbonyl)-*N*'-(1,2-dithiolane-3-pentanoyl)-4,4'-oxydianiline (10); 5-(1,2-Dithiolan-3-yl)-N-(6-hydroxy-2,2,5,7,8-pentamethyl-3,4-dihydro-2H-chromen-4-yl) pentanamide (12); 5-(1,2-Dithiolan-3-yl)-N-(6-methoxy-2,2, 5,7,8-pentamethyl-3,4-dihydro-2H-chromen-5-yl)pentanamide (13a).

3.7.2. Coupling of N-(Lipoyloxy)succinimide to Trolox Aminoamides, (6-Hydroxy-2,5,7,8-tetramethyl-3,4-dihydro-2H-benzopyran-2-yl) methylamine and Quinolinone Carboxamides (11, 19)

The appropriate amides (1 eq) were mixed with *N*-(lipoyloxy) succinimide (1 eq) in dichloromethane and the mixture was stirred and light-protected overnight. CH_2Cl_2 and saturated aqueous Na_2CO_3 were added. The organic layer was washed with saturated aqueous NaCl, dried with Na_2SO_4, and the solvent was evaporated in vacuo. The crude products were purified by flash column chromatography (*see* Note 5). The following compounds were obtained according this procedure: *N*-(3,4-Dihydro-6-hydroxy-2,5,7,8-tetramethyl-2*H*-1-benzopyran-2-carbonyl)-*N*'-(1,2-dithiolane-3-pentanoyl)-ethylene-diamine (1); *N*-(3,4-Dihydro-6-hydroxy-2,5,7,8-tetramethyl-2*H*-1-benzopyran-2-carbonyl)-*N*'-(1,2-dithiolane-3-pentanoyl)-hexamethylenediamine (2); *N*-{2-[(2-{[5-(1,2-Dithiolan-3-yl)pentanoyl]-amino}ethyl)amino] ethyl}-6-hydroxy-2,5,7,8-tetra-methyl-2-chromanecarboxamide (4); 5-(1,2-Dithiolan-3-yl)-N-[(6-hydroxy-2,5,7,8-tetramethyl-3,4-dihydro-2H-chromen-2-yl)methyl]pentanamide (11); 5-(1,2-Dithiolan-3-yl)-N-[(6-methoxy-2,2,7,8-tetramethyl-3,4-dihydro-2H-chromen-5-yl)methyl]-pentanamide (14a)-N-(1,2-Dihydro-4-hydroxy-1-methyl-2-oxo-3-quinoline-carbonyl)-*N*'-(1,2-dithiolane-3-pentanoyl)-ethylenediamine(15); N-(1,2-Dihydro-4-hydroxy-2-oxo-1-phenyl-3-quinolinecarbonyl)-*N*'-(1,2-dithiolane-3-pentanoyl)-ethylenediamine (16); N-(1,2-Dihydro-4-hydroxy-1-methyl-2-oxo-3-quinolinecarbonyl)-*N*'-(1,2-dithiolane-3-pentanoyl)-hexamethylenediamine (17); N-(1,2-Dihydro-4-hydroxy-2-oxo-1-phenyl-3-quinolinecarbonyl)-*N*'-(1,2-dithiolane-3-pentanoyl)-hexamethylenediamine (18).

3.7.3. Synthesis of Compounds of Category C: 1,2-Dithiolane-Catechol Hybrids (18)

1. *1-(5-(1,2-dithiolan-3-yl)pentyl)-4-(3,4-dimethoxyphenyl)-1H-1,2,3-triazole (19a)*. To a solution of 3-(5-azidopentyl)-1,2-dithiolane (0.040 g, 0.17 mmol) in 2.5 mL t-BuOH/H_2O 4:1, 3,4-dimethoxyphenylacetylene (0.027 g, 0.17 mmol), $CuSO_4 \cdot 5H_2O$ (0.009 mmol) and sodium ascorbate (0.017 mmol) were added and the mixture was stirred at ambient temperature overnight. The crude product was taken up with ethyl acetate and water. The organic layer was washed with saturated aqueous NaCl, dried over sodium sulfate and the solvent was evaporated in vacuo. The residue was purified by flash-column chromatography (Petroleum ether/Ethyl acetate 50:50). Yield: 0.020 g (33 %), yellow viscous oil.

2. *N'-(3-(3,4-dimethoxyphenyl)propanoyl)-5-(1,2-dithiolan-3-yl) pentane hydrazide.* To a solution of 3-(3,4-dimethoxyphenyl)

propanehydrazide (0.050 g, 0.22 mmol) in 3 mL dry THF was added dropwise a solution of lipoic acid activated N-hydroxysuccinimide (0.068 g, 0.22 mmol) in 3 mL dry THF and the mixture was stirred at ambient temperature for 2 days. The reaction was quenched with saturated aqueous NaHCO₃ and extracted with CH₂Cl₂. The organic layer was washed with saturated aqueous NaCl, dried, and the solvent was evaporated in vacuo. The crude product was purified by flash-column chromatography (CH₂Cl₂/CH₃OH 95:5). Yield: 0.060 g (67 %), beige solid.

3. *2-(4-(1,2-dithiolan-3-yl)butyl)-5-(3,4-dimethoxyphenethyl)-1,3,4-oxadiazole (20a)*. To 40 mg of N'-(3-(3,4-dimethoxy-phenyl)propanoyl)-5-(1,2-dithiolan-3-yl)-pentanehydrazide was added dry POCl₃ (0.5 mL) and the mixture was refluxed for 3 h. The reaction was quenched with cold water and extracted with CH₂Cl₂. The organic layer was washed with saturated aqueous NaCl, dried and the solvent was evaporated in vacuo. The crude product was purified by flash-column chromatography (CH₂Cl₂/CH₃OH 97:3). Yield: 0.023 g (59 %), yellow waxy solid.

4. *N'-(5-(1,2-dithiolan-3-yl)pentanoyloxy)-3,4-dimethoxybenzim-idamide*. To a solution of N'-hydroxy-3,4-dimethoxybenzimi-damide (0.080 g, 0.41 mmol) in 4 mL dry CH₂Cl₂, lipoic acid (0.084 g, 0.41 mmol) and DCC (0.093 g, 0.45 mmol) were added and the reaction mixture was stirred at ambient tem-perature overnight. The mixture was extracted with CH₂Cl₂ and water. The organic layer was washed with saturated aque-ous NaCl, dried over sodium sulfate, filtered and the solvent was evaporated in vacuo. The residue was purified by flash-column chromatography (CH₂Cl₂/CH₃OH 97:3). Yield: 0.120 g (76 %), yellow-green solid.

5. *5-(4-(1,2-dithiolan-3-yl)butyl)-3-(3,4-dimethoxyphenyl)-1,2,4-oxadiazole (21a)*. To a solution of N'-(5-(1,2-dithiolan-3-yl)pentanoyloxy)-3,4-dimethoxy-benzimidamide (0.040 g, 0.1 mmol) in 4 mL anhydrous THF, n-tetrabutylammonium fluoride (0.1 mL) was added and the reaction mixture was stirred at ambient temperature. After 1 h it was extracted with ethyl acetate and water. The organic layer was washed with saturated aqueous NaCl, dried over sodium sulfate, filtered, and the solvent was evaporated in vacuo. The resi-due was purified by flash-column chromatography (Petroleum ether/Ethyl acetate 75:25). Yield: 0.035 g (94 %), yellow waxy solid.

6. *N-(3,4-dimethoxyphenethyl)-5-(1,2-dithiolan-3-yl)pentanethio-amide*. To a solution of N-(3,4-dimethoxyphenethyl)-5-(1,2-dithiolan-3-yl)pentanamide (0.110 g, 0.3 mmol) in 3 mL anh.

THF, Lawesson's reagent (0.120 g, 0.3 mmol) was added and the mixture was refluxed at 70 C overnight. After evaporation of the solvent under argon the residue was purified by flash-column chromatography (CH_2Cl_2/CH_3OH 97:3). Yield: 0.090 g (78 %), yellow waxy solid.

7. *5-(4-(1,2-dithiolan-3-yl)butyl)-1-(3,4-dimethoxyphenethyl)-1H-tetrazole (22a).* To a solution of N-(3,4-dimethoxyphenethyl)-5-(1,2-dithiolan-3-yl)pentanethioamide (0.090 g, 0.23 mmol) in 1.7 mL anh. THF were added diisopropylazodicarboxylate (0.07 mL, 0.35 mmol) and triphenylphosphine (0.092 g, 0.35 mmol) and after stirring for 5 min trimethylsilylazide (0.05 mL, 0.35 mmol) was added. The reaction mixture was stirred at 40 C for 3 h and then the solvent was evaporated in vacuo. The residue was purified by flash-column chromatography (Petroleum Ether/Ethyl Acetate 50:50). Yield: 0.070 g (78 %), white solid.

8. *2-(4-(1,2-dithiolan-3-yl)butyl)-4-(3,4-dimethoxyphenyl)thiazole (23a).* To a solution of 2-bromo-1-(3,4-dimethoxyphenyl)ethanone (0.047 g, 0.18 mmol), in 3 mL abs. EtOH was added a solution of lipothioamide (0.040 g, 0.18 mmol) in 3 mL abs EtOH and the resulting mixture was refluxed for 2 h. After evaporating EtOH, the residue was extracted with ethyl acetate and water. The organic layer was washed with saturated aqueous $NaHCO_3$ and saturated aqueous NaCl, dried over sodium sulfate, and the solvent was evaporated in vacuo. The crude product was purified by flash-column chromatography (CH_2Cl_2/CH_3OH 97:3). Yield: 0.020 g (29%), waxy solid.

9. *General method for the deprotection of the methoxy group.* To a solution of the appropriate protected catechol (1 eq) in CH_2Cl_2 was added $BF_3 \cdot S(Me)_2$ (20 eq) and the mixture was stirred at ambient temperature overnight. The solvent was evaporated under argon and the residue was extracted with ethyl acetate and water. The organic layer was washed with saturated aqueous NaCl, dried, and the solvent was evaporated in vacuo. The following compounds were obtained according this procedure *(22)*: 5-(1,2-Dithiolan-3-yl)-N-(6-hydroxy-2,2,5,7,8-pentamethyl-3,4-dihydro-2H-chromen-5-yl)pentanamide (13b); 5-(1,2-Dithiolan-3-yl)-N-[(6-hydroxy-2,2,7,8-tetramethyl-3,4-dihydro-2H-chromen-5-yl)methyl]pentanamide (14b); 1-(5-(1,2-dithiolan-3-yl)pentyl)-4-(3,4-dihydroxyphenyl)-1H-1,2,3-triazole (19b); 2-(4-(1,2-dithiolan-3-yl)butyl)-5-(3,4-dihydroxyphenethyl)-1,3,4-oxadiazole (20b); 5-(4-(1,2-dithiolan-3-yl)butyl)-3-(3,4-dihydroxyphenyl)-1,2,4-oxadiazole (21b); 5-(4-(1,2-dithiolan-3-yl)butyl)-1-(3,4-dihydroxyphenethyl)-1H-tetrazole (22b); 2-(4-(1,2-dithiolan-3-yl)butyl)-4-(3,4-dihydroxyphenyl)thiazole (23b).

3.8. Results

Tables 21.1–21.5 show representative results of the biological activity of selected compounds.

3.8.1. Trolox–lipoic acid hybrids (Compounds of Category A)

The antiarrhythmic and antioxidant activity of compounds 1–14 (Tables 21.1 and 21.2) was evaluated on isolated heart preparations using the Krebs perfused Langerdorff model. The tissue malondialdehyde (MDA) levels reflect the antioxidant capacity of the analogs under study (10, 11).

3.8.2. Quinolinone–Lipoic Acid Hybrids (Compounds of Category B)

The *in vitro* antioxidant activity of the quinolinone–lipoic acid hybrids was assessed by examination of the interaction of the tested compounds with the stable free radical DPPH (indicating the radical scavenging ability of the compounds in an iron-free system) and of the competition of the quinolinone analogs with DMSO for the hydroxyl radical (HO•), generated by the $Fe^{3+}/$ascorbic acid system, expressed as percent inhibition of formaldehyde production (Tables 21.3) (19). The *in vivo* anti-inflammatory effects of compounds 15–17 were assessed by using the functional model of carrageenin induced rat paw oedema and are presented in Table 21.4, as percent inhibition of induced rat paw oedema (19).

3.8.3. 1,2-Dithiolane-Catechol hybrids (Compounds of Category C)

The neuroprotective activity of compounds 19a,b-23a,b was evaluated using glutamate-challenged hippocampal HT22 cells (Table 21.5) (18).

Table 21.1
Antiarrhythmic activity is presented as incidence of premature beats: % of total heart beats during the first 10 min of reperfusion, for each treatment. Antioxidant activity is expressed as malondialdehyde content (nmoles/g wet tissue), at the end of reperfusion

Compound (5 μM)	Premature beats (%)	MDA
None	15 ± 3.4	2.11 ± 0.3
Lipoic acid	6.3 ± 2.1^{a}	1.03 ± 0.02^{b}
Trolox	3.15 ± 1.7^{a}	1.07 ± 0.05^{b}
1	0	1.10 ± 0.1^{b}
3	16 ± 4.1	1.19 ± 0.09^{b}
6	0	0.99 ± 0.06^{b}

$n = 2–4$
[a]$p < 0.001$ vs. control
[b]$p < 0.001$ vs. control

Table 21.2
Antiarrhythmic activity is presented as incidence of premature beats: % of total heart beats during the first 10 min of reperfusion, for each treatment Antioxidant activity is expressed as malondialdehyde content (nmoles/gr wet tissue), at the end of reperfusion

Compound (1 μM)	Premature beats (%)	MDA
None	13.7 ± 3.6	0.23 ± 0.03
4	0.53 ± 0.2^a	0.12 ± 0.01^a
5	0.95 ± 0.05^a	0.15 ± 0.02^b
11	0.85 ± 0.15^a	0.16 ± 0.008^b
12	5.2 ± 0.05^b	0.19 ± 0.017
13	0.95 ± 0.05^a	0.18 ± 0.007
14	2.1 ± 0.7^a	0.112 ± 0.016^a
Lipoic acid + trolox	6.45 ± 1.3^b	0.16 ± 0.03

$n = 3\text{–}4$
[a] $p < 0.01$ vs. control
[b] $p < 0.05$ vs. control

Table 21.3
Interaction % with DPPH (RA %); Competition % with DMSO for hydroxyl radical (·OH%)

Compound	0.1 (mM)[a] RA% 20 min	0.1 (mM)[a] RA% 60 min	0.5 (mM)[a] RA% 20 min	0.5 (mM)[a] RA% 60 min	·OH(%) 0.1 (mM)[a]
15	10.9	3.4	28.8	40.1	67.9
16	6.4	7.8	no	14.1	0
17	9.1	13.0	28.4	38.8	28.6
18	6.2	10.2	23.9	29.5	no[c]
Lipoic acid	20.3	27.1	5.2	6.8	38.6
NDGA (Nordihydroguaiaretic acid)	81	82.6	96.5	98	nt[b]
Trolox	nt[b]	nt[b]	nt[b]	nt[b]	88.2

[a] SD < 10%
[b] not tested
[c] no action under the reported experimental conditions

Table 21.4
Inhibition % of induced carrageenin rat paw edema (CPE %) at 0.01 mmol/kg body weight

Compound	CPE%[a]
15	48.2*
16	45.5**
17	53*
18	50*
Lipoic acid	29.6**
Indomethacin	47*

[a]Statistical studies were done with student's T-test, $*p < 0.05$, $**p < 0.01$

Table 21.5
Neuroprotective activity of compounds 19a,b-23a,b

Compound	EC$_{50}$ (μM) (mean ± SEM)	compound	EC$_{50}$ (μM)[a] (mean ± SEM)
19a	>10	21a	>10
19b	6.23 ± 0.97	21b	3.63 ± 0.33
20a	>10	22a	>10
20b	4.21 ± 0.40	22b	2.99 ± 0.14
		23a	>10

[a]EC$_{50}$ values are compound concentrations required to secure a viability in the glutamate-exposed cells equal to 50% of that of the non-exposed cells calculated as described in the Experimental Section. Values are mean ± SEM of at least three independent experiments

4. Notes

1. Lipoic acid chloride is a yellow oily compound, which is air-sensitive and hydrolyzes easily upon standing, therefore it requires special handling: it should be kept in the refrigerator, under nitrogen atmosphere. In any case, optimum results are obtained when lipoic acid chloride is freshly synthesized.

2. N-(lipoyloxy)succinimide is a yellowish solid which is easily stored, in the refridgerator and is very stable for a long time.

3. 3-(5-Azidopentyl)-1,2-dithiolane is a sensitive compound, which should be stored in the refrigerator and should be

freshly prepared prior to use. In addition, this compound is used as produced from the reaction without further purification.

4. The volume ratio THF/H$_2$O = 3:2 is critical for the reaction, since it ensures the homogeneity of the mixture of these solvents. In the case of compound 13a the reaction was carried out in THF using triethylamine as a base.

5. In the case of quinolinone–lipoic acid hybrids, drops of DMF were added to the dichloromethane solution in order to facilitate dissolution of the reactants. Moreover, the work-up of the reaction was slightly modified as follows: H$_2$O (10 mL) was added to the mixture and afterward it was extracted with dichloromethane (3–10 mL) and washed repeatedly with H$_2$O. The organic extracts were combined, dried (Na$_2$SO$_4$), and concentrated in vacuo which were triturated with diethylether and filtered off. Purification of the lipoic acid hybrids was performed using flash column chromatography. In some cases, purification with flash column chromatography was not adequate to obtain the analytically pure compound; therefore further purification was required (recrystallization).

References

1. Packer L, Witt EH, Tritschler HJ (1995) Alphalipoic acid as a biological antioxidant. *Free Radic Biol Med* **19**:227–250

2. Moini H, Packer L, Saris N-EL (2002) Antioxidant and prooxidant activities of R-lipoic acid and dihydrolipoic acid. *Toxicol Appl Pharmacol* **182**:84–90

3. Navari-Izzo F, Quartacci MF, Sgherri C (2002) Lipoic acid: A unique antioxidant in the detoxification of activated oxygen species. *Plant Physiol Biochem* **40**:463–470

4. Freisleben HJ (2000) Lipoic acid reduces ischemia-reperfusion injury in animal models. *Toxicology* **148**:159–171

5. Packer L, Tritschler HJ, Wessel K (1997) Neuroprotection by the metabolic antioxidant alpha-lipoic acid. *Free Radic Biol Med* **22**:359–378

6. Pirlich M, Kiok K, Sandig G, Lochs H, Grune T (2002) *a*-Lipoic acid prevents ethanolinduced protein oxidation in mouse hippocampal HT22 cells. *Neurosci Lett* **328**:93–96

7. Ha H, Lee J-H, Kim H-N, Kim H-M, Kwak HB, Lee S, Kim H-H, Lee ZH (2006) alphaLipoic acid inhibits inflammatory bone resorption by suppressing prostaglandin E2 synthesis. *J Immunol* **176**:111–117

8. Henriksen EJ (2006) Exercise training and the antioxidant *a*-lipoic acid in the treatment of insulin resistance and type 2 diabetes. *Free Radic Biol Med* **40**:3–12

9. Meunier B (2008) Hybrid molecules with a dual mode of action: dream or reality? *Acc Chem Res* **41**:69–77

10. Koufaki M, Calogeropoulou T, Detsi A, Roditis A, Kourounakis AP, Papazafiri P, Tsiakitzis K, Gaitanaki C, Beis I, Kourounakis PN (2001) Novel potent inhibitors of lipid peroxidation with protective effects against reperfusion arrhythmias. *J Med Chem* **44**:4300–4303

11. Koufaki M, Detsi A, Theodorou E, Kiziridi C, Calogeropoulou T, Vassilopoulos A, Kourounakis AP, Rekka E, Kourounakis PN, Gaitanaki C, Papazafiri P (2004) Synthesis of chroman analogues of lipoic acid and evaluation of their activity against reperfusion arrhythmias. *Bioorg Med Chem* **12**:4835–4841

12. Harnett JJ, Auguet M, Viossat I, Dolo C, Bigg D, Chabrier P-E (2002) Novel lipoic acid analogues that inhibit nitric oxide synthase. *Bioorg Med Chem Lett* **12**:1439–1442

13. Durand G, Polidori A, Salles JP, Prost M, Durand P, Pucci B (2003) Synthesis and antioxidant efficiency of a new amphiphilic spin-trap derived from PBN and lipoic acid. *Bioorg Med ChemLett* **13**:2673–2676

14. Antonello A, Hrelia P, Leonardi A, Marucci G, Rosini M, Tarozzi A, Tumiatti V, Melchiorre C (2005) Design, synthesis, and biological evaluation of prazosin-related derivatives as multipotent compounds. *J Med Chem* **48**:28–31

15. Rosini M, Andrisano V, Bartolini M, Bolognesi ML, Hrelia P, Minarini A, Tarozzi A, Melchiorre C (2005) Rational approach to discover multipotent anti-Alzheimer drugs. *J Med Chem* **48**:360–363

16. Antonello A, Tarozzi A, Morroni F, Cavalli A, Rosini M, Hrelia P, Bolognesi ML, Melchiorre C (2006) Multitarget-directed drug design strategy: a novel molecule designed to block epidermal growth factor receptor (EGFR) and to exert proapoptotic effects. *J Med Chem* **49**:6642–6645

17. Venkatachalam SR, Salaskar A, Chattopadhyay A, Barik A, Mishra B, Gangabhagirathic R, Priyadarsini KI (2006) Synthesis, pulse radiolysis, and in vitro radioprotection studies of melatoninolipoamide, a novel conjugate of melatonin and a-lipoic acid. *Bioorg Med Chem* **14**:6414–6419

18. Koufaki M, Kiziridi C, Nikoloudaki F, Alexis MN (2007) Design and synthesis of 1, 2-dithiolane derivatives and evaluation of their neuroprotective activity. *Bioorg Med Chem Lett* **17**:4223–4227

19. Detsi A, Bouloumbasi D, Prousis KC, Koufaki M, Athanasellis G, Melagraki G, Afantitis A, Igglessi-Markopoulou O, Kontogiorgis C, Hadjipavlou-Litina DJ (2007) Design and synthesis of novel quinolinone-3-aminoamides and their a-lipoic acid adducts as antioxidant and anti-inflammatory agents. *J Med Chem* **50**:2450–2458

20. Decker M, Kraus B, Heilmann J (2008) Design, synthesis and pharmacological evaluation of hybrid molecules out of quinazolinimines and lipoic acid lead to highly potent and selective butyrylcholinesterase inhibitors with antioxidant properties. *Bioorg Med Chem* **16**:4252–4261

21. Wagner AF, Walton E, Boxer GE, Pruss MP, Holly FW, Folkers K (1956) Properties and derivatives of α-lipoic acid. *J Am Chem Soc* **78**:5079–5081

22. Compton B, Sheng S, Carlson K, Rebacz N, Lee I, Katzenellenbeogen B, Katzenellenbeogen J (2004) Pyrazolo [1, 5-a] pyrimidines: estrogen receptor ligands possessing estrogen receptor b antagonist activity. *J Med Chem* **47**:5872–5893

Chapter 22

Characterization of the Antioxidant Properties of Pentaerithrityl Tetranitrate (PETN)-Induction of the Intrinsic Antioxidative System Heme Oxygenase-1 (HO-1)

Andreas Daiber and Thomas Münzel

Abstract

Organic nitrates are among the oldest and yet most commonly employed drugs in the chronic therapy of coronary artery disease and congestive heart failure. While they have long been used in clinical practise, our understanding of their mechanism of action and of their side effects remains incomplete. To date, the most commonly employed nitrates are isosorbide mononitrate (ISMN), isosorbide dinitrate (ISDN), and nitroglycerin (GTN). Another nitrate, pentaerithrityl tetranitrate (PETN), has long been employed in eastern European countries and is currently being reintroduced also in western countries. So far, PETN is the only organic nitrate in clinical use, which is devoid of induction of oxidative stress and related side-effects such as endothelial dysfunction and nitrate tolerance. Some of these effects are related to special pharmacokinetics of PETN, but upon chronic administration, PETN also induces antioxidative pathways at the genomic level, resulting in increased expression of heme oxygenase-1 (HO-1) and ferritin, both possessing highly protective properties. There is good experimental evidence that at least part of the beneficial profile of long-term PETN treatment is based on activation of the heme oxygenase-1/ferritin system.

Key words: Nitrate tolerance, Endothelial dysfunction, Mitochondrial aldehyde dehydrogenase (ALDH-2), Heme oxygenase-1 (HO-1), Reactive oxygen species, Vascular oxidative stress

1. Introduction

Previous studies have demonstrated that nitrate tolerance in response to GTN in vivo treatment is a multifactorial phenomenon (Fig. 22.1) (1). The "oxidative stress concept" in the setting of nitrate tolerance was established by Münzel et al. (2) and refined during the last couple of years (3, 4). In essence, the concept consists of increased superoxide formation in response to nitrate treatment, which decreases NO bioavailability, leads

D. Armstrong (ed.), *Advanced Protocols in Oxidative Stress II*, Methods in Molecular Biology, vol. 594
DOI 10.1007/978-1-60761-411-1_22, © Humana Press, a part of Springer Science + Business Media, LLC 2010

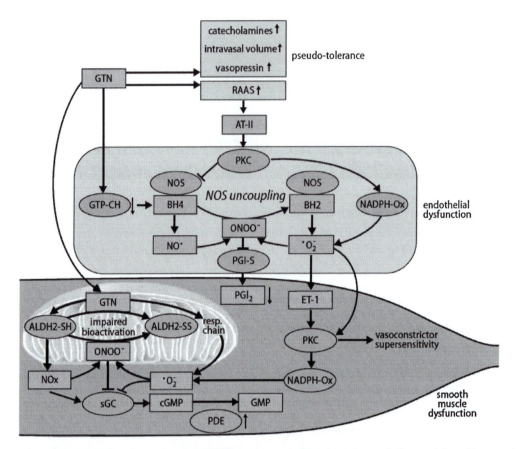

Fig. 22.1. Within 1 day of continuous low dose GTN therapy, neurohormonal counter-regulation consisting of increased catecholamine and vasopressin plasma levels, increased intravasal volume, and activation of the renin–angiotensin–aldosteron system (RAAS) reduces therapeutic efficacy (pseudo-tolerance). After 3 days, endothelial and smooth muscle dysfunction develops (vascular tolerance and cross-tolerance) by different mechanisms: (1) Increased endothelial and smooth muscle superoxide formation from NADPH oxidase activation by protein kinase C (PKC) and from the mitochondria, (2) Direct inhibition of NOS activation by PKC, (3) Uncoupling of endothelial NOS due to limited tetrahydrobiopterin (BH$_4$) availability caused by peroxynitrite (ONOO⁻)-induced oxidation of BH$_4$ and reduced expression of GTP-cyclohydrolase (GTP-CH), (4) Vasoconstrictor supersensitivity due to increased smooth muscle PKC activity, (5) Impaired bioactivation of GTN because of inhibition of ALDH-2, (6) Inhibition of smooth muscle soluble guanylyl cyclase (sGC) by superoxide and peroxynitrite, (7) Increased inactivation of cyclic GMP (cGMP) by phosphodiesterases (PDE), (8) Inhibition of prostacyclin synthase (PGI$_2$-S) by peroxynitrite, leading to reduced prostacyclin (PGI$_2$) formation. For the sake of clarity, tolerance-induced radical generation in endothelial mitochondria was omitted from the scheme. Adopted from (1). Copyright by Lippincott Williams & Wilkins.

to peroxynitrite formation, NOS uncoupling, and impairs NO/cGMP signaling (3). Moreover, oxidative inhibition of prostacyclin synthase (5) as well as mitochondrial ALDH activity (6) may present other key events in the development of nitrate tolerance. For PETN an NO-dependent pathway has been suggested to be responsible for the antiatherosclerotic actions of chronic PETN therapy and the decrease in oxidative stress in this model (7). PETN could even reverse endothelial dysfunction in established atherosclerosis (8). In contrast to GTN, platelet activation and increased ROS formation were not observed upon chronic PETN treatment (9). Also, in vascular cell culture, GTN but not PETN significantly increased ROS production (10). Taken together, there is a growing body of evidence that PETN mediated tolerance-devoid anti-ischemic protection is associated with or is a consequence of the antioxidative profile of this organic nitrate.

Excessive formation of peroxynitrite has been suggested to be a major determinant for the development of nitrate tolerance (5, 11). Peroxynitrite leads to uncoupling of endothelial NOS via oxidative depletion of BH_4 (12), which would best explain the cross tolerance to endothelium-dependent vasodilators such as acetylcholine (ACh) (3). However, this concept was questioned by studies in eNOS knockout mice (13). Since these eNOS deficient mice still developed nitrate tolerance in response to GTN treatment, a causal role of eNOS uncoupling for the development of tolerance was ruled out. Nevertheless, GTN treatment of cells led to eNOS uncoupling, which probably contributes to the development of clinical nitrate and cross tolerance (1, 14), and, more importantly, eNOS dysfunction is a major determinant for endothelial dysfunction. Therefore, eNOS uncoupling induced by chronic organic nitrate therapy could lead to endothelial dysfunction in patients as observed for isosorbide mononitrate (ISMN) (15) and dinitrate (ISDN) (16) as well as nitroglycerin (GTN) (3). The contribution of NADPH oxidases to ROS production and oxidative damage in response to GTN therapy was established by a number of independent cell culture and animal studies (11, 17). Moreover, it was demonstrated that GTN-induced activation of PKC mediates late preconditioning, a protective effect that is probably based on initial ROS formation and mild oxidative damage (18). Additional studies showed that statins (19) and ACE inhibition (20) positively influence the development of tolerance, and it was hypothesized that this protective effect is mediated via inhibition of NADPH oxidase dependent ROS formation. For PETN, it was described that it does not activate NADPH oxidase activity (21).

In contrast to other long-acting nitrates, PETN induces tolerance-free vasodilation in humans (22, 23). In contrast to all other organic nitrates, PETN and its metabolite pentaerythrityl trinitrate (PETriN) induce the antioxidant defense protein heme oxygenase-1 (HO-1), also known as a chaperone, heat shock protein 32 (hsp32), and increase the formation of the antioxidant molecule bilirubin and the vasodilator carbon monoxide (CO) (24). In addition, PETN and PETriN led to a marked increase in protein expression of a second antioxidant protein, ferritin, via the HO-1-dependent release of free iron from endogenous heme sources (25). In addition, we could show recently that PETN and PETriN in contrast to GTN did not affect the nitrate esterase activity of ALDH-2, and did not elicit ROS formation with isolated arteries and mitochondria, adding a further mechanism to explain the lack of tolerance development to PETN (6). Altogether, these defense mechanisms protected endothelial cells from hydrogen peroxide-induced toxicity, and might explain the previously observed antiatherogenic actions of PETN in vivo (Fig. 22.2) (8, 26). The concept of PETN-mediated activation of intrinsic anti-oxidative pathways at the genomic level was further supported by recent findings that acute (high bolus) challenges with PETN resulted in similar tolerance development as compared to GTN (6, 27). Therefore, PETN triggered protective effects

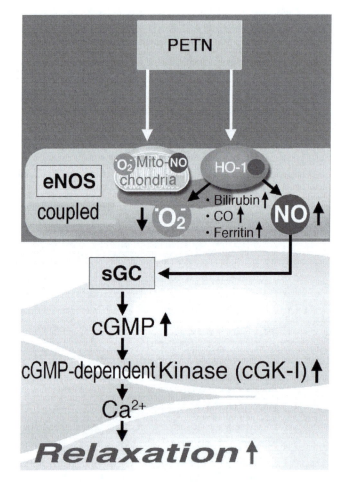

Fig. 22.2. In contrast to GTN, PETN stimulates the expression of the protective protein heme oxygenase-1 (HO-1), and thus triggers formation of the potent antioxidant bilirubin as well as the weak, antiatherosclerotic activator of soluble guanylyl cyclase (sGC), carbon monoxide (CO). As a consequence of HO-1 dependent degradation of metallo-porphyrins and increase in free iron, the expression of ferritin is increased providing protection against iron-induced Fenton toxicity. Activation of these intrinsic antioxidative pathways prevents eNOS uncoupling, mitochondrial oxidative stress, inactivation of ALDH-2, thereby maintaining the vasodilatory action of PETN as well as endothelial function. Adopted from Münzel et al., *Dtsch. Med. Wochenschr.* 2008.

requires at least 6 h of continuous treatment, allowing HO-1 and ferritin induction.

However, until recently, this explanation of protective effects of PETN at a molecular level was restricted to cell culture-based data. In a recent study, it was demonstrated that PETN also in response to chronic treatment in animals up-regulates heme oxygenase-1 and ferritin, thereby increasing bilirubin levels (28). In this study, GTN but not PETN induced nitrate tolerance and endothelial dysfunction, increased vascular and mitochondrial ROS production, and impaired vascular and mitochondrial ALDH-2 activity. It was also important to demonstrate that GTN-induced tolerance, and all related side-effects were completely normalized by cotherapy with the heme oxygenase-1 inducer hemin. On the contrary, cotreatment with the heme oxygenase-1 suppressor apigenin induced a tolerance-like phenomenon in the PETN-treated rats, clearly indicating that the heme oxygenase-1 defense system is highly potent and may successfully prevent nitrate-induced tolerance. Bilirubin has been demonstrated to be a highly efficient peroxynitrite scavenger (29), and it also decreased GTN-induced oxidative stress in isolated mitochondria (Fig. 22.3) (28, 30). Preliminary data in heme oxygenase-1 knockout mice support these previous findings. Finally, ongoing studies in our laboratory demonstrate that PETN cotreatment is able to improve angiotensin-II induced hypertension in Wistar rats. Therefore, we think that characterization of the antioxidative properties of PETN is of clinical interest and pharmacologic induction of intrinsic antioxidative systems such as heme oxygenase-1 and ferritin might be future therapeutic strategies for the treatment of cardiovascular diseases.

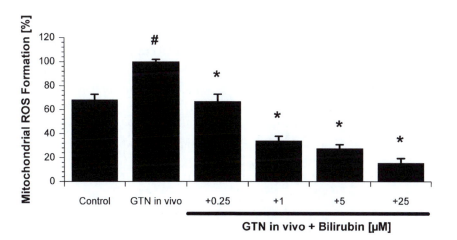

Fig. 22.3. Effects of bilirubin on GTN (in vivo)-induced mitochondrial ROS formation. Bilirubin efficiently decreased mitochondrial ROS (L-012 ECL) in response to GTN in vivo treatment of Wistar rats. Data are mean $\pm$ SEM of 6–11 independent experiments. #$P < 0.05$ vs. control, *$P < 0.05$ vs. GTN in vivo group. Adopted from ref. 30.

2. Materials

2.1. In Vivo Treatment with Organic Nitrates

1. GTN was purchased as an ethanolic stock solution (102 g/l, 450 mM) from UNIKEM (Copenhagen, Denmark) (*see* Note 1).

2. PETN (with 33% (w/w) water) was from Dottikom (Switzerland) (*see* Note 1). For in vivo treatment stock solutions of 450 mM, PETN in DMSO were freshly prepared.

3. Male Wistar rats (200 g) were purchased from Charles River (Sulzfeld, Germany).

4. Chronic infusion of organic nitrates in Wistar rats was performed by micro osmotic minipumps (Alzet, Cupertino, CA) model 2001 with infusion rates of 1 μl/h for 7 days (*see* Notes 2 and 3).

2.2. Induction of Heme Oxygenase-1 by PETN

1. Bradford reagent was purchased from BioRad (Hercules, CA, USA).

2. *Hg-buffer (mM in water)*. 20 Tris–HCl, 250 sucrose, 3 EGTA, 20 EDTA. The pH was adjusted to 7.5. A protease inhibitor cocktail tablet (Protease inhibitor complete, Roche Diagnostics, Mannheim, Germany) was added to 10 ml of this buffer. Finally, 1% (v/v) Triton was added to the buffer.

3. The rabbit monoclonal α-actinin antibody was purchased from Sigma (Steinheim, Switzerland) (*see* Note 4).

4. The rabbit polyclonal heme oxygenase-1 antibody was from Stressgen (San Diego, CA, USA).

5. Laemmli buffer contained 1 ml Tris–HCl (1 M), 3-ml sodium dodecyl sulfate (SDS, 20 % (w/w)), 3-ml glycerol, 300-μl bromophenol blue (1 %), 320-μl aqua dest. Aliquots (850 μl) were stored at –20°C and were thawed just before use, and 100 μl 2-mercaptoethanol were added to the aliquot.

6. Separation gel (9%) contained 3-ml acrylamide/bisacrylamide (30/0.8 % in aqua dest.) (Roth, Germany), 2.5-ml lower Tris, 100-μl ammoniumpersulfate (10%, APS), 10-ml aqua dest.

7. Loading gel (3%) contained 0.5-ml acrylamide/bisacrylamide (30/0.8% in aqua dest.) (Roth, Germany), 1.25-ml upper Tris, 0-μl ammoniumpersulfate (10 %, APS), 5-ml aqua dest.

8. Electrophoresis buffer (g/l water): 3.03 Tris, 14.4 glycin, 1 SDS.

9. Blotting buffer (g): 3.03 Tris, 14.4 glycin in 750-ml water and 250-ml methanol.

10. K-phosphate buffer (mM in water): 99.01 NaCl, 4.69 KCl, 2.5 $CaCl_2$, 1.2 $MgSO_4$, 25 $NaHCO_3$, 1.03 K_2HPO_4, 20 sodium HEPES, 11.1 d-glucose. The pH was adjusted to 7.35.

11. Enhanced chemiluminescence (ECL) kit was from Pierce (Rockford, IL, USA).

2.3. Oxidative Stress in Response to Organic Nitrate Treatment

1. L-012 (8-amino-5-chloro-7-phenylpyrido(3,4-d)pyridazine-1,4-(2H,3H)dione sodium salt) was from Wako Pure Chemical Industries (Osaka, Japan) (*see* Note 5). L-012 stocks (50 mM in DMSO) were stored at –20°C.

2. Succinate stock solutions were 500 mM in water and kept at 4°C (*see* Note 6).

3. Mitochondria isolation buffer I: 50-mM HEPES, 70-mM sucrose, 220-mM mannitol, 1-mM EGTA, and 0.033-mM bovine serum albumin.

4. Mitochondria isolation buffer II: 10-mM Tris, 340-mM sucrose, 100-mM KCl, and 1-mM EDTA.

3. Methods

3.1. In Vivo Treatment with Organic Nitrates

1. This section describes how to treat rats with organic nitrates for the induction of nitrate tolerance or determination of the antioxidative properties of PETN.

2. Male Wistar rats were anaesthetized with isoflurane and equipped with micro osmotic pumps containing 450 mM GTN (in ethanol) or PETN (in DMSO) or the solvent, and infusion was maintained for 3 days at 1 μl/h.

3. The resulting doses were 10.5 and 6.6 μg/kg/min for PETN and GTN, respectively. Implantation was performed according to a previously described procedure (Fig. 22.4) (31).

3.2. Induction of Heme Oxygenase-1 by PETN

This section describes how to assess the protective antioxidative effects of PETN as compared to GTN by induction of heme oxygenase-1. The Western blot technique for quantification of heme oxygenase-1 protein is described in detail. Assessment of HO-1 and ferritin mRNA expression by RT-PCR and of HO-1 activity by measurement of bilirubin is only briefly presented.

3.2.1. RT-PCR

1. mRNA expression of HO-1 and ferritin (heavy-chain) was analyzed with quantitative real-time RT-PCR using an iCyclerTM iQ system (Bio-Rad Laboratories, Munich, Germany) (Fig. 22.5) (28).

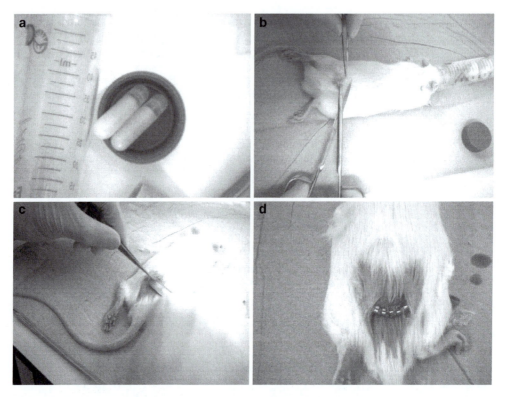

Fig. 22.4. (**a**) Osmotic pumps are prepared for implantation. (**b**) After shaving at the hind quarters and disinfection, the skin was opened with a scissor, and a pocket was formed between corium and connective tissue. (**c**) The osmotic pump was inserted into the pocket. (**d**) The wound was closed with brackets and disinfected.

2. Briefly, total RNA from rat aorta and heart was isolated according to the manufacturer's protocol (RNeasy Fibrous Tissue Mini Kit; Qiagen, Hilden, Germany).

3. 0.5 µg of total RNA was used for real-time RT-PCR analysis with the QuantiTect™ Probe RT-PCR kit (Qiagen).

4. TaqMan® Gene Expression assays (Applied Biosystems, Foster City, CA) for HO-1, ferritin, and GAPDH were purchased as probe and primer sets.

5. The comparative Ct method was used for relative mRNA quantification (32).

6. Gene expression was normalized to the endogenous control, GAPDH mRNA, and the amount of target gene mRNA expression in each sample was expressed relative to that of control.

7. We have also published a detailed protocol for quantification of HO-1 and ferritin mRNA in EA.hy cells (30).

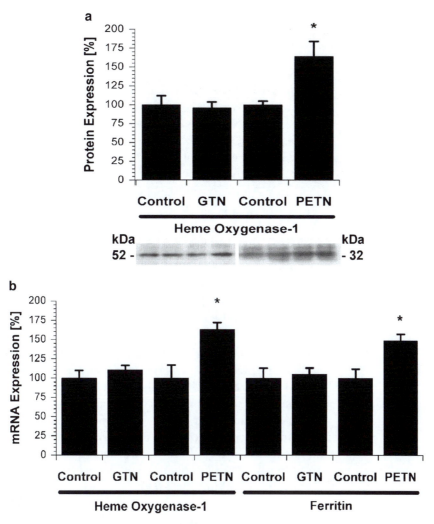

Fig. 22.5. (**a**) Protein expression of HO-1 in isolated aortic vessel segments of rats upon treatment with PETN or GTN. Below the densitometric data, the representative original blots are shown. Data are mean ± SEM of n = 4-6 independent experiments. *P < 0.05 vs. solvent control. (**b**) mRNA expression of HO-1 and ferritin (heavy chain) in aorta in response to chronic treatment with PETN or GTN. Data are mean ± SEM of n = 6–10 independent experiments. *P < 0.05 vs. solvent control. Adopted from ref. 28. Copyright by Lippincott Williams & Wilkins.

3.2.2. Measurement of Heme Oxygenase-1 Activity

Heme oxygenase activity was determined by measuring serum levels of bilirubin photometrically by a Hitachi 917 R Analyzer (Boehringer Mannheim, Germany) in the Department of Clinical Chemistry, University Hospital, Mainz, according to a standard procedure (2,5-dichlorophenyl diazonium (DPD) method) (33) (*see* Note 7).

3.2.3. Measurement of Heme Oxygenase-1 Protein Induction

1. Vascular or cardiac tissue was homogenized with a mortar in liquid nitrogen, and the frozen tissue powder was resuspended in Hg-buffer.

2. Protein concentration was determined using Bradford reagent according to the manufacturer's instructions and normalized by a BSA standard curve. The absorbance was read at 595 nm.

3. SDS-PAGE was performed with a gel electrophoresis system Mini Protean II (BioRad). A 1-mm thick and 5-cm separation gel was prepared. To generate a horizontal upper edge, an isopropanol phase was added on top of the gel. After polymerization of the gel, the isopropanol was removed, and a 1-cm long loading gel was prepared on top of the separation gel and a 12-well comb for the samples was inserted.

4. The tissue homogenates were mixed with Laemmli-buffer, incubated for 5 min at 95°C, and then loaded (20 µg/pocket). Prestained weight markers were loaded in the right and left pocket of the gel (10–120 kDa at 10 kDa-steps, Gibco-BRL).

5. SDS-PAGE was performed with electrophoresis buffer at 60 V until the samples were completely loaded to the loading gel. After this step, the power was increased to 200 V for approximately 1.5 h until the desired separation was achieved (envisaged by the prestained markers).

6. Blotting of the proteins from the gel to a nitrocellulose membrane (BioRad) was performed with a wet-blot system Mini Trans-Blot Cell (BioRad). The gel was covered by a nitrocellulose membrane from Schleicher & Schuell (Dassel, Germany); the lower and upper layers were 1.5-mm thick Mini Trans-Blot filter papers (BioRad). After positioning of the gel-membrane assembly in the wet-blot tank that was filled with blotting-buffer, the proteins were blotted to the membrane by a current of 250 mA for 2 h.

7. The transfer was controlled by staining the membrane with Ponceau S solution (Sigma), and the membrane was washed afterwards to remove the Ponceau S.

8. The membrane was divided into two parts (<60 kDa and >60 kDa), and both were incubated over night in blocking-buffer (5% (w/w) milk powder in K-phosphate buffer) at 4°C. Under constant shaking, the membrane (<60 kDa) was then incubated for 1.5 h with a heme oxygenase-1 (32 kDa) antibody (1:1,000) in 3% milk at room temperature. The >60 kDa part of the membrane was incubated 1.5 h with a α-actinin (100 kDa) antibody (1:2,000) in 3% milk at room temperature.

9. After 5 × 5 min washing steps with PBS-T (Dulbecco's phosphate buffered solution containing 0.1% (v/v) Tween 20), the membranes were incubated 1.5 h with the secondary antibody (peroxidase-labeled goat-anti-rabbit (GAR-Pox) or goat-anti-mouse (GAM-Pox) both from Vector Labs, Burlingame, CA, USA) at a dilution of 1:10,000 in 3% milk at room temperature.

10. After 5×5 min washing steps with PBS-T, the membranes were incubated with the two components of the ECL Kit (luminol and hydrogen peroxide) yielding the chemiluminescence together with the peroxidase at the secondary antibody. By this way, the first antibody and accordingly the target proteins (actinin and HO-1) cause staining on a film (Kodak, BioMax MR), resulting in the typical "bands" upon conventional film development (Fig. 22.5). The heme oxygenase-1 content in each band is normalized to the α-actinin loading control by commercial scanning and densitometry software (Gel Pro Analyzer, Media Cybernetics, Bethesda, MD, USA) (*see* Note 8).

3.3. Suppression of Oxidative Stress by PETN

This section describes how to measure mitochondrial oxidative stress in response to GTN or PETN in vivo treatment by an enhanced chemiluminescence-based assay using L-012, a luminol analog with better signal to noise ratio. The isolation of cardiac mitochondria and measurement of reactive oxygen and nitrogen species (ROS and RNS) are described in detail.

1. Hearts from Wistar rats treated with either GTN or PETN were homogenized in mitochondria isolation buffer I and centrifuged at $1,500 \times g$ (10 min at 4°C) and $2,000 \times g$ for 5 min (the pellets were discarded). The supernatant was then centrifuged at $20,000 \times g$ for 20 min, and the pellet was resuspended in 1 ml of mitochondria isolation buffer I.

2. The latter step was repeated and the pellet resuspended in 1 ml of mitochondria isolation buffer I. The mitochondrial fraction (total protein approximately 5–10 mg/ml) was kept on ice and diluted to approximately 1 mg/ml protein in 0.25 ml of PBS (*see* Note 9).

3. For detection of mitochondrial oxidative stress (ROS and RNS), the mitochondrial stocks were diluted to a final protein concentration of 0.1 mg/ml in PBS containing L-012 (100 µM) (*see* Note 10).

4. Finally, the complex II substrate succinate (5 mM, 5 µl from the stock) was added, and the ROS and RNS were detected using a chemiluminescence counter, Lumat LB 9507 (Berthold Techn., Bad Wildbad, Germany)

5. The chemiluminescence was detected in 30 s intervals for 5 min and the last value expressed as counts/min (*see* Notes 11 and 12) (Fig. 22.6).

6. For detection of vascular oxidative stress, thoracic aorta were freshly obtained from the rats, cut into segments of 0.5-cm length, and kept in Krebs-Hepes (KH) buffer until use.

7. The segment was equilibrated for 20 min in KH-buffer at 37°C and then placed in the chemiluminometer, a single

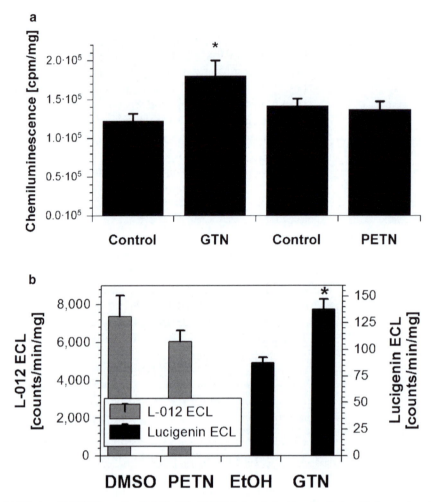

Fig. 22.6. (**a**) Mitochondrial ROS formation (L-012, 100 μM ECL) in isolated rat heart mitochondria from rats upon chronic treatment with PETN or GTN. Data are mean ± SEM of 11–20 independent experiments. *$P < 0.05$ vs. solvent control. (**b**) Vascular ROS formation (L-012, 100 μM or lucigenin, 5 μM ECL) in isolated aortic vessel segments from rats upon chronic treatment with PETN (*gray*, L-012) or GTN (black, lucigenin). Data are mean ± SEM of 11–15 independent experiments. *$P < 0.05$ vs. solvent control. Data were published in table format by ref. 28. Copyright by Lippincott Williams & Wilkins.

photon counter (Lumat, Berthold Techn.) in phosphate-buffered saline (PBS) containing the luminol analog L-012 (100 μM).

8. The chemiluminescence was detected in 60 s intervals for 20 min and the last value expressed as counts/min (*see* Note 13) (Fig. 22.6). This procedure was adapted from previous studies (4, 5).

9. The aortic vessel segments are then dried, and the vascular counts are normalized on dry weight of the vessel segments.

4. Notes

1. GTN and PETN are explosives and should be handled carefully. Both organic nitrates are provided in a save form: GTN is dissolved in ethanol, and PETN contains 33% water. But PETN preparation should never be dried, and GTN solutions should not be concentrated. Never evaporate GTN solutions or PETN preparations.

2. All animal treatment was in accordance with the Guide for the Care and Use of Laboratory Animals as adopted and promulgated by the U.S. National Institutes of Health and was granted by the Ethics Committee of the University Hospital Eppendorf and the University Hospital Mainz.

3. Osmotic pumps were activated prior to subcutaneous implantation to ensure the maximal infusion rate from the beginning.

4. α-actinin was used for normalization of the HO-1 signal and served as the loading control.

5. L-012 is a luminol analog, which is slightly more lipophilic than luminol.

6. The pH of the succinate stock solution should be controlled and adjusted to pH 7.4.

7. A more sensitive method for the detection of tissue bilirubin levels is based on the optical detection of the conversion of biliverdin to bilirubin by recombinant purified biliverdin reductase (Roman Klemz et al., unpublished, Charité, Berlin Germany).

8. Meanwhile, we use a Western blot imaging system Chemi-Lux from Intas (Göttingen, Germany) with integrated densitometry software (Gel Pro Family). This system allows high-throughput analysis and saves money since no films and developer chemicals are required.

9. The protein concentration was determined by a standard DC (Lowry) protein assay from BioRad.

10. L-012 has an increased selectivity and sensitivity for peroxynitrite and/or –derived free radicals (hydroxyl and nitrogen dioxide radicals) as previously demonstrated (34). This could be the reason why L-012 is the most sensitive dye to detect organic nitrate-induced oxidative stress. Organic nitrates trigger superoxide formation but also release nitric oxide, which react in a diffusion-controlled reaction and form peroxynitrite which is detected by L-012. Hydralazin is a highly efficient peroxynitrite scavenger but also suppresses nitroglycerin-induced mitochondrial

(supporting GTN-triggered peroxynitrite formation) and vascular oxidative stress as well as peroxynitrite-dependent protein tyrosine nitration (35).

11. The detection is based on the counting of photons, which are emitted by the chemiluminescence dye upon reaction with free radicals and reactive oxygen and nitrogen species. The Lumat LB 9507 is one of the most sensitive chemiluminescence counters and even detects basal oxidative stress from isolated aortic vessel segments without any stimulation.

12. Attention should be paid to the determination of the protein content since the ECL signal largely depends on the protein concentration.

13. The aortic rings should be cleaned from fatty tissue and blood since these impurities will cause artificial chemiluminescence signal, thereby leading to false-positive detection of oxidative stress.

Acknowledgments

We thank the German Research Foundation for continuous funding of our ongoing research on nitrate tolerance (SFB 553-C17 to T.M. and A.D.) and the University Hospital Mainz for financial support (MAIFOR to A.D.).

References

1. Munzel T, Daiber A, Mulsch A (2005) Explaining the phenomenon of nitrate tolerance. Circ Res 97:618–28

2. Munzel T, Sayegh H, Freeman BA, Tarpey MM, Harrison DG (1995) Evidence for enhanced vascular superoxide anion production in nitrate tolerance. A novel mechanism underlying tolerance and cross-tolerance. J Clin Invest 95:187–94

3. Schulz E, Tsilimingas N, Rinze R, Reiter B, Wendt M, Oelze M, Woelken-Weckmuller S, Walter U, Reichenspurner H, Meinertz T, Munzel T (2002) Functional and biochemical analysis of endothelial (dys)function and NO/cGMP signaling in human blood vessels with and without nitroglycerin pretreatment. Circulation 105:1170–5

4. Daiber A, Oelze M, Sulyok S, Coldewey M, Schulz E, Treiber N, Hink U, Mulsch A, Scharffetter-Kochanek K, Munzel T (2005) Heterozygous deficiency of manganese superoxide dismutase in mice (Mn-SOD+/−): a novel approach to assess the role of oxidative stress for the development of nitrate tolerance. Mol Pharmacol 68:579–88

5. Hink U, Oelze M, Kolb P, Bachschmid M, Zou MH, Daiber A, Mollnau H, August M, Baldus S, Tsilimingas N, Walter U, Ullrich V, Munzel T (2003) Role for peroxynitrite in the inhibition of prostacyclin synthase in nitrate tolerance. J Am Coll Cardiol 42:1826–34

6. Daiber A, Oelze M, Coldewey M, Bachschmid M, Wenzel P, Sydow K, Wendt M, Kleschyov AL, Stalleicken D, Ullrich V, Mulsch A, Munzel T (2004) Oxidative stress and mitochondrial aldehyde dehydrogenase activity: a comparison of pentaerythritol tetranitrate with other organic nitrates. Mol Pharmacol 66:1372–82

7. Kojda G, Hacker A, Noack E (1998) Effects of nonintermittent treatment of rabbits with pentaerythritol tetranitrate on vascular reactivity and superoxide production. Eur J Pharmacol 355:23–31

8. Hacker A, Muller S, Meyer W, Kojda G (2001) The nitric oxide donor pentaerythritol tetranitrate can preserve endothelial function in established atherosclerosis. Br J Pharmacol 132:1707–14

9. Fink B, Bassenge E (1997) Unexpected, tolerance-devoid vasomotor and platelet actions of pentaerythrityl tetranitrate. J Cardiovasc Pharmacol 30:831–6

10. Dikalov S, Fink B, Skatchkov M, Stalleicken D, Bassenge E (1998) Formation of reactive oxygen species by pentaerithrityltetranitrate and glyceryl trinitrate in vitro and development of nitrate tolerance. J Pharmacol Exp Ther 286:938–44

11. Abou-Mohamed G, Johnson JA, Jin L, El-Remessy AB, Do K, Kaesemeyer WH, Caldwell RB, Caldwell RW (2004) Roles of superoxide, peroxynitrite, and protein kinase C in the development of tolerance to nitroglycerin. J Pharmacol Exp Ther 308:289–99

12. Kuzkaya N, Weissmann N, Harrison DG, Dikalov S (2003) Interactions of peroxynitrite, tetrahydrobiopterin, ascorbic acid, and thiols: implications for uncoupling endothelial nitric-oxide synthase. J Biol Chem 278:22546–54

13. Wang EQ, Lee WI, Fung HL (2002) Lack of critical involvement of endothelial nitric oxide synthase in vascular nitrate tolerance in mice. Br J Pharmacol 135:299–302

14. Abou-Mohamed G, Kaesemeyer WH, Caldwell RB, Caldwell RW (2000) Role of l-arginine in the vascular actions and development of tolerance to nitroglycerin. Br J Pharmacol 130:211–8

15. Thomas GR, DiFabio JM, Tommaso T, Parker JD (2007) Once daily therapy with isosorbide-5-mononitrate causes endothelial dysfunction in humans: evidence of a free radical mediated mechanism. J Am Coll Cardiol 49:1289–1295

16. Sekiya M, Sato M, Funada J, Ohtani T, Akutsu H, Watanabe K (2005) Effects of the long-term administration of nicorandil on vascular endothelial function and the progression of arteriosclerosis. J Cardiovasc Pharmacol 46:63–7

17. Munzel T, Kurz S, Rajagopalan S, Thoenes M, Berrington WR, Thompson JA, Freeman BA, Harrison DG (1996) Hydralazine prevents nitroglycerin tolerance by inhibiting activation of a membrane-bound NADH oxidase. A new action for an old drug. J Clin Invest 98:1465–70

18. Banerjee S, Tang XL, Qiu Y, Takano H, Manchikalapudi S, Dawn B, Shirk G, Bolli R (1999) Nitroglycerin induces late preconditioning against myocardial stunning via a PKC-dependent pathway. Am J Physiol 277: H2488–94

19. Otto A, Fontaine D, Fontaine J, Berkenboom G (2005) Rosuvastatin treatment protects against nitrate-induced oxidative stress. J Cardiovasc Pharmacol 46:177–84

20. Salvemini D, Pistelli A, Mollace V (1993) Release of nitric oxide from glyceryl trinitrate by captopril but not enalaprilat: in vitro and in vivo studies. Br J Pharmacol 109:430–6

21. Schwemmer M, Bassenge E (2003) New approaches to overcome tolerance to nitrates. Cardiovasc Drugs Ther 17:159–73

22. Gori T, Al-Hesayen A, Jolliffe C, Parker JD (2003) Comparison of the effects of pentaerythritol tetranitrate and nitroglycerin on endothelium-dependent vasorelaxation in male volunteers. Am J Cardiol 91:1392–4

23. Jurt U, Gori T, Ravandi A, Babaei S, Zeman P, Parker JD (2001) Differential effects of pentaerythritol tetranitrate and nitroglycerin on the development of tolerance and evidence of lipid peroxidation: a human in vivo study. J Am Coll Cardiol 38:854–9

24. Oberle S, Abate A, Grosser N, Hemmerle A, Vreman HJ, Dennery PA, Schneider HT, Stalleicken D, Schroder H (2003) Endothelial protection by pentaerithrityl trinitrate: bilirubin and carbon monoxide as possible mediators. Exp Biol Med (Maywood) 228:529–34

25. Oberle S, Schwartz P, Abate A, Schroder H (1999) The antioxidant defense protein ferritin is a novel and specific target for pentaerithrityl tetranitrate in endothelial cells. Biochem Biophys Res Commun 261:28–34

26. Grosser N, Schroder H (2004) Therapy with NO donors-antiatherogenic and antioxidant actions. Herz 29:116–22

27. Koenig A, Lange K, Konter J, Daiber A, Stalleicken D, Glusa E, Lehmann J (2007) Potency and in vitro tolerance of organic nitrates: partially denitrated metabolites contribute to the tolerance-devoid activity of pentaerythrityl tetranitrate. J Cardiovasc Pharmacol 50:68–74

28. Wenzel P, Oelze M, Coldewey M, Hortmann M, Seeling A, Hink U, Mollnau H, Stalleicken D, Weiner H, Lehmann J, Li H, Forstermann U, Munzel T, Daiber A (2007) Heme oxygenase-1: a novel key player in the development of tolerance in response to organic nitrates. Arterioscler Thromb Vasc Biol 27:1729–35

29. Minetti M, Mallozzi C, Di Stasi AM, Pietraforte D (1998) Bilirubin is an effective

antioxidant of peroxynitrite-mediated protein oxidation in human blood plasma. Arch Biochem Biophys 352:165–74

30. Mollnau H, Wenzel P, Oelze M, Treiber N, Pautz A, Schulz E, Schuhmacher S, Reifenberg K, Stalleicken D, Scharffetter-Kochanek K, Kleinert H, Munzel T, Daiber A (2006) Mitochondrial oxidative stress and nitrate tolerance – comparison of nitroglycerin and pentaerithrityl tetranitrate in Mn-SOD+/– mice. BMC Cardiovasc Disord 6:44

31. Sydow K, Daiber A, Oelze M, Chen Z, August M, Wendt M, Ullrich V, Mulsch A, Schulz E, Keaney JF Jr, Stamler JS, Munzel T (2004) Central role of mitochondrial aldehyde dehydrogenase and reactive oxygen species in nitroglycerin tolerance and cross-tolerance. J Clin Invest 113:482–9

32. Livak KJ, Schmittgen TD (2001) Analysis of relative gene expression data using real-time quantitative PCR and the 2(-Delta Delta C(T)) Method. Methods 25:402–8

33. Parviainen MT (1997) A modification of the acid diazo coupling method (Malloy-Evelyn) for the determination of serum total bilirubin. Scand J Clin Lab Invest 57:275–9

34. Daiber A, Oelze M, August M, Wendt M, Sydow K, Wieboldt H, Kleschyov AL, Munzel T (2004) Detection of superoxide and peroxynitrite in model systems and mitochondria by the luminol analogue L-012. Free Radic Res 38:259–69

35. Daiber A, Oelze M, Coldewey M, Kaiser K, Huth C, Schildknecht S, Bachschmid M, Nazirisadeh Y, Ullrich V, Mulsch A, Munzel T, Tsilimingas N (2005) Hydralazine is a powerful inhibitor of peroxynitrite formation as a possible explanation for its beneficial effects on prognosis in patients with congestive heart failure. Biochem Biophys Res Commun 338:1865–74

Chapter 23

Direct Determination of Tissue Aminothiol, Disulfide, and Thioether Levels Using HPLC-ECD with a Novel Stable Boron-Doped Diamond Working Electrode

Bruce Bailey, John Waraska, and Ian Acworth

Abstract

This chapter describes two different methods using reversed-phase HPLC with electrochemical detection on a boron-doped diamond (BDD) working electrode for the direct, routine, sensitive and simultaneous measurement of a number of aminothiols, disulfides, and thioethers, in either plasma or tissue homogenates.

Key words: Thiol, Disulfide, Glutathione, Glutathione disulfide, Cysteine, Cystine, S-nitrosoglutathione, Liver, Brain, Plasma, Boron-doped diamond, Electrochemical detection, HPLC

1. Introduction

Aminothiols (e.g., cysteine, glutathione [GSH], homocysteine), disulfides (e.g., cystine, glutathione disulfide [GSSG], homocystine), and thioethers (e.g., methionine) play many important biochemical roles in vivo (reviewed in (1)). GSH is not only a crucial antioxidant per se, but is involved in many aspects of cellular protection: it can regenerate other antioxidants (e.g., ascorbate); it acts as a cofactor for antioxidant enzymes (e.g., GSH peroxidase); and, through its involvement in the mercapturic acid pathway, is critical for detoxification of xenobiotics. Furthermore, maintenance of a high [GSH]/[GSSG] ratio (i.e., reductive environment) is essential for normal cellular functioning (see (1) and references therein). Methionine, an essential amino acid, is an intermediate in the biosynthesis of several key biomolecules including cysteine, carnitine, taurine, lecithin, and phosphatidylcholine. Its derivative, S-adenosyl methionine (SAM), serves as a methyl donor.

D. Armstrong (ed.), *Advanced Protocols in Oxidative Stress II,* Methods in Molecular Biology, vol. 594
DOI 10.1007/978-1-60761-411-1_23, © Humana Press, a part of Springer Science+Business Media, LLC 2010

A variety of HPLC techniques have been developed for measurement of thiols, disulfides, and thioethers (1). As these compounds do not possess an active chromophore, they must first be derivatized for measurement by either HPLC-UV (e.g., with Sanger's reagent) (2) or HPLC-fluorescence detection after derivative formation with OPA (3, 4)), monobromobimane (5)) or dansyl chloride (6). HPLC-UV often does not possess sufficient sensitivity for the measure of thiols and disulfides in some biological tissues or fluids.

HPLC with electrochemical detection (HPLC-ECD) is a sensitive approach that can be used for the *direct* measurement of thiols, disulfides, and thioethers, without the problems of sample dilution, reaction kinetics, and reaction efficiency often associated with derivatization procedures (1). Although numerous HPLC-ECD methods using different working electrode materials exist (e.g., mercury–gold amalgam and dropping mercury), these analytes are often measured using carbon-based (amperometric, thin layer glassy carbon, or coulometric, flow-through graphite) working electrodes (7, 8). Unfortunately, although these approaches are well published they often suffer from a number of problems that adversely affect method performance and reliability. The following lists possible problems, and their potential resolution:

- Continual loss of sensitivity due to absorption/poisoning of the working electrode surface by oxidation products. This requires routine maintenance to physically clean (polish) amperometric electrodes or the use of electrochemical conditioning to reactivate the surface of coulometric electrodes.

- Loss of analyte due to metal-induced auto-oxidation of labile thiols. Obviated by routine passivation of the HPLC system, or the use of a metal-reduced or metal-free system.

- Decreased sensitivity resulting from the increased noise associated with electrolysis of the mobile phase (typically water) by the high potentials required for the measurement of disulfides. The impact can be lessened by using fresh, degassed mobile phase, and by careful choice of applied potential.

- Increased chance of chromatographic co-elution due to the high potentials required for disulfide measurement. This can be addressed by sample preparation procedures, separation chemistry, appropriate choice of applied potential, and the use of a screening electrode (7).

- Chromatographic ghost peaks and baseline perturbations due to elution of highly retained hydrophobic materials on the analytical column – a consequence of using low % organic mobile phases. This requires the analytical column to be frequently washed with highly organic solvents.

Another carbon-based working electrode, the boron-doped diamond (BDD) electrode, shows unique physical and electrochemical properties (9, 10), and recently became commercially available (11). The BDD overcomes many of the issues described for other carbon-based working electrodes (above). The BDD electrode is highly inert, and free from fouling. It is stable, does not show changes in its surface characteristics over time, and shows low background currents and noise, even at high potentials. Furthermore, the electrolysis of the mobile phase occurs at a much higher potential relative to the oxidation of analytes of interest, thereby enabling the electrochemical measurement of analytes that is not possible with conventional working electrodes.

Presented in this chapter are two reversed-phase HPLC-ECD methods using a BDD working electrode for the measurement of thiols, disulfides, and thioethers in deproteinized plasma extracts, and in brain and liver homogenates.

2. Materials

2.1. Analytical HPLC System

All from ESA Biosciences, Inc.

1. Pump – Model 584 – EC compatible.
2. Autosampler – Model 540, or 542 – biocompatible with tray cooling.
3. Electrochemical detector: Coulochem® – Model 5300 – 2-channel system; or CoulArray® – Model 5600.
4. Analytical cell (single amperometric) – Model 5040.
5. BDD target and 25 μm gasket.
6. Guard cell (single coulometric electrode) – Model 5020. Placed prior to autosampler.
7. Thermal organizer.
8. PEEK pulse damper.
9. Software – EZChrom Elite™.
10. Column – Tissue analysis: Shiseido Capcell Pak C8 DD (3×250 mm; $3\,\mu$m) with guard column - Shiseido Capcell Pak C18 MG (4×20 mm; $3\,\mu$m). Plasma analysis: Interesil ODS C18 (3×150 mm; $3\,\mu$m) with guard column – Shiseido Capcell Pak C18 MG (4×20 mm; $3\,\mu$m).
11. Filters – post pump & pre-column, $0.5\,\mu$m PEEK.

2.2. Miscellaneous Equipment

1. Vortex-Genie® 2 Mixer (or equivalent).
2. Microcentrifuge, accuSpin, Micro AR (Fisher).

3. pH Meter, Accumet AB15 (Fisher), and appropriate pH linearity calibration solutions (e.g., pH 1 and pH 5).

4. Positive displacement microliter pipettes: 20 µL, 50 µL, 100 µL, 200 µL, 500 µL and replacement tips.

5. 2-L Glass beaker.

6. Microfilter centrifuge tubes – polysulfone 30K MW cut off. VectraSpin.

7. Microcentrifuge tubes, 2.0 mL (Fisher or equivalent).

8. Microcentrifuge tube caps – (Fisher or equivalent).

9. 2L Graduated cylinder.

10. Mobile phase filtering equipment. Consisting of: vacuum pump, 2L Erlenmeyer glass flask with side arm, solvent-inert connecting tubing (e.g., Teflon), sintered glass filtration unit (Alltech), and 0.2 µm nylon 66 filters (Varian).

11. Ultrasonic bath (Branson).

12. Sep-Pak® C18 SPE cartridges (Waters Inc) for water polishing.

13. Centrifuge accuSpin (Fisher).

14. Magnetic stirrer and stirring bar.

15. Tissue disruptor and sonic probe. Misonix (Fisher).

16. 2L Glass mobile phase bottle.

17. EDTA Vaccutainer® (Becton, Dickinson Co).

18. Limited Volume autosampler vials, polypropylene, 12×32 mm, 750 µL (e.g., ESA Vial kit 70-1695).

2.3. Reagents

1. De-ionized water: 18.2 MΩ-cm.

2. Acetonitrile (EMD, Omnisolve).

3. Sodium dihydrogen phosphate (EM Science).

4. 1-Octanesulfonic acid sodium salt (JT Baker, HPLC grade).

5. Phosphoric acid (Fisher Scientific, HPLC grade).

6. Perchloric acid (PCA), 70% (GFS Chemicals, Superior Reagent grade, 2477).

7. Ethylenediamine tetraacetic acid (EDTA) (Sigma, Ultra). Make 100 µM solution (3.7 mg in 100-mL water).

8. Tris[2-carboxyethyl phosphine] (TCEP) (Pierce). The TCEP reducing solution contains 50-mg TCEP/mL water.

9. Standards were obtained from Sigma. Cystathionine [CYSTATHIONINE] (C7505), cysteine [CYS] (C4022), cysteinyl-glycine [CYSGLY] (C0166), cystine [CYS2] (C8755), glutathione [GSH], glutathione disulfide [GSSG], homocysteine [HCYS], homocystine [HCYS2] (H0501);

methionine [METHIONINE] (M6039), *N*-acetylcysteine [NAC] (A7250), and *S*-Nitrosoglutathione (N4148).

10. Isotonic saline – 0.9% w/v aq sodium chloride.

3. Methods

3.1. Samples
(See Note 1)

3.1.1. Plasma (Human)

Collect whole blood into a 7-mL Vacutainer®-type tube containing EDTA as the anticoagulant. Mix by inversion a few times then immediately centrifuge the samples for 10 min, 15,000 rpm/17,860×G, at 4°C. Transfer plasma to a capped microcentrifuge tube and store at –80°C until analyzed. Do not subject samples to additional freeze-thaw cycles.

3.1.2. Brain and Liver Tissue (Rat)

Tissue samples were obtained from a local university. Male Sprague-Dawley rats (200–250 g) were sacrificed by decapitation. The brain was rapidly excised and then immediately washed in ice-cold isotonic saline prior to dissection into several gross regions (e.g., cortex, cerebellum, hippocampus etc), on ice. Samples were blotted dry and immediately stored in capped microcentrifuge tubes at –80°C until analysis. Similarly, a portion of liver was removed, washed in ice-cold saline, blotted dry, and stored in a microcentrifuge tube at –80°C until analysis.

3.2. Standards
(See Note 1)

1. Individual stock solutions were prepared in water at 1.0 mg/mL and stored at 4°C prior to use.

2. Working standards were prepared in cold mobile phase on the day of the analysis.

3. Working standards were kept at 5°C on the autosampler, throughout the analytical run.

3.3. Sample Preparation

3.3.1. Plasma

1. A 100 μL volume of plasma was placed in a 1.5-mL microcentrifuge tube on ice.

2. Add 100 μL of deionized water. (Control for TCEP protocol).

3. Vortex for 0.5 min.

4. Refrigerate. Wait for 10 min (Control for TCEP protocol).

5. Add 200 μL 0.1N PCA (0.86 mL 70% PCA+99.14-mL water).

6. Vortex for 0.5 min.

7. Transfer to microfilter centrifuge tube.

8. Centrifuge at 15,000 rpm/17,860×G for 30 min, at 5°C.

9. Transfer filtrate into an autosampler vial and cap.

10. Place on autosampler at 5°C.

3.3.2. Brain and Liver

1. Thaw tissue sample on ice.

2. Weigh potion of tissue in a microcentrifuge tube.

3. Make sure weight is <200 mg.

4. Add 0.4N PCA/100 nM EDTA solution (3.44 mL 70% + 95.56 mL water + 1.0 mL 100 µM EDTA solution), 1/10 w/v.

5. Sonicate 3 × 15 s to disrupt tissue.

6. Centrifuge 15,000 rpm/17,860×G, 15 min, at 5°C.

7. Transfer 100-µL supernatant to an autosampler vial.

8. Add 100 µL of mobile phase.

9. Mix.

10. Place on autosampler at 5°C.

3.4. Plasma TCEP Reduction Procedure

1. A 100 µL volume of plasma was placed in a 1.5-mL microcentrifuge tube on ice.

2. Add 75 µL of deionized water and 25 µL of TCEP solution. Vortex for 0.5 min.

3. Refrigerate. Wait for 10 min for reduction of disulfides.

4. Add 200 µL 0.1N PCA (0.86 mL 70% PCA + 99.14 mL water).

5. Vortex for 0.5 min.

6. Transfer to microfilter centrifuge tube.

7. Centrifuge at 15,000 rpm/17,860×G for 30 min, at 5°C.

8. Transfer filtrate into an autosampler vial and cap.

9. Place on autosampler at 5°C.

3.5. HPLC Analysis

1. The system, in sequence, consists of: pump, pulse dampener, PEEK in-line filter, guard cell, autosampler, PEEK in-line filter, column (in thermal organizer), and analytical cell (in thermal organizer). *See* Note 2.

2. Prepare mobile phase: *See* Note 3

   (a) Weigh the monobasic phosphate, monohydrate (6.9 g/2L – 25 mM), and place into a 2-L glass beaker.

   (b) Add about 1,000 mL of polished water (see ESA Biosciences Technical Note (70-4821) "Guidelines for mobile phase selection and preparation" for further details) and dissolve using a magnetic stirrer and stirrer bar.

(c) Filter the dissolved phosphate using the mobile phase filtering equipment. *See* Note 4.

(d) Weigh and add the OSA (0.6 g/2L – 1.4 mM).

(e) Add the acetonitrile (120 mL/2L – 6% v/v).

(f) Add water to about 1,900 mL.

(g) Adjust the pH to 2.65 with phosphoric acid

(h) Transfer to 2-L graduated cylinder and bring to mark with water.

(i) Pour into 2-L glass mobile phase bottle

(j) Degas prior to use. Place bottle in sonicator. Connect bottle to vacuum pump via Teflon tubing and rubber stopper). Draw vacuum, and then sonicate for 5 min. Turn off sonicator, and then vacuum pump.

(k) Place mobile phase on the HPLC system.

(l) Purge pump with new phase.

3. Pass fresh mobile phase through the system for at least 1 h, prior to analysis.

4. HPLC conditions:

(a) Flow rate: 0.75 mL/min (plasma) or 0.6 mL/min (tissue).

(b) Column oven temperature: 35°C.

(c) Injection volume: 10 μL.

(d) Autosampler tray temperature: 4°C.

(e) Applied potentials (mV vs. Pd reference): *See* Note 5

• Guard cell = +900 mV

• BDD Analytical cell = +1,500 mV

3.6. Results

The hydrodynamic voltammograms (HDVs) (sometimes referred to as current-voltage curve) for aminothiols, thioethers, and disulfides are shown in Fig. 23.1. This graph illustrates that the optimal applied potential for the detection of these compounds on BDD electrode is approximately +1,500 mV (vs. palladium reference electrode). *See* Note 6.

The analysis of various concentrations of aminothiols using HPLC with a BDD electrode produced a linear response curve as shown in Fig. 23.2. The limit of detection (LOD) was ~500 pg (on column).

The analysis of a typical plasma sample, using the plasma chromatography method is presented in Fig. 23.3. The authenticity of thiols was established by exposing the plasma sample to TCEP, which effectively reduces all disulfides to their corresponding thiols. This can be seen by the reduction of CYS2 (4.3 min)

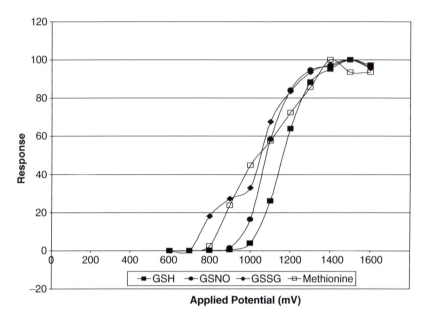

Fig. 23.1. HDVs of a representative thiol, disulfide, thioether, and nitrosothiol (10 μg/mL mixture on column). The potential applied to the BDD electrode started at +1,500 mV, and was decreased by 100 mV with each subsequent injection. The signal (current) produced for each analyte was plotted as a function of applied potential. The oxidation potential of 1,500 mV was chosen for all further experimentation. For HDVs of all analytes, see ref. 11

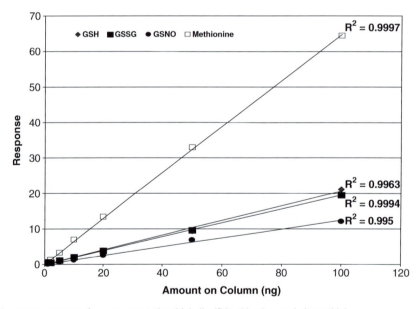

Fig. 23.2. The response curves for a representative thiol, disulfide, thioether, and nitrosothiol

to CYS (4.8 min) in Fig. 23.4. The response for GSH and GSSG were stable for at least 65 h of repetitive plasma analysis (%RSD were 1.5% and 6.5% for GSH and GSSG, respectively).

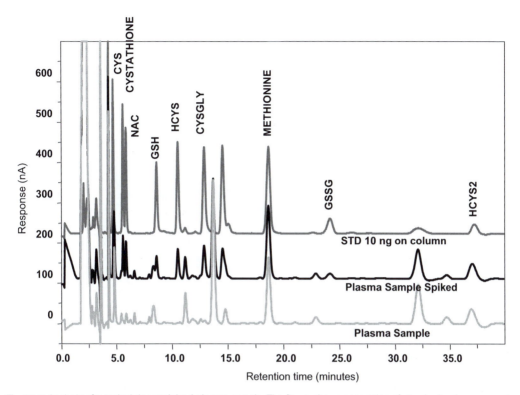

Fig. 23.3. Analysis of a typical deproteinized plasma sample. The figure shows separation of standards, plasma sample, and the same plasma sample spiked with external standards (20 ng into 100 µL sample). NAC co-eluted with an endogenous metabolite. Reproduced with permission from American Laboratory

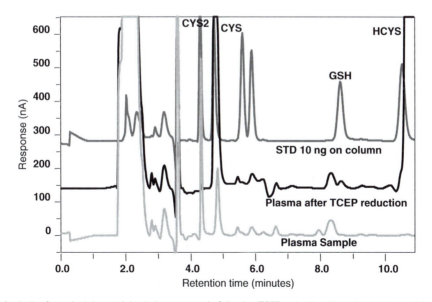

Fig. 23.4. Analysis of a typical deproteinized plasma sample following TCEP reduction. Note the decrease of the disulfide CYS2, and the increase in thiols CYS, and GSH, following TCEP treatment

The plasma levels of GSH and GSSG were found to be 1.39 ng/mL and 0.36 ng/mL, respectively. These were within the range of previous published values (1).

The analysis of GSH, GSSG, and methionine in rat brain and liver samples are presented in Figs. 23.5 and 23.6, respectively. *See* Note 7.

The levels of GSH and GSSG are presented in Table 23.1 and are in agreement with those published in literature (1).

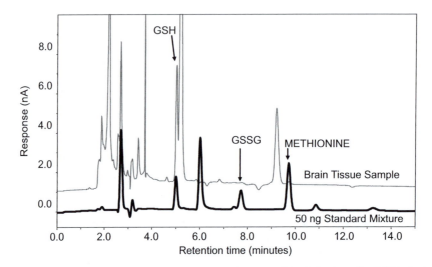

Fig. 23.5. Analysis of rat brain cortical tissue. Note there is an unknown analyte eluting just after GSH in the brain tissue

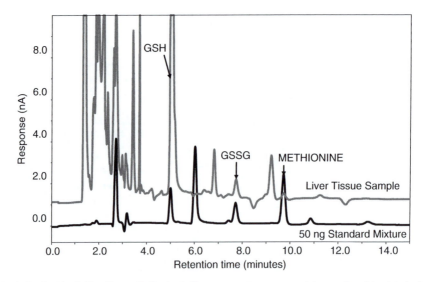

Fig. 23.6. Analysis of rat brain liver tissue. Unlike brain tissue, no endogenous analyte was found to elute just after GSH

Table 23.1
GSH and GSSG tissue levels

		GSH	GSSG
Sample	Region	µmol/g	nmol/g
1	Cortex	0.216	10.85
2	Cortex	0.204	9.04
3	Liver	3.659	35.68
4	Liver	4.609	49.61
5	Liver	4.152	45.24

4. Notes

1. Thiol standards show varying degrees of stability dependent on their chemical structure. Oxidation of thiols to disulfides can be minimized by cooling, lowering the pH of the standard solution to <2.00, and by the inclusion of EDTA to prevent transition metal-induced auto-oxidation. Thiols in blood can be rapidly lost due to mixed disulfide formation and through protein binding (e.g., mixed disulfide with cysteinyl thiols of albumin). Blood should be collected into EDTA vacutainers and centrifuged at 4°C immediately. Other precautions for GSH determination include the use of serine borate to inhibit γ-glutamyltransferase activity, not rupturing red blood cells (filled with GSH), and sampling at a specific time of day to avoid possible circadian rhythms (1, 12). Plasma should be either frozen at –80°C or rapidly deproteinized, filtered (centrifugation through microfilter), and analyzed immediately (2). Tissue must be rapidly prepared using acid deproteinization (above).

2. All HPLC components should be EC compatibility and be either metal free or contain minimum amounts of metal in order to prevent transition metal-induced oxidation of thiols to disulfides or formation of thiol–metal complexes. Routine passivation of the HPLC pump and autosampler, and inclusion of EDTA in the mobile phase can help keep transition metal levels to a minimum.

3. All chemicals used in the sample preparation solutions and in the mobile phase must be of the highest quality with minimum amounts of transition metals. See ESA Biosciences Technical Note (70-4821) "Guidelines for mobile phase selection and preparation" for further details.

4. It is usually only necessary to filter the sodium phosphate solution, as other chemicals rarely contain particulates.

5. A "clean cell command" (1,900 mV for 30 s) is usually applied at the end of each run. This helps to maintain the electrode. A 5-min re-equilibration step is necessary before the next run begins.

6. It is interesting to note that the oxidation of disulfides on BDD occurs at a lower potential relative to that for the thiols. This is the opposite to their electrochemical behavior on glassy carbon amperometric and graphite coulometric working electrodes (1, 8), suggesting that a very different electrochemical oxidation mechanism is taking place.

7. Modification of tissue method (flow rate = 0.7 mL/min; column temp = 32°C) can also be used to measure the biologically important nitrosothiol, S-nitrosoglutathione (GSNO) (13). GSNO elutes at ~12 min. Although the external standard is well resolved, and spiking into the tissue homogenate showed there to be no interferences, endogenous GSNO could not be detected in brain homogenates. With an LOD of ~500 pg (on column) this suggests that GSNO was not present in these tissues, was lost during sample collection/preparation due to its liability, or if present, was below detection limits.

References

1. Acworth IN (2003) The handbook of redox biochemistry. CD-ROM. ESA Biosciences. Part number: 70-6090

2. Fariss MW, Reed DJ (1987) High-performance liquid chromatography of thiols and disulfides: Dinitrophenol derivatives. Methods Enzymol 143:101–109

3. Keller DA, Menzel DB (1985) Picomole analysis of glutathione, glutathione disulfide, glutathione S-sulfonate, and cysteine S-sulfonate by high-performance liquid chromatography. Anal Biochem 151:418–423

4. Michelet F, Gueguen R, Leroy P, Wellman M, Nicolas A, Siest G (1995) Blood and plasma glutathione measured in healthy subjects by HPLC: Relation to sex, aging, biological variables, and life habits. Clin Chem 41:1509–1517

5. Fahey RC, Newton GL (1987) Determination of low-molecular-weight thiols using monobromobimane fluorescent labeling and high-performance liquid chromatography. Methods Enzymol 143:85–96

6. Martin J, White IN (1991) Fluorimetric determination of oxidised and reduced glutathione in cells and tissues by high-performance liquid chromatography following derivatization with dansyl chloride. J Chromatogr 568:219–225

7. Acworth IN, Bowers M (1997) An introduction to HPLC-based electrochemical detection: From single electrode to multi-electrode arrays. In: Acworth IN, Naoi M, Parvez H, Parvez S (eds) Coulometric electrode array detectors for HPLC. Progress in HPLC-HPCE, vol 6. VSP, Utrecht, The Netherlands, pp 3–50

8. Flanagan RJ, Perrett D, Whelpton R (eds) (2005) Thiols, disulfides and related compounds. In: Smith RM (Series ed) Electrochemical detection in HPLC: Analysis of drugs and poisons. RSC chromatography monographs, Athenaeum, UK, pp79–103

9. Xu J, Granger M, Chen Q, Strojek JW, Lister TE, Swain GM (1997) Boron-Doped Diamond Thin-Film Electrodes: Diamond thin films could be an electrochemist's best friend. Anal Chem 69:591A–597A

10. Rao T, Fujishima A, Angus J (2005) In: Fujishima A, Einaga Y, Rao T, Tryk D (eds) Historical survey of diamond electrodes in dia-

mond electrochemistry. Elsevier, Amsterdam, pp 1–10

11. Waraska J, Acworth IN (2007) A novel electrode material that extends the utility of HPLC-electrochemical detection. American Lab 39(14): 38–41

12. Jones DP, Carlson JL, Samiec PS, Sternberg P, Mody VC, Reed RL, Brown LA (1998) Glutathione measurement in human plasma. Evaluation of sample collection, storage and derivatization conditions for analysis of derivatives by HPLC. Clin Chim Acta 275: 175–184

13. Zhang Y, Hogg N (2005) S-Nitrosothiols: Cellular formation and transport. Free Radic Biol Med 38:831–838

Chapter 24

Activation of Erythrocyte Plasma Membrane Redox System Provides a Useful Method to Evaluate Antioxidant Potential of Plant Polyphenols

Syed Ibrahim Rizvi, Rashmi Jha, and Kanti Bhooshan Pandey

Abstract

Plant polyphenols are known to possess antioxidant acitivities. In recent years, there have been numerous reports confirming the efficacy of these compounds to improve plasma antioxidant capacity in humans. Current methods to evaluate the antioxidant potential of polyphenols are based on in vitro assay procedures (TEAC, ORAC, FRAP, DPPH). However, the antioxidant potential assessed by these methods does not correlate with the biological activity observed in vivo. Eukaryotic cells display a plasma membrane redox system (PMRS) that transfers electrons from intracellular substrates to extracellular electron acceptors. Here, we describe a method to evaluate the antioxidant potential of plant polyphenols based on their ability to enter the erythrocytes and donate electrons to PMRS. We also present results to show the potentiating effect of quercetin, EGCG, EC and catechin on erythrocyte PMRS activity.

Key words: Erythrocyte, PMRS, Polyphenols, Antioxidant potential

1. Introduction

Polyphenols are common constituents of food that are main sources of dietary polyphenols. An important activity ascribed to plant polyphenols is their strong antioxidant activity. As antioxidants, polyphenols may protect cell constituents against oxidative damage and, therefore, limit the risk of various degenerative diseases associated with oxidative stress (1). Several reports confirm that consumption of a polyphenol-rich diet increases plasma antioxidant capacity (2–4), although the correlation between antioxidant capacity in vitro and that observed under biological conditions is difficult to predict.

There are several methods to evaluate the antioxidant potential of polyphenols in vitro. Most of these methods are based on

D. Armstrong (ed.), *Advanced Protocols in Oxidative Stress II,* Methods in Molecular Biology, vol. 594
DOI 10.1007/978-1-60761-411-1_24, © Humana Press, a part of Springer Science+Business Media, LLC 2010

the ability of these classes of compounds to trap free radicals and reduce chemicals. The antioxidant potency is frequently compared to that of a reference substance, usually Trolox (a water soluble derivative of vitamin E), gallic acid, or catechins. The common methods to evaluate antioxidant potential of polyphenols are listed below:

1. TEAC (Trolox Equivalent Antioxidant Capacity) method: The method is based on the colorimetric determination of colored ABTS⁺ radical, formed from ABTS (2,2-azinobis 3-ethyl-benzothiazoline-6-sulphonic acid) (5). This method is applicable for both lipophilic and hydrophilic compounds.

2. ORAC (Oxygen Radical Absorbance Capacity) or TRAP (Total Radical-trapping Antioxidant Capacity) method: The method is based on the formation of radicals by heating AAPH (2,2-azobis 2-amidopropane hydrochloride) and monitoring the degradation of phycoerythrin by fluorimetry (6, 7).

3. FRAP (Ferric Reducing Ability of Plasma) method: It is based on the reduction of ferric ions (8).

4. DPPH radical scavenging assay: When DPPH· reacts with an antioxidant compound, which can donate hydrogen a radical, DPPH radical is reduced and a change in color is measured at 515 nm (9).

Frequently, results obtained from these methods are not comparable (10, 11). It is therefore not surprising that no correlation between the antioxidant potency of various polyphenols measured in vitro and their biological activity determined in vivo or at the cellular level has been observed in a large number of studies.

Eukaryotic cells display a plasma membrane redox system (PMRS) that transfers electrons from intracellular substrates to extracellular electron acceptors (12, 13). An important role of PMRS is in the maintenance of extracellular concentration of reduced ascorbic acid (ASC) (14). This function of PMRS is especially important in erythrocytes where it enables the cell to respond to changes in both intra- and extracellular redox environments. Recently, it has been shown that the activity of erythrocyte PMRS correlates with plasma antioxidant potential and that its activity increases as a function of human age (15), the increase in PMRS has been hypothesized to be a defense mechanism of the body activated to maintain plasma ascorbic acid level and antioxidant status (Fig. 24.1). This role of PMRS may assume significance in all disease conditions where there is a decrease in plasma antioxidant potential.

The predominant electron acceptor used to investigate erythrocyte PMRS is the membrane-impermeant oxidant, ferricyanide. The reduction of ferricyanide to ferrocyanide occurs at the cell membrane and can be monitored spectrophotometrically. Initial studies showed the reduction of ferricyanide due to the transfer of

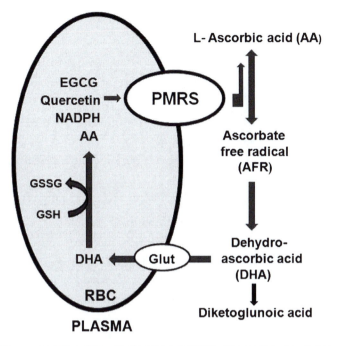

Fig. 24.1. Diagrammatic representation of the role of erythrocyte PMRS with ascorbate recycling in plasma. *Glut* glucose transporter; *GSG* reduced glutathione; *GSSG* oxidized glutathione

reducing equivalents from intracellular NADH (16). However, later it was found that intracellular ascorbate was the main source of reducing equivalents (17). It is now known that some polyphenols can also donate electrons to erythrocyte PMRS (18); this may play an important role in the maintenance of plasma reduced ascorbic acid level, which is the primary antioxidant.

The following sections describe a method to evaluate the antioxidant potential of plant polyphenols based on their ability to enter erythrocytes donating electrons to PMRS. The PMRS through its ascorbate free radical (AFR) reductase activity (19) uses the electrons to reduce extracellular AFR to ASC. The relative ability of individual polyphenols to donate electrons to PMRS provides a good parameter to assess their antioxidant potential under in vivo conditions.

2. Materials

2.1. Equipments

1. pH meter.
2. Incubator.
3. UV-VIS Spectrophotometer (Elico, India).
4. Cooling centrifuge (REMI, India, model C-24).

5. PC with software for statistical analysis (Graphpad Prism Version 5).

2.2. Reagents

1. Phosphate Buffered Saline (PBS): 0.154 M NaCl, 10 mM Na_2HPO_4 dissolve in double distilled water and pH adjusted to 7.4.

2. Potassium ferricyanide solution: 20 mM, freshly prepared in PBS.

3. Glucose Phosphate Buffer Saline (GPBS): PBS containing 5 mM glucose.

4. Sodium acetate: 3 M in H_2O, pH ranges 6–6.5.

5. Citric acid: 0.2 M solution in H_2O.

6. Acetic acid: 0.1 N.

7. Ferric chloride: 3.3 mM solution in 0.1 N acetic acid (*see* Note 1).

8. Bathophenanthroline disulfonic acid (DPI): 6.2 mM in H_2O (*see* Note 1).

3. Methods

3.1. Blood Collection and Isolation of Erythrocytes

1. Human venous blood from different healthy volunteers was collected by venipuncture in heparin containing tubes (10 units/mL). Collected blood was stored at 4°C and processed as fast as possible (*see* Note 2).

2. Blood samples were centrifuged at $1,800 \times g$ for 10 min at 4°C. Plasma, buffy coat and upper 15% of the packed red blood cells was taken by aspiration and the RBC were washed twice with cold PBS (0.9% NaCl, 10 mM Na_2HPO_4, pH 7.4). The packed red blood cells (pRBC) were thus obtained.

3.2. Incubation of Erythrocytes with Polyphenols

To 0.3 mL of packed red blood cells (pRBC) add 2.4 mL of PBS (containing 5 mM glucose) followed by 0.3 mL of polyphenols (10^{-5} M final concentration). Incubate the suspension under mild magnetic stirring at 37°C for 30 min (*see* Note 3). After incubation, centrifuge the suspension and wash once with PBS. The pRBC obtained to be used for PMRS assay.

3.3. Determination of the Activity of Erythrocyte PMRS

1. The procedure for measuring the activity of erythrocyte PMRS is same as described by Avron and Shavit (20).

2. Suspend 0.2 mL pRBC in PBS containing 5 mM glucose and freshly prepared potassium ferricyanide. Incubate the suspension for 30 min at 37°C and then centrifuge at $1,800 \times g$ at 4°C.

3. Collect the supernatant for assay of ferrocyanide content. 0.2 mL of supernatant is diluted to a final volume of 1.4 mL with water. After which 0.2 mL of sodium acetate, 0.2 mL of citric acid and 0.1 mL of ferric chloride is added to the solution.

4. Lastly 0.1 mL of DPI is added to the reaction mixture, making it to 2 mL final solution.

5. A pink color is developed; the absorbance is measured after 5 min at 535 nm and calculations are done by using molar absorption coefficient (ε) 20,500/M/cm.

6. Results are expressed in micromole ferrocyanide/mL PRBC/30 min. This represents the erythrocyte PMRS activity.

3.4. Results

1. Figure 24.2 shows the result of the activation of erythrocyte PMRS activity by quercetin and tea catechins (epigallocatechin gallate, epicatechin and catechin) at a 10^{-5} M final concentration (*see* Note 3). Our results show that quercetin significantly ($p<0.001$) increases the activity of erythrocyte PMRS confirming previous studies (21). EGCG also has a significant (86%) activating effect on erythrocyte PMRS activity. The effect of EGCG is followed by catechin and (–)epicatechin.

2. It has been established that the electron-donating ability of flavonoids depends on the position and degree of hydroxylation (22), the reducing ability being enhanced by the presence

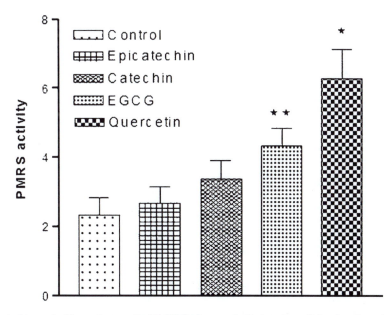

Fig. 24.2. Effect of tea catechins and quercetin (10^{-5} M final concentration) on the activity of erythrocyte PMRS activity. *$p<0.001$ and **$p<0.01$ compared with control

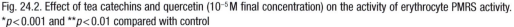

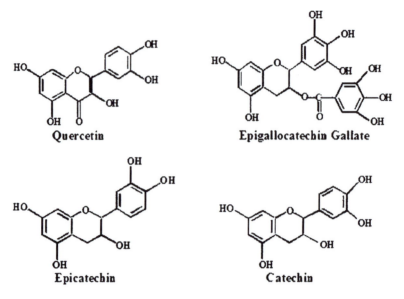

Fig. 24.3   Structures of different polyphenols used in the study

of hydroxyl groups at 3 and 5 positions. Our results do not substantiate this assumption. We observe a wide variation in the electron-donating ability between EGCG, catechin and (–)epicatechin, although all these have –OH group at positions 3 and 5. The structures of quercetin, EGCG, EC and catechins are shown in Fig. 24.3

3. The erythrocyte PMRS can play a defensive role in combating extracellular oxidative stress. During aging process, its activation provides a clue to the physiological role of PMRS in the maintenance of plasma ascorbate level (15) (*see* Note 4). The fact that some flavonoids display the ability to activate erythrocyte PMRS suggests a possible mechanism for the antioxidant effect of flavonoids.

4. The activation of erythrocyte PMRS by polyphenols gives a better assessment of their in vivo antioxidant potential of polyphenols compared to in vitro methods (TEAC, ORAC, FRAP, DPPH) and possibly provides a better correlation with their biologic activity.

4. Notes

1. Solutions ferric chloride and DPI should be freshly prepared.

2. Care should be taken to avoid hemolysis of erythrocytes during collection of blood and isolation of erythrocytes. Very

gentle pressure must be applied to expel blood from the syringe into the heparin tubes.

3. Erythrocyte should be very mildly stirred (magnetic stirring) during incubation with flavonoids to avoid hemolysis.

4. Since the activity of erythrocyte PMRS is age-dependent, in comparative studies only age-matched subjects should be taken.

Acknowledgement

The work was supported by a research project grant to S I Rizvi from University Grants Commission, New Delhi, India.

References

1. Scalbert A, Manach C, Morand C, Remesy C (2005) Dietary polyphenols and the prevention of diseases. Crit Rev Food Sci Nutr 45:287–306

2. Leenen R, Roodenburg AJ, Tijburg LB, Wiseman SA (2000) A single dose of tea with or without milk increases plasma antioxidant activity in humans. Eur J Clin Nutr 54: 87–92

3. Fuhrman B, Lavy A, Aviram M (1995) Consumption of red wine with meals reduces the susceptibility of human plasma and low-density lipoprotein to lipid peroxidation. Am J Clin Nutr 61:549–554

4. Maxwell S, Cruickshank A, Thorpe G (1994) Red wine and antioxidant activity in serum. Lancet 344:193–194

5. Rice-Evans C, Miller NJ (1994) Total antioxidant status in plasma and body fluids. Methods Enzymol 234:279–293

6. Cao G, Russell RM, Lischner M, Prior RL (1998) Serum antioxidant capacity is increased by consumption of strawberries, spinach, red wine or vitamin C in elderly women. J Nutr 128:2383–2390

7. Ghiselli A, Serafini M, Maiani G, Aazzini E, Ferro-Luzzi A (1995) A fluroresence-based method for measuring total plasma antioxidant capacity. Free Radic Biol Med 18:29–36

8. Benzie IF, Strain JJ (1996) The ferric reducing ability of plasma (FRAP) as a measure of "antioxidant power": the FRAP assay. Anal Biochem 239:70–76

9. Ratty AK, Sunamoto J, Das NP (1988) Interaction of flavonoids with 1, 1-diphenyl 2-picrylhydrazyl free radical, liposomal membranes and soybean lipoxygenase-1. Biochem Pharmacol 37:989–995

10. Rice-Evans CA, Miller NJ, Bolwell PG, Bramley PM, Pridham JB (1995) The relative antioxidant activities of plant-derived polyphenolic flavonoids. Free Radical Res 22: 375–383

11. Guo C, Cao G, Sofic E, Prior RL (1997) High-performance liquid chromatography coupled with coulometric array detection of electro active components in fruits and vegetables: relationship to oxygen radical absorbance capacity. J Agric Food Chem 45:1787–1796

12. May JM, Qu ZC (1999) Ascorbate-dependent electron transfer across the human erythrocyte membrane. Biochem Biophys Acta 1421:19–31

13. Crane FI, Low H (1976) NADH oxidation in liver and fat cell plasma membranes. FEBS Lett 68:153–156

14. May JM, Qu ZC, Cobb CE (2000) Extracellular reduction of ascorbate free radical by human erythrocytes. Biochem Biophys Res Commun 267:118–123

15. Rizvi SI, Jha R, Maurya PK (2006) Erythrocyte plasma membrane redox system in human aging. Rejuvenation Res 9(4):470–474

16. Zamudio I, Canessa M (1966) Nicotinamide–adenine dinucliotide dehydyrogenase activity of human eryhtrocyte membranes. Biochim Biophys Acta 120:165–169

17. May JM, Qu ZC, Cobb CE (2004) Human erythrocyte recycling of ascorbic acid relative

contributions from the ascorbate free radical and dehydroascorbic acid. J Biol Chem 279:14975–14982

18. Fiorani M, Accorsi A (2005) Dietary flavonoids as intracellular substrates for an erythrocyte trans-plasma membrane oxidoreductase activity. Br J Nutr 94:338–345

19. Su D, May JM, Koury MJ, Asard H (2006) Human erythrocyte membranes contain a cytochrome b561 that may be involved in extracellular ascorbate recycling. J Biol Chem 281:39852–39859

20. Avron M, Shavit N (1963) A sensitive and simple method for determination of ferrocyanide. Anal Biochem 6:549–554

21. Fironi M, Sanctis RD, Bellis RD, Dacha M (2002) Intracellular flavonoids as electron donors for extracelluar ferricynide reduction in human erythroytes. Free Radic Biol Med 32(1):64–72

22. Rice Evans C, Miller NJ, Paganga G (1996) Structure-antioxidant activity relationships of flavonoids and phenolic acids. Free Radic Biol Med 20:933–956

Chapter 25

Antioxidant Activity of Biotransformed Sex Hormones Facilitated by *Bacillus Stearothermophilus*

Mohammad Afzal, Sameera Al-Awadi, and Sosamma Oommen

Abstract

Bacillus stearothermophilus, a thermophilic bacterium isolated from Kuwaiti desert, when incubated with exogenous progesterone for 10 days at 65°C produced two monohydroxylated, two dihydroxy isomers of progesterone and a B-Seco compound. These metabolites were purified by TLC and HPLC followed by their identification through ^1H, ^{13}C NMR and other spectroscopic data. Microbial hydroxylation of 17β-estradiol resulted in the production of estrone. The effect of some inducers resulted in the production of two metabolites from 17β-estradiol one of which was identified as 3,6β,17β-trihydroxyestra-1,3,5,14(10)-tetrene and the other metabolite remained unidentified. The transformation products were identified through their spectral data and comparison with reference to compounds. Antioxidant activities of progesterone transformed mixture and purified metabolites of 17β-estradiol were studied by linoleic acid/β-carotene assay. An enhanced antioxidant activity for progesterone transformation products was observed, when compared to progesterone. A comparison of antioxidant activity of progesterone and 17β-estradiol transformation products is reported.

Key words: Antioxidant activity, Steroids, Biotransformation, *Bacillus stearothermophilus*

1. Introduction

Oxidative stress is a principal cause of aging and chronic diseases such as inflammation, infection, cancer, and cardiovascular disorders (1, 2). Exogenous or endogenous sources of oxidative stress and weakened antioxidative defenses can damage macromolecules such as DNA, lipids, and proteins.

Estrogens are powerful antioxidants, and through their receptor status, estrogen/progesterone play an important role, in development, growth, and the differentiation of both male and female secondary sex characteristics (3) and protection against neurodegeneration and oxidative stress (4). In addition, estrogen

D. Armstrong (ed.), *Advanced Protocols in Oxidative Stress II,* Methods in Molecular Biology, vol. 594
DOI 10.1007/978-1-60761-411-1_25, © Humana Press, a part of Springer Science + Business Media, LLC 2010

has favorable effect on lipid profile, endothelial cell function, vascular reactivity and haemostatic factors. Thus estrogens, such as 17β-estradiol (E_2), have protective effect on oxidative stress mediated by estrogen receptor ∴α(ERα) (5) and may also protect from cardiovascular disease and hepatic fibrosis (6). Exogenous estrogen is known to protect against atherosclerosis by modulating low-density lipoprotein oxidation, binding free radicals and lowering plasma cholesterol (7).

Steroids are used as progestational, anabolic, antitumor agents and oral contraceptives as well as sedatives. Hydroxylation of steroids is valuable due to thier physiological and clinical bearing. Thus hydroxylated estrogens are important therapeutic agents that are used in the treatment of breast tumors etc. However, chemical synthesis of these new estrogens is difficult to achieve. We have investigated a bacterial transformation of these hormones, and have achieved many new hydroxylated estrogens identified through their spectral data. This paper reviews the newly biotransformed sex hormones by a thermophilic bacteria *B. stearothermophilus* isolated from oil polluted desert soil around the Kuwait oil fields. We have made a study on biotransformation of progesterone, testosterone and 17β-estradiol. Transformation of progesterone for 24 h resulted in three hydroxylated progesterone derivatives 20-hydroxyprogesterone, 6β-hydroxyprogesterone and the rare 6α-hydroxyprogesterone. A new biotransformed metabolite 9,10-seco-4-pregnene-3,9,20-trione was also purified and identified (8). Prolonged incubation of progesterone resulted in dihydroxy and 5α-progesterone along with B-seco-progesterone (9). Some of the products identified from testosterone biotransformation are 6α-hydroxyandrost-4-en-3,17-dione, 6β-hydroxyandrost-4-en-3,17-dione, 6α-hydroxytestosterone, 6β-hydroxytestosterone and androst-4-en-3,17-dione (10). The metabolism of 17β-estradiol resulted in the formation of estrone, 3,6β,17β-trihydroxyestra-1,3,5,14(10)-tetrene and an unidentified compound.

The steroidogenic capacity and oxidative stress-related parameters of the human corpus luteum (CL) at different stages of the luteal phase have been studied under basal and human chorionic gonadotropin (hCG)-stimulated conditions (11). Oxidative stress is known to inhibit ovarian and testicular steroidogenesis and progesterone receptor modulators with antioxidant effects which have been investigated as possible treatment for endometriosis (12). Stereochemical synthesis of hydroxylated steroids is not only difficult, but also results in enantiomeric-mixtures posing yet another problem for purification. Biotrsnformation of steroids by thermophilic bacteria offers a facile approach for obtaining stereospecific hyroxylated sex hormones with enhanced antioxidant activity. In this chapter, we discuss biotrsnformation of progesterone and a comparative study of the antioxidant activity of the biotransformed hydroxylated products.

2. Materials

2.1. Equipment

1. Analytical HPLC system (GBC Australia)
2. UV detector (GBC Model LC 1205 UV/visible detector).
3. Winchrome chromatography software
4. HPLC column (Waters C_{18} symmetry column 4.6×150 mm).
5. ^{1}H and ^{13}C NMR spectroscopy techniques (400 MHz NMR – Bruker AC 400).
6. Infrared data (Perkin Elmer System 2000 FTIR).
7. Mass spectra were determined using a V.G. Analytical Ltd, (Manchester, England) model 305 mass spectrometer-2025 attached with a library data system.

2.2. Reagents

1. Culture media were purchased from Fluka Riedel-deHaën (Germany) and Scharlau (Barcelona, Spain).
2. All chemicals were of analytical grade and were supplied by Merck (Darmstadt, Germany) and Scharlau.
3. Kieselgel-60 F_{254} fluorescent thin layer chromatographic plates (TLC) were obtained from Merck (Darmstadt, Germany).
4. Progesterone, 17β-estradiol and their derivatives for the identification of metabolites were obtained from Sigma-Aldrich Co. UK.

3. Methods

3.1. Biotransformation of 17β-Estradiol

Starter cultures (50 mL) of *B. stearothermophilus* were grown overnight on TYE medium at 65°C in a shaker incubator (*see* Note 1). The culture was transferred to 500 mL of fresh TYE media in a 1 L flask and was kept under the same condition until the end log phase of the growth (4–4.5 h). Cells were collected by centrifugation and were washed with 0.05 M sodium phosphate buffer pH 7 and mixed with 10 mL of the same buffer and transferred to 500 mL of the phosphate buffer containing 10% Castenholtz mineral salt solution with 0.1% ascorbic acid (*see* Note 2). A final concentration of 0.3 mg/mL of 17β-estradiol was added and the cultures were re-incubated at 65°C for 72 h (*see* Note 3). Inducers for the transformation of 17β-estradiol like chloramphenicol (50 μg/mL), cyclodextrin (2 moles/mole of substrate) and riboflavin (1 mg/50 mL) were individually studied. The metabolites were extracted with ethyl acetate and the solvent was evaporated under vacuum. The residue, redissolved in methanol, was resolved on silica gel TLC plates (Merck 60 F_{254}) using benzene:dichloromethane:acetone:ethylacetate (22:22:3:3 v/v/v) as a mobile phase. The purified metabolites were identified from their respective spectral data including ^{1}H, ^{13}C NMR, mass spectroscopy and FT infrared spectrophotometry.

3.2. Biotransformation of Progesterone

Cells of *B. stearothermophilus* were collected after incubation as described for 17β-estradiol. Cell suspension was transferred to phosphate buffer containing 10% Castenholtz mineral salt solution (*see* Note 4). An ethanolic stock solution of progesterone (*see* Note 5) was added to this flask to achieve a final concentration of 50 μg/mL and was re-incubated for a further period of 10 days.

Progesterone metabolites were extracted by shaking the buffer with an equal volume of chloroform. Solvent from the dried extract was evaporated under reduced pressure and the residue was redissolved in methanol and chromatographed using 20 × 20 cm Kieselgel-60 F_{254} fluorescent TLC plates, developed in diethylether:toluene:ethylacetate (3:5:2 v/v/v) as a mobile phase. The metabolites were viewed under UV light (*see* Note 6), marked with a pencil and edge of the TLC chromatogram was stained with an anisaldehyde reagent, which produced a spectrum of different colors. An isocratic elution of these compounds using acetonitrile:water mixed with 0.1% acetic acid (31:69 v/v; *see* Note 7) as a mobile phase in HPLC separation, resulted in good resolution of the dihydroxy compounds from the corresponding monohydroxy-progesterone derivatives (Fig. 25.1).

3.3. Results

3.3.1. Estradiol metabolites

Metabolism of 17β-estradiol resulted in the formation of estrone which is commonly reported as conversion product. Biotransformation was also studied by an addition of inducers such as cyclodextrin, riboflavin and chloramphenicol, that resulted in the production of other two additional metabolites one of which was identified as 3,6β,17β-trihydroxyestra-1,3,5,14(10)-tetrene and the other remained unidentified. The molecular structures of these metabolites are shown in Fig. 25.2.

3.3.2. Progesterone Metabolites

Thus thermophilic bacillus was capable of site-selective hydroxylation of progesterone at position C-6, producing a rare microbial transformation product 6α-hydroxyprogesterone and its congener 6β-hydroxy progesterone. In addition, two other metabolites were also produced, one of which was identified as a common biotransformation isomer 20-hydroxyprogesterone and the other 9,10-seco-4-pregnene-3,9,20-trione (B-seco compound) which arose from rare B ring cleavage of progesterone molecule. With a prolonged incubation of up to 10 days, dihydroxylated products, 6α,20-dihydroxyprogesterone and

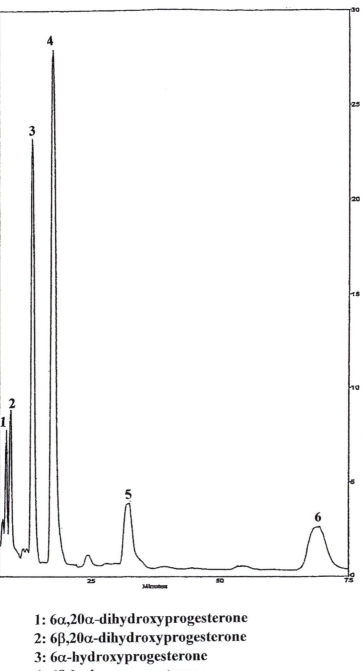

1: 6α,20α-dihydroxyprogesterone
2: 6β,20α-dihydroxyprogesterone
3: 6α-hydroxyprogesterone
4: 6β-hydroxyprogesterone
5: 4-Pregnene 3,9,20 trione (B-Seco compound)
6: 20α-hydroxyprogesterone.

Fig. 25.1. HPLC chromatogram of 10 day progesterone transformation by *B. stearothermophilus*. From *left* to *right*, Peak 1, 6α,20-dihydroxyprogesterone; Peak 2, 6β,20-dihydroxyprogesterone; Peak 3, 6α-hydroxyprogesterone; Peak 4, 6β-hydroxyprogesterone; Peak 5, B-Seco compound; Peak 6, 20α-hydroxyprogesterone.

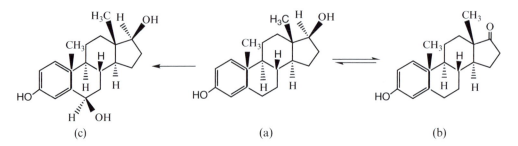

Fig. 25.2. The structures of 17β-estradiol and their transformation products. (a) 17β-estradiol, (b) Estrone, (c) 3,6β,17β-trihydroxyestra-1,3,5,14 (10)-tetrene.

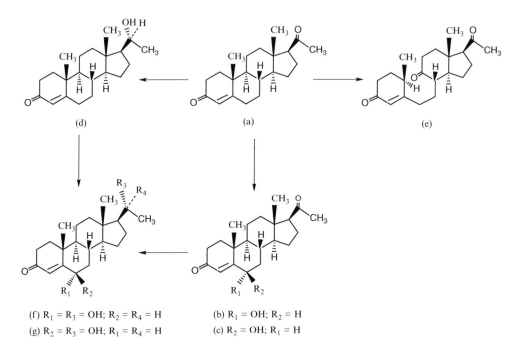

(f) $R_1 = R_3 = OH; R_2 = R_4 = H$ (b) $R_1 = OH; R_2 = H$

(g) $R_2 = R_3 = OH; R_1 = R_4 = H$ (c) $R_2 = OH; R_1 = H$

Fig. 25.3. Molecular structures of progesterone and its metabolites. (a) Progesterone (b) 6α-hydroxyprogesterone (c) 6β-hydroxyprogesterone (d) 20α-hydroxyprogesterone (e) 9,10-Seco-4-pregnene-3,9,20-trione (f) 6α,20-dihydroxyprogesterone (g) 6β,20-dihydroxyprogesterone.

6β,20α- dihydroxyprogesterone were also formed. The structures of the metabolites and its formation are shown in Fig. 25.3.

3.3.3. Antioxidant Studies of the Biotransformed Products

Antioxidant activities of progesterone transformed mixture and purified metabolites of 17β-estradiol were studied by linoleic acid/ β-carotene assay and are shown in Fig. 25.4a, b. Transformation products of 17β-estradiol, the estrone and unidentified product showed little enhancement in their antioxidant properties since these metabolites were mainly oxidized products of the parent sterol (Fig 25.4a). However, progesterone transformation products mixture with their multiple hydroxyl

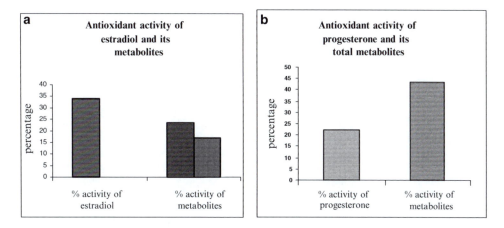

Fig. 25.4. Antioxidant activity of estradiol and its hydroxy metabolites (**a**), progesterone and its hydroxy metabolites (**b**)

groups, showed enhanced antioxidant activity as shown in Fig. 25.4b. Thus biotransformed progesterone metabolites carry improved antioxidant activity that may have clinical implications.

4. Notes

1. As we deal with thermophiles, it is preferred to autoclave media and buffer at higher temperature for a longer period of time.

2. Extreme care should be taken to avoid the contamination of the strains. Contamination can result in unexpected metabolites.

3. Increase in time of incubation can enhance the number and quantity of the biotransformed metabolites.

4. Transformation experiment in buffer with minerals is preferred rather than media to maximize the biotransformation.

5. Stock solution of steroids should be freshly prepared to prevent decomposition during storage in solvents

6. Care should be taken while viewing with UV lamp

7. Acetic acid in the HPLC mobile phase helps in getting sharp peaks and a good separation of hydroxysteroid metabolites

Acknowledgement

This work was supported by a Kuwait University grant # SB040 and SL02/02 for which the authors are thankful.

References

1. Yoshishige U, Yoshito I, Hiroaki M, Shinji G, Takehiko K, Junji Y, Satoshi I, Takahito K (2006) 17β-Estradiol protects against oxidative stress-induced cell death through the glutathione/glutaredoxin-dependent redox regulation of Akt in myocardiac H9c2 cells. J Biol Chem 218(19):13092–13102

2. Finkel T, Holbrook NJ (2000) Oxidants, oxidative stress and the biology of aging. Nature 408:239–247

3. Yang SH, Liu R, Perez EJ, Wen Y, Stevens SM Jr, Valencia T, Brun-Zinkernagel AM, Prokai L, Will Y, Dykens J, Koulen P, Simpkins JW (2004) Mitochondrial localization of estrogen receptors β. Proc Natl Acad Sci U S A 101: 4130–4135

4. Noriko S, Kenji T, Masahiro A, Junya A, Masayoshi H, Katsuya I, Seungbum K, Yi-Qiang L, Yumiko O, Tokumitsu W, Ieharu Y, Masao Y, Masato E, Yasuyoshi O (2001) Estrogen prevents oxidative stress–induced endothelial cell apoptosis in rats. Circulation 103:724–731

5. Baba T, Shimizu T, Suzuki Y, Ogawara M, Isono K, Koseki H, Kurosawa H, Shirakawa T (2005) Estrogen, insulin, and dietary signals cooperatively regulate longevity signals to enhance resistance to oxidative stress in mice. J Biol Chem 280:16417–16426

6. Yan L, Ji Z, Yan L, Xiao-Song G (2002) Protective effect of estradiol on hepatocytic oxidative damage. Gastroenterology 8(2): 363–366

7. Kuhl H (1994) Cardiovascular effects and estrogen/gestagen substitution therapy. Ther Umsch 51:748–754

8. Al-Awadi S, Afzal M, Oommen S (2001) Studies on *Bacillus stearothermophilus*. Part 1. Transformation of progesterone to a new metabolite 9,10-seco-4-pregnene-3,9,20-trione. J Steroid Biochem Mol Biol 78(5): 493–498

9. Al-Awadi S, Afzal M, Oommen S (2002) Studies on *Bacillus stearothermophilus*. Part II. Transformation of progesterone. J Steroid Biochem Mol Biol 82(2–3):251–256

10. Al-Awadi S, Afzal M, Oommen S (2003) Studies on *Bacillus stearothermophilus*. Part III. Transformation of testosterone. Appl Microbiol Biotechnol 62:48–52

11. Vga M, Castillo T, Retamales I, Heras JL, Devoto L, Videla LA (1994) Steroidogenic capacity and oxidative stress-related parameters in human luteal cell regression. Free Radic Biol Med 17(6):493–499

12. Gupta S, Agarwal A, Krajcir N, Alvarez JG (2006) Role of oxidative stress in endometriosis. Reprod Biomed 13(1):126–134

Chapter 26

Separation of Phenylpropanoids and Evaluation of Their Antioxidant Activity

Sammer Yousuf, M. Iqbal Choudhary, and Atta Ur Rahman

Abstract

Phenylpropanoids are a group of natural products with a wide range of biological and pharmacological importance. They have been isolated from a large number of plants by utilizing a diverse range of chromatographic techniques. We describe here the utilization of normal, reverse phase, and high pressure liquid column chromatographic techniques for the purification of phenylpropanoids from the crude plant extracts. Most important of them is the recycling reverse phase HPLC effectively utilized to purify the highly oxygenated phenylpropanoids glycosides from the butanolic and water extracts. The antioxidant activity of the different phenylpropanoids is also reported on the basis of DPPH, superoxide anion scavenging and Fe^{2+}-chelating assays, along with electron spin resonance (ESR) method.

Key words: Phenylpropanoids, Chromatographic separation, Antioxidant activity, DPPH scavenging activity, Superoxide anion-scavenging activity, Fe^{2+}-chelating assays, ESR

1. Introduction

Phenylpropanoids are the bioactive secondary metabolites with a three-carbon side chain, attached to a phenyl ring. They are produced in nature from the amino acid, phenylalanine (1).

Phenylpropanoids have been reported from the plants of the families Acanthaceae, Asterraceae, Araliaceae, Boranginaceae, Buddlejaceae, Bignoniaceae, Crassulaceae, Icacinaceae, Lamiaceae, Martyniaceae, Magnoliaceae, Oleaceae, Orobanchaceae, Pedaliaceae, Plantaginaceae, Scrophulariaceae, Salicaceae, Thymelaeaceae and Verbenaceae (2, 3).

Although phenylpropanoids are known as a large group of biologically active compounds, still no acceptable classification exits. The literature, however, suggests the following categories of phenylpropanoids:

D. Armstrong (ed.), *Advanced Protocols in Oxidative Stress II,* Methods in Molecular Biology, vol. 594
DOI 10.1007/978-1-60761-411-1_26, © Humana Press, a part of Springer Science+Business Media, LLC 2010

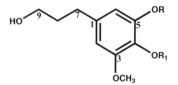

3-[7-(Methyl)-butyroyloxy)-6-hydroxy, 8-methoxy]-phenyl-propanol (1), R=H, R₁=

3-[6-(Methyl)-hydroxy-7-butyroyloxy), 8-methoxy]-phenyl-propanol (2), R=

1. Cinnamyl alcohol.

2. Cinnamic acid and their derivatives, such as esters, glycosides, amides, aldehydes.

3. Complex oxidative coupling products such as lignans, neolignans, and lignane glycosides are also reported from this group (2, 3).

Phenylpropanoids show a wide range of biological activities such as antimicrobial (4), antibacterial (2, 5–7), anti fungal (5–7) anti viral (8), adaptogenic (9), neuro-tropic, and immunostimulating properties (10). Some of the phenylpropanoids possess hepato-protective activity against the galactosamine intoxication, induced by hepatitis virus in man. They also showed significant hepato-protective activity against the CCl₄-induced toxicity (11). Some of these compounds have also exhibited agonist activity for the transient receptor potential vanilloid 1 (TRPVI), which is a Ca²⁺ permeable cation channel. The activity was evaluated by conducting an analysis of the calcium concentration in TRPVI-expressing HEK293 cells (12). Some of them also have anticancer properties (2). The anti-ulcer activity in water-immersed rats showed that many phenylpropanoids inhibit stress-induced gastric ulceration (13). A neolignan, isolated from *Zizyphus jujuba* leaves, was reported as an endogenous prostaglandin inducer (14).

Phenylpropanoids can also be useful as anti-inflammatory agents, as they significantly inhibit the histamine release from rat basophilic leukemia cells (RBL-2H3), stimulated with A23187. Furthermore, these compounds also cause a decline in TNF-alpha levels in culture supernatants of RBL-2H3 cells, followed by treatment with A23187 (15).

The major phenylpropanoids from *Marrubium vulgare* exhibited an inhibitory potency against the cyclooxygenase-2. This explains the traditional uses of the phenylpropanoids containing plants as anti-inflammatory drugs (16). Phenylpropanoids are also found to possess antiproliferative and proptotic activities on malignant melanoma cells (17).

The most appreciated property of phenylpropanoids is their antioxidant and free radical scavenging activity (18–26). It has been proved by a number of studies worldwide that free radicals, so called "reactive oxygen species (ROS)," are involved in various diseases of living systems. The positive aspect of these radicals in living systems is cell signaling and protection, but nevertheless their adverse reactions are very damaging. ROS are produced by human body in many physiological processes during the mitochondrial oxidation, inflammation in response to various stimuli, in peroxisomes, during oxidation of fatty acids, and during the neutralization of different xenobiotics. These reactive species are found to be involved in various life-threatening diseases, such as cancers, cardiovascular disorders, immune system decline, and diabetes. Among various ROS, superoxide anion radicals are extensively studied. They have an important role in the progression and complications of diabetes.

There is an increasing interest in antioxidants, particularly in those which prevent the presumed deleterious effects of free radicals in the human body, and to prevent the deterioration of foodstuffs. In both cases, there is a preference for antioxidants from natural, rather than from synthetic sources. There is a parallel development in the use of methods for estimating the efficiency of such substances as antioxidants. The most widely used methods to evaluate the antioxidant activities of the phenylpropanoids are DPPH free radical scavenging assay, the superoxide scavenging, metal (Fe^{2+})-chelating assays, and electron spin resonance (ESR) methods (19–29).

2. Materials

2.1. Equipments

1. HPLC LC-10AS chromatograph, Shimadzu (Compounds 3 and 4).
2. HPLC pump with Rheodyne 7510 injector and a UV detector, Milton Roy (Compounds 9 and 10).
3. Prep. HPLC (Lichrospher RP 18, 7 μm, 250×10 mm) (Compounds 9 and 10).
4. Recycling Preparative HPLC: JAI, LC-908W (Compounds 11–29).
5. Microtiter Plate Reader, SpectraMax Plus 384 (DPPH and Fe^{2+}-chelating assays).
6. Eliza Micro plate reader (DPPH and Fe^{2+}-chelating assays).
7. Vortex Mixer (DPPH and Fe^{2+}-chelating assays).
8. JEOL JES-FR30 ESR method.

2.2. Reagents

1. Highly purified DDH_2O (Compounds 3 and 4).

2. Antioxidant DPPH radical scavenging and Fe^{2+}-chelating assays.

   (a) 1,1-Diphenyl-2-picrylhydrazyl radical (DPPH), Sigma Aldrich.

   (b) Phenazine methosulfate (PMS), Sigma.

   (c) β-Nicotinamide adenine dinucleotide, reduced, Disodium salt, trihydrate (NADH), Research Organics.

   (d) Nitro Blue tetrazolium Salt (NBT), Sigma.

   (e) Dimethyl sulphoxide (DMSO), Merck.

   (f) Ferrozine, ICN, Biomedicals.

   (g) Ferrous sulfate, Merck.

   (h) 0.1 M Phosphate buffer (pH = 7.40), prepared by mixing Na_2HPO_4, Merck and NaH_2PO_4, Sigma.

3. Antioxidant – ESR

   (a) Diethylenetriamine penta acetic acid (DETAPAC) and phosphate buffer, Wako pure chemical industries, Osaka, Japan.

   (b) 5,5-Dimethyl-1-pyrroline N-oxide (DMPO), Dojin Chemicals, Kumamoto, Japan.

   (c) Hypoxanthine, Sigma-Aldrich.

   (d) Xanthine oxidase (XOD), Roche, Inc, USA.

2.3. Plant Material

Phenylpropanoids are reported from almost all parts of the plants of Acanthaceae, Asterraceae, Araliaceae, Boranginaceae, Buddlejaceae, Bignoniaceae, Crassulaceae, Icacinaceae, Lamiaceae, Martyniaceae, Magnoliaceae, Oleaceae, Orobanchaceae, Pedaliaceae, Plantaginaceae, Scrophulariaceae, Salicaceae, Thymelaeaceae, and Verbenaceae families. The whole plants or their required parts were dried under the shade, followed by soaking in different organic solvents.

2.4. Supplies

1. Silica gel column (9.4×16.5 cm) (Compounds 1 and 2).

2. Columns YMC ODS H-80 or L-80 (4 μm) (250×20 mm) (YMC, Japan) (Compounds 11–29).

3. Silica gel BW-200 (Compounds 3 and 4).

4. Chromatorex ODS DM1020T (Compounds 3 and 4).

5. Precoated TLC-plates with silica gel $60F_{254}$ (Merck, 0.25 mm) (ordinary phase) and silica gel $60F_{254}$ (Merck, 0.25 mm) (reverse phase) (Compounds 3 and 4).

6. Silica gel 60H, Merck (Compounds 5–8).

7. Sephadex LH-20, Fluka (Compounds 5–8).

8. Silica gel (Kieselgel MN 60, Macherey Nagel) (Compounds 9 and 10).

9. Cellulose F precoated TLC plates, Merck (Compounds 9 and 10).

10. Polyamide, Woelm (Compounds 9 and 10).

11. Fluka silica gel (70–230 mesh size) (Compounds 11–29).

12. Diaion HP-20, SUPELCO (Compounds 11–29).

13. Polyamide-6-powder, SERVA (Compounds 11–29).

14. Analytical grade MeOH and H_2O, Fluka (Compounds 11–29).

3. Methods

For the isolation and purification of phenylpropanoids from the plant extracts (air dried arial parts, flowers, fruits, roots), a broad range of chromatographic techniques have been utilized. Most of the phenylpropanoids are non-crystalline in nature and, therefore, require a complete chromatographic examination and separation by using solvent–solvent extraction, followed by column chromatography and high performance liquid chromatographic techniques.

3.1. Extraction

Phenylpropanoids have been reported from the ethyl acetate and butanolic fractions of various parts of the plants; however, there is a slight difference in the extraction procedures to obtain pure constituents from these two mentioned extracts.

The most widely used procedure is to dry the plant material under the shade. This dried plant material is then crushed to powder and soaked in methanol or in ethanol either in pure solvent or in different proportions with water for a couple of weeks. The soaking of dry plant material in hexane and acetone or extraction with hot water has also been reported. The soaking procedure to obtain the crude extracts is summarized in Scheme 26.1.

Extracts thus obtained were then evaporated to gums under vacuum. This gummy material was suspended in water and defatted with pet. ether or hexane or directly subjected to solvent–solvent extraction with EtOAc and BuOH (Scheme 26.2). The extraction of alcoholic extracts with diethyl ether or dichloromethane, followed by EtOAc and BuOH (Scheme 26.2).

3.2. Separation and Purification

Silica gel, Sephadex LH-20, diaion HP-20, and polyamide are the most common sorbents utilized for the fractionation and purification

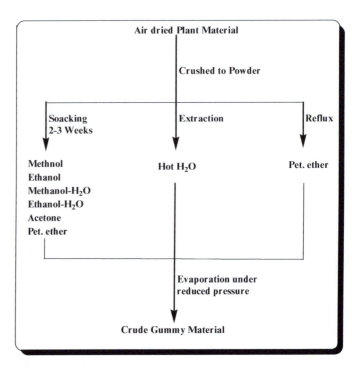

Scheme 26.1. Extraction of crude plant materials.

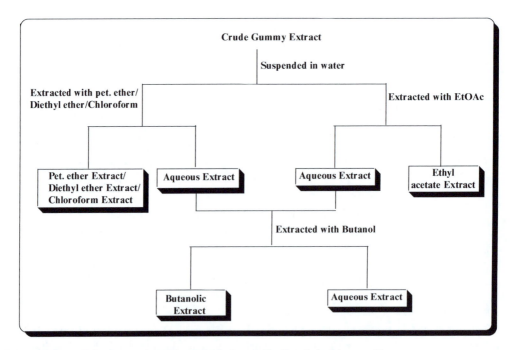

Scheme 26.2. Fractionation in various solvents for further purification of phenylpropanoids.

of plant extracts for the preliminary isolation of phenylpropanoids during different chromatographic techniques.

The ordinary-phase silica gel (*see* Note 1) column chromatography is an effective technique which is widely utilized to purify the less oxygenated phenylpropanoids such as two acylated phenylpropanoids named 3-[7-(α-methyl-butyroyloxy)-6-hydroxy-8-methoxy]-phenylpropanol (1) and 3-[7-(α-methyl-butyroyloxy)-6-hydroxy-8-methoxy]-phenyl-propanol (2) from the acetone extract of the *Schkuhria schkuhrioides*. The acetone extract was loaded on to a silica gel column and eluted with gradient *n*-hexane-EtOAc to afford different fractions. The compounds 1 and 2 obtained from fraction 1 after repeated column chromatography (eluting with gradient *n*-hexane-EtOAc) and TLC, developed with *n*-hexane-EtOAc, 7:3 (*see* Notes 1 and 2).

As the number of polar substitutents (–OH, –COOH, etc.) increase on the basic skeleton of phenylpropanoids, the combined use of silica gel and reverse phase medium become an effective tool to purify them, such as *trans-O*-coumaric acid (3) an and scopoletin (4) obtained from the ethyl acetate extract of the leaves of *Artemisia argyi*. The aforementioned plant extract was first subjected to the ordinary-phase silica gel column chromatography to obtain six fractions (gradient *n*-hexane-EtOAc). Fraction 4 was purified by ordinary-phase (gradient *n*-hexane-EtOAc) and reverse phase silica gel column chromatographies to obtain compounds 3 and 4 (*see* Note 3).

The glycosides of natural products are large size molecules with polar sugar moieties. The combination of gel filtration on Sephadex LH-20 (*see* Note 4) and silica gel column chromatography

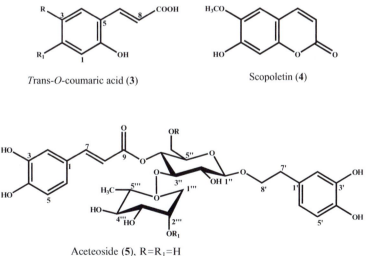

Trans-O-coumaric acid (3)   Scopoletin (4)

Aceteoside (5), R=R$_1$=H
Forsythoside B (6), R=β-D-apiose, R$_1$=H
Arenarioside (7), R=β-D-xylose, R$_1$=H
Ballotetroside (8), R=β-D-apiose, R1=α-L-arabinose

is considered to be an important tool to purify the large and highly oxygenated molecules. Phenylpropanoid glycosides such as acetoside (5), forsythoside B (6), arenarioside (7), and ballotetroside (8) from *M. vulgare* were purified by gel filtration on Sephadex LH-20 (MeOH–H$_2$O, 1:1) followed by column chromatography on silica gel 60H (gradient mixture of MeOH–EtOAc–H$_2$O).

Another effective combination for the purification of highly polar and oxygenated phenylpropanoids is gel filtration by Sephadex LH-20 and polyamide column chromatography (*see* Note 5). Echinacoside (9) and 6-*O*-caffeoylechinacoside (10) from the butanolic extract of the *Echinacea pallida* were purified by gel filtration by Sephadex LH-20 (H$_2$O–MeOH, 1:1) followed by the purification on polyamide column chromatography (gradient H$_2$O–MeOH mixture).

In addition to above-mentioned techniques to purify the phenylpropanoids with the polar functional groups and monosaccharides, HPLC with reverse phase column is also considered to be an effective technique to purify the polar substituents and both monosaccharides and polysaccharides in high yields (*see* Note 6).

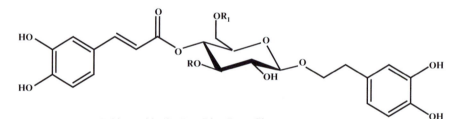

Echinacoside (**9**), R = Rha, R$_1$ = Glu
6-*O*-Caffeoylechinacoside (**10**), R = Rha, R$_1$= 6-Caffeoyl-Glu

p-Coumeric acid (**11**), R=R$_2$=H, R$_1$=OH, R$_3$=COOH
Ferulic acid (**12**), R=OMe, R$_1$=OH, R$_2$=H, R$_3$=COOH

4-Hydroxy-N-{4-[(*E*)-3-(4-hydroxy-3-methoxyphenyl)prop-2-enamido]butyl}benzamide (**19**),
R=OMe, R$_1$=OH, R$_2$=H, R$_3$=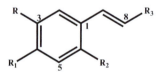

2-[3-Hydroxy-4-(4-hydroxyphenoxy)phenyl]-1-(methoxycarbonyl)ethyl (*E*)-3-(3,4-dihydroxyphenyl)prop-2-enote (**20**),
R=R$_1$=OH, R$_2$=H, R$_3$=

HPLC is effectively utilized to purify phenylpropanoids from the butanolic and water extracts, which contain glycosides with many sugars attached to aglycon and polar functionalities. The preliminary fractionation of these crude plant extracts was carried out on normal-phase silica gel, Sephadex LH-20 or polyamide column chromatography, followed by purification by HPLC on ODS H-80 or L-80 columns. The above-mentioned combination was employed to purify the phenylpropanoids from the ethyl acetate extract of the *Lindelofia stylosa* (Scheme 26.3). The EtOAc-soluble fraction was subjected to silica gel column chromatography and eluted with gradient mixtures of EtOAc/MeOH to obtain subfractions, LE-1 to LE-6. The fraction LE-3 was further fractionated into subfractions A–C by using acetone:CHCl$_3$ (1:1) as eluent. These subfractions were further subjected to preparative recycling HPLC (H$_2$O/MeOH 1:1) on L-80 and H-80 reverse phase column to obtain pure compounds 11 (Rt. 40 min), 12 (Rt. 36 min), 13 (Rt. 38 min), 14 (Rt.

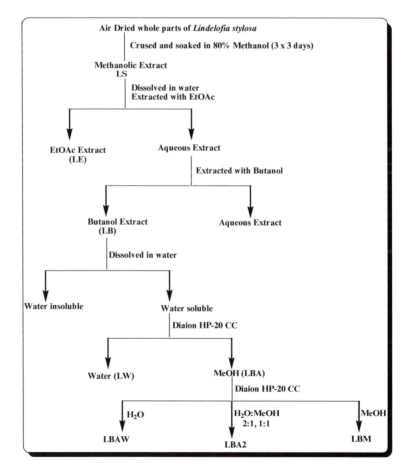

Scheme 26.3. General scheme for the extraction and fractionation of *Lindelofia stylosa*, a phenylpropanoids containing plant.

56 min). and 15 (Rt. 56 min), respectively, at flow rate of 4 mL/min (fraction A and C) and 3 mL/min (fraction B). Similarly, fraction LE-4 yielded compound 16 (Rt.30 min) under the same conditions. The fraction LE-5, which was eluted at EtOAc–MeOH (1:1) from silica gel column was passed through a polyamide column and eluted with $CHCl_3$–MeOH (1:1) to furnish fraction Fr-2, which was purified by preparative recycling HPLC (H_2O/MeOH 1:1) by using ODS H-80, reverse phase column to obtain pure compounds 17 (Rt. 52 min) and 18 (Rt. 40 min) with a flow rate of 3 mL/min. The purification procedure is summarized in Scheme 26.4.

Fraction LE-6 (Obtained from EtOAc fraction) was further subfractionated to Fr-1 to Fr-4 by loading on a silica gel column

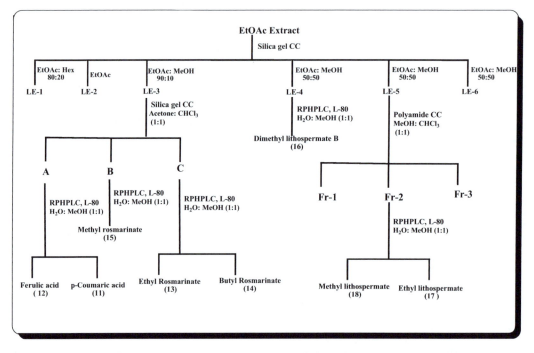

Scheme 26.4. Isolation of phenylpropanoids from butanolic extract of *Lindelofia stylosa*, a phenylpropanoids containing plant.

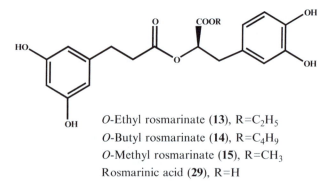

O-Ethyl rosmarinate (**13**), R=C_2H_5
O-Butyl rosmarinate (**14**), R=C_4H_9
O-Methyl rosmarinate (**15**), R=CH_3
Rosmarinic acid (**29**), R=H

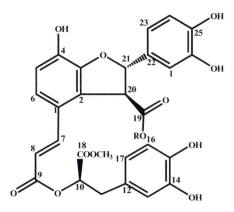

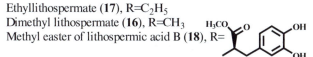

Ethyllithospermate (**17**), R=C$_2$H$_5$
Dimethyl lithospermate (**16**), R=CH$_3$
Methyl easter of lithospermic acid B (**18**), R=

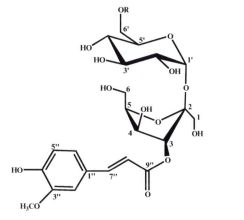

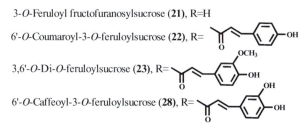

3-*O*-Feruloyl fructofuranosylsucrose (**21**), R=H

6'-*O*-Coumaroyl-3-*O*-feruloylsucrose (**22**), R=

3,6'-*O*-Di-*O*-feruloylsucrose (**23**), R=

6'-*O*-Caffeoyl-3-*O*-feruloylsucrose (**28**), R=

followed by polyamide column chromatography, by using acetone–CHCl$_3$ and MeOH–CHCl$_3$ as eluent, respectively. Fractions Fr-1 and Fr-4 were rechromatographed (silica gel column, MeOH–CHCl$_3$, 10:90) to obtain pure compounds 19 and 20. Fractions Fr-2 and 3 were further subjected to recycling HPLC (H$_2$O/MeOH, 1:2 and 1:1, respectively) on ODS H-80, preparative reverse phase column, with a flow rate of 4 mL and 3.5 mL/min to obtain compounds 21 (Rt. 30 min), 22 (Rt. 32 min), and 23 (Rt. 30 min). The purification procedure is shown in Scheme 26.5.

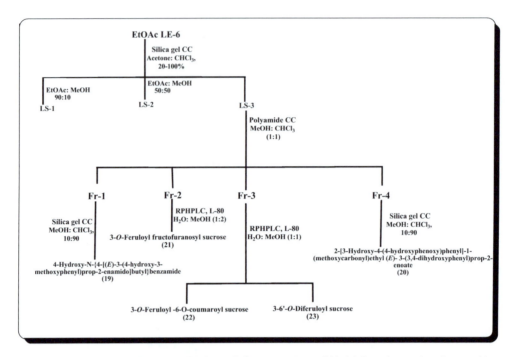

Scheme 26.5. Purification of phenylpropanoids from ethyl acetate extract of *Lindelofia stylosa*, a phenylpropanoids containing plant.

Many phenylpropanoids have been reported from the butanolic and water fractions of plant extracts. The isolation was initiated with diaion HP-20 column by using MeOH–H$_2$O gradient mixtures as eluent. The MeOH–H$_2$O eluent were further subfractionated by silica gel column chromatography, polyamide, or Sephadex LH-20, either individually or in combinations. These subfractions were finally purified by analytical or preparative high pressure liquid chromatography on reverse phase column to obtain different glycosides of phenylpropanoids. For example, the water soluble part of the butanolic extract of *L. stylosa* was passed through a diaion HP-20 column and eluted with H$_2$O, H$_2$O–MeOH (1:1, 1:2) and MeOH to obtain fractions LBAW, LBA2, and LBM, respectively. Fraction LBA2 was again loaded on a polyamide column and eluted with chloroform with an increasing proportion of methanol to obtained fractions LBA2a (CHCl$_3$, 100%), LBA2b (CHCl$_3$–MeOH, 85:15), LBA2c (CHCl$_3$–MeOH, 80:20/70:30), LBA2d (CHCl$_3$–MeOH, 50:50). Fraction LBA2b was then purified by RHPLC on an H-80 column by using 1:1 H$_2$O–MeOH as mobile phase with a flow rate of 3.5 mL/min to give compounds 24 (Rt. 40 min), 25 (Rt. 45 min), and 26 (Rt. 30 min), respectively. Fraction LBA2c was chromatographed on Diaion HP-20 and Sephadex LH-20 columns to obtain fractions LBACB and LBACC. These fractions yielded compounds 27 (Rt. 34 min) and 28 (Rt. 20 min) by recycling HPLC on an H-80 column by using 1:1 H_2O–MeOH as mobile phase with a flow rate of 3.5 mL/min. Fraction LBA2d was passed through the Sephadex LH-20 column, followed by purification on recycling HPLC by using ODS L-80 column under the same solvent system as described above to obtain rosmarinic acid (29) (Rt. 18 min, flow rate 4 mL/min) (Scheme 26.6).

3.3. Antioxidant Activities of Phenylpropanoids

3.3.1. DDPH Radical Scavenging Assay

The method developed by Lee et al. (27) is use to determine the antioxidant activity of compounds. Test samples (5 μL dissolved in DMSO) were mixed with DPPH (95 μL in ethanol). The mixtures were dispersed in a 96-well plate and incubated at 37°C for half an hour. After incubation, decrease in absorption was measured at 515 nm by using a multiplate reader (SpectraMAX-340) and percent radical scavenging activity of samples was determined in comparison with a DMSO-treated control (3-*t*-butyl-4-hydroxyanisole).

3.3.2 . Superoxide Anion Radical Scavenging Assay

The reaction mixture was prepared by mixing 280 μM β-nicotinamide adenine dinucleotide (NADH, reduced form; 40 μL), 80 μM nitro blue tetrazolium (NBT, 40 μL), 8 μM phenazine methosulfate (PMS, 20 μL), 10 μL of 1 mM sample and 90 μL of 0.1 M phosphate buffer (pH 7.4). The NBT, NADPH, and PMS solutions were prepared in the same buffer and sample was dissolved in DMSO. The reaction was performed in 96-well microtitre plate at room temperature and absorbance was measured at

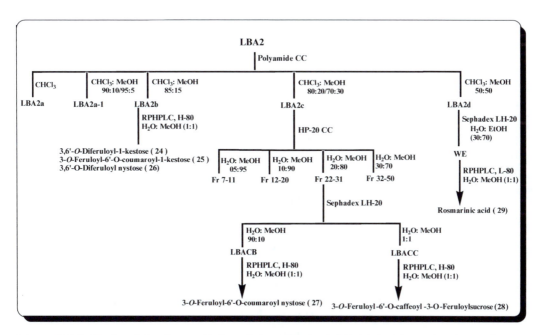

Scheme 26.6. Purification of phenylpropanoids from ethyl acetate extract of *Lindelofia stylosa*, a phenylpropanoids containing plant.

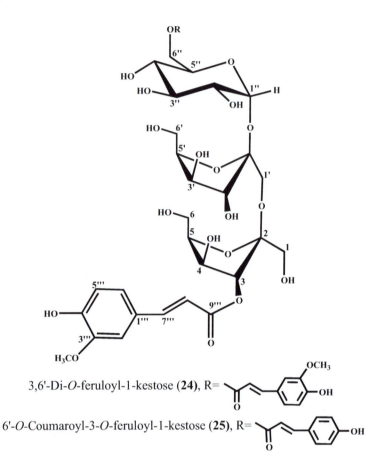

3,6'-Di-O-feruloyl-1-kestose (**24**), R=

6'-O-Coumaroyl-3-O-feruloyl-1-kestose (**25**), R=

560 nm by a microplate reader. The formation of superoxide is monitored by measuring the formation of water-soluble blue Formazan dye. A lower absorbance of reaction mixture indicates a higher scavenging activity of the sample (*28*).

3.3.3. Fe²⁺-Chelating Assay

The Fe^{2+}-chelating ability was determined according to the modified method of Decker and Welch (*28*). The concentration of Fe^{2+} was monitored by measuring the formation of ferrous ion-ferrozine complex. Pure compound in DMSO (0.5 μL) was mixed with 400 μM $FeCl_2$ (35 μL) and 1,000 μM ferrozine (60 μL). The mixture was shaken and left at room temperature for 10 min. The absorbance of resulting mixture was measured at 562 nm. A lower absorbance of reaction mixture indicated a higher Fe^{2+}-chelating ability.

The IC_{50} values in all assays were determined at different concentrations of test samples using by the kinetic program Ez-Fit

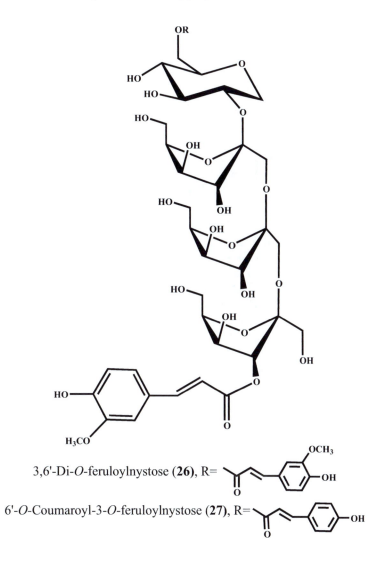

3,6'-Di-*O*-feruloylnystose (**26**), R=

6'-*O*-Coumaroyl-3-*O*-feruloylnystose (**27**), R=

and the % radical scavenging activity was calculated by using the following formula:

$$\% RSA = 100 - \{(OD \text{ test compound} / OD \text{ control}) \times 100\}$$

3.3.4. ESR Spin-Trapping for Measuring O$_2$

Hosoya et al. in 2008 (18) utilized the ESR methods to evaluate the antioxidant activities of compounds. Thirty five microliters of 5.5 mmol/L DETAPAC, 15 μL of 9.2 mmol DMPO, 50 μL of 2 mmol/L HPX, and 50 μL of each of the compounds were mixed in a test tube. The mixture was transferred to the ESR spectrometer cell and DMPO-O$_2$· spin adduct was quantified for 45 s after the addition of 50 μL of 0.4 unit/mL XOD from cow's milk.

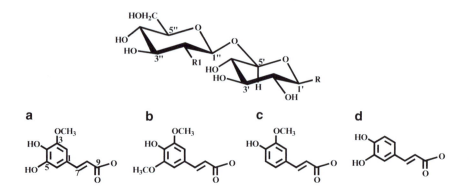

a   b   c   d

1-(3",4"-Dihydroxy-5"-methoxy)-*O*-*trans*-cinnamoyl-2'-*O*-trans-sinapoyl gentiobiose (**30**), R=a, R$_1$=b
1-*O*-*Trans*-caffeoyl-2'-O-trans-sinapoyl gentiobiose (**31**), R=d, R$_1$=b
1',2'-Di-(3",4"-dihydroxy-5"-methoxy)-*O*-*trans*-cinnamoyl gentiobiose (**32**), R=R$_1$=a
1-(3",4"-Dihydroxy-5"-methoxy)-*O*-*trans*-cinnamoyl-2'-*O*-trans-feruloyl gentiobiose (**33**), R=a, R$_1$=c

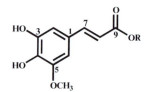

3,4-Dihydroxy-5methoxy-trans-cinammic acid (**34**), R=H
3,4-Dihydroxy-5-methoxy-trans-cinammic acid methyl ester (**35**), R=CH$_3$

The signal intensities were evaluated in terms of the peak height of the first signal of the DMPO-O$_2$· spin adduct.

Radical scavenging were recorded on JEOL JES-FR30 by the conditions, magnetic field $335.5 \pm mT$, Power: 4 mW, 9 GHZ, Sweep time: 1 min, Modulation: 100 kHz, Mod. Wid: 0.1 mT, Amplitude: 160, Time constant: 0.1 s., Temperature: 293K.

3.4. Results

The phenylpropanoids, obtained during the phytochemical investigation of *L. stylosa*, 4-hydroxy-*N*-{4-[(*E*)-3-(4-hydroxy-3-methoxyphenyl)prop-2-enamido]butyl}-benzamide (**19**), 2-[3-hydroxy-4-(4-hydroxyphenoxy)phenyl]-1-(methoxycarbonyl)-ethyl (*E*)-3-(3,4-dihydroxyphenyl)-prop-2-enoate (**20**), were tested by DPPH radical scavenging assay, and both of them showed significant activity. These two compounds were found to be more active than the standards used in superoxide anion scavenging assay. 2-[3-Hydroxy-4-(4-hydroxyphenoxy)-phenyl]-1-(methoxycarbonyl)ethyl (*E*)-3-(3,4-dihydroxyphenyl)-prop-2-enoate (**20**) was found to be more potent than the standard in Fe^{2+}-chelating assay. The chelation of compound with Fe^{2+} depends on the catechol group. It was found that derivatives of lithospermic acid, named dimethyllithospermate (16) together with O-methyl derivative of rosmarinic acid (15) showed significant antioxidant activity in Fe^{2+} chelating and super anion scavenging assays. The IC_{50} values in three assays are listed in Table 26.1–26.3.

Table 26.4 shows the antioxidant activity of 1-(3″,4″-dihydroxy-5″-methoxy)-O-$trans$-cinnamoyl-2′-O-$trans$-sinapoyl gentiobiose (30), 1-O-$trans$-caffeoyl-2′-O-$trans$-sinapoyl gentiobiose (31), 1′,2′-di-(3″,4″-dihydroxy-5″-methoxy)-O-$trans$-cinnamoyl gentio biose (32), 1-(3″,4″-dihydroxy-5″-methoxy)-O-$trans$-cinnamoyl-2′-O-$trans$-feruloyl gentiobiose (33), and cinnamic acid derivatives (34) and (35) against the super anion radicals by using ESR method. The results showed that the scavenging ability of hydroxyl cinnamic acid was largely dependent on the number of hydroxyl groups on the benzene moiety and $ortho$ substitutions with the electron donor methoxy group.

Table 26.1
DDPH radical scavenging activity of some phenylpropanoids

Compound	DDPH radical scavenging activity IC_{50} mM
p-Coumaric acid	0.0630
Ferulic acid	0.0560
4-Hydroxy-N-{4-[(E)-3-(4-hydroxy-3-methoxyphenyl)prop-2-enamido] butyl}-benzamide	0.2383
2-[3-hydroxy-4-(4-hydroxyphenoxy)phenyl]-1-(methoxycarbonyl)ethyl (E)-3-(3,4-dihydroxyphenyl)-prop-2-enoate	0.2202
Methyl lithospermate	0.1028
Citanoside F	0.1197
Dimethyl lithospermate	0.0506
Ethyl rosmariante	0.0801
Butyl rosmarinate	0.1456
2-(4-Hydroxy-3-methoxyphenyl) ethyl O-α-L-rahmnopyranosyl-(1-3)-O-β-D-glucopyranoside	0.0412
α-L-rahmnopyranosyl-(1-3)-O-(4-O-caffeoyl)-D-glucopyranoside	0.2706
3-(Ter-Butyl)-4-hydroxyanisole[a]	0.0442
Propyl gallate[a]	0.030

[a]Standard reference compounds

Table 26.2
Superoxide anion scavenging activity of some phenylpropanoids

Compound	Superoxide anion scavenging activity IC_{50} mM
p-Coumaric acid	0.0415
Ferulic acid	0.0492
4-Hydroxy-N-{4-[(E)-3-(4-hydroxy-3-methoxyphenyl)prop-2-enamido]butyl}-benzamide	Inactive
2-[3-Hydroxy-4-(4-hydroxyphenoxy)phenyl]-1-(methoxycarbonyl) ethyl (E)-3-(3,4-dihydroxyphenyl)-prop-2-enoate	Inactive
Methyl lithospermate	Inactive
Citanoside F	Inactive
Dimethyl lithospermate	0.113
Ethyl rosmariante	0.282
Butyl rosmarinate	0.443
2-(4-Hydroxy-3-methoxyphenyl) ethyl O-α-L-rahmnopyranosyl-(1-3)-O-β-D-glucopyranoside	Inactive
α-L-Rahmnopyranosyl-(1-3)-O-(4-O-caffeoyl)-D-glucopyranoside	Inactive
3-(Ter-Butyl)-4-hydroxyanisole[a]	0.096
Propyl gallate[a]	0.106

[a]Standard reference compound

---

4. Notes

1. Silica gel is a stationary phase composed of polymers of silicon dioxide and is involved in strong interaction with those compounds having polar compounds such as hydroxyl and carboxylic acid. As a result non polar compounds are eluted first. Because of this reason it is called normal phase chromatography.

2. Different combinations of organic solvents such as $CHCl_3$–MeOH, n-hexane-$(CH_3)_2CO$, n-hexane-EtOAc, EtOAC–MeOH, benzene-$(CH_3)_2CO$, $CHCl_3$–MeOH–H_2O, and MeOH–EtOAc–H_2O have been widely used in gradient manner for fractionation and purification of phenylpropanoids by normal phase silica gel and preparative thick layer chromatography.

Table 26.3

Antioxidant activities of some phenylpropanoids by Fe²⁺-Chelating assay

Compound	Fe²⁺-chelating assay IC$_{50}$ mM
p-Coumaric acid	Inactive
Ferulic acid	0.0320
4-Hydroxy-*N*-{4-[(*E*)-3-(4-hydroxy-3-methoxyphenyl)prop-2-enamido]butyl}-benzamide	Inactive
2-[3-Hydroxy-4-(4-hydroxyphenoxy)phenyl]-1-(methoxycarbonyl) ethyl (*E*)-3-(3,4-dihydroxyphenyl)-prop-2-enoate	Inactive
Methyl lithospermate	Inactive
Citanoside F	Inactive
Dimethyl lithospermate	0.053
Ethyl rosmarinate	0.034
Butyl rosmarinate	0.092
2-(4-Hydroxy-3-methoxyphenyl) ethyl *O*-α-L-rahmnopyranosyl-(1-3)-*O*-β-D-glucopyranoside	Inactive
α-L-Rahmnopyranosyl-(1-3)-*O*-(4-*O*-caffeoyl)-D-glucopyranoside	Inactive
Propyl gallate[a]	0.064

[a]Standard reference compound

Table 26.4

Superoxide anion scavenging activity of some phenylpropanoids by using ESR

Compound	Superoxide anion scavenging activity by using ESR IC$_{50}$ μM
1-(3″,4″-Dihydroxy-5″-methoxy)-*O*-*trans*-cinnamoyl-2′-*O*-*trans*-sinapoyl gentiobiose	28.5
1-*O*-*Trans*-caffeoyl-2′-*O*-*trans*-sinapoyl gentibiose	84.5
1′,2′-Di-(3″,4″-dihydroxy-5″-methoxy)-*O*-*trans*-cinnamoyl gentibiose	8.4
1-(3″,4″-Dihydroxy-5″-methoxy)-*O*-*trans*-cinnamoyl-2′-*O*-*trans*-feruloyl gentiobiose	17.1
3,4-Dihydroxy-5-methoxy-*trans*-cinnamic acid	36.0
3,4-Dihydroxy-5-methoxy-*trans*-cinnamic acid methyl ester	31.3
Ascorbic acid[a]	140

[a]Standard reference compound

3. As normal-phase silica gel column chromatography results in strong adsorption of the polar compounds, reverse phase chromatographic methods were used to purify the polar components. Reverse phase adsorbents are the modified silica gel with a long alkyl chain, such as C_{18} attached. As a result they are hydrophobic in nature. The long alkyl chain attached to the normal phase silica gel reverses the elution order and polar components elute first from the column. Gradient mixture of H_2O–MeOH is a widely used eluent in reverse phase columns.

4. Sephadex is a highly specialized size exclusion gel filtration media for the separation of large molecules from smaller ones. The adsorbent has the largest pore size in water and as a result the large molecules elute first. The method is generally applied in combination with other chromatographic techniques like normal-phase silica gel, HPLC etc. The most commonly used eluents for the size exclusion method are H_2O–MeOH (1:1/1:2).

5. Polyamide is a chromatographic material used for the detection and purification of acidic compounds. The coordination between the acidic/polar functionality and the amino group of polyamide is the key in the mechanism of separation. Different combinations of the solvents such as $CHCl_3$–MeOH and MeOH–H_2O in gradient can be used as eluent during the separation.

6. High pressure liquid chromatography (HPLC) is the preferred chromatographic technique for the purification and quantification of components of a mixture on the basis that different substances have different migration rates on the column. The efficiency of the column is high because of the pressure and the detection is more precise because of the wide range of detectors available for the system. HPLC can be used both for analytical and preparative purposes, depending upon the size of the column and the amount of sample loaded on it. Preparative recycling HPLC allows the sample to be recycled in the column and each cycle improve the purity and separation of the sample.

References

1. Hahlbrock K, Scheel D (1989) Physiology and molecular biology of phenylpropanoids metabolism. Ann Rev Plant Physiol Plant Mol Biol 40:347–369

2. Kurkin VA (2003) Phenylpropanoids from medicinal plants: Distribution, classification, structural analysis, and biological activity. Chem Nat Compd 39(2):123–153

3. Cometa E, Tomassini L, Nicolletti M, Piretti S (1992) Phenylpropanoid glycosides. Distribution and pharmacological activity. Fitopterapia LXIV:197–217

4. Salatino A, Teixeira EW, Negri G, Message D (2005) Original and chemical variations of Brazilian Propolis. eCAM 2(1):33–38

5. Ravn H, Brimer L (1988) Structure and antibacterial activity of plantamajoside, A caffeic acid sugar ester from *Plantago major* subsp. *major*. Phytochemistry 27(11):3433–3437

6. Canno E, Veiga M, Jimenez C, Riguera R (1990) Pharmacological effects of three phenylpropanoid glycosides from *Mussatia*. Planta Med 56:24–26

7. Salomao K, Pereira PRS, Campus LC, Borba CM, Cabello PH, Marcucci MC, Castro SLD (2008) Brazilian Propolis: Correlation between chemical composition and antimicrobial activity. eCAM 5(3):317–324

8. Whiting DA (1987) Lignans neolignans and related compounds. Nat Prod Rep 4:499–525

9. Duh CY, Phoebe CH Jr, Pezzuto JM, Kinghorn AD, Fransworth NR (1986) Plant anticancer agents, XLII, cytotoxic constituents from *Wikstroemia eliptica*. J Nat Prod 49:706

10. Kurkin VA (1998) 2nd International electronic conference on synthetic organic chemistry (ECSOC-2) http://www.mdpi.org/ September 1–30

11. Lee EJ, Kim SR, Kim J, Kim YC (2002) Hepatoprotective phenylpropanoids from Scrophularia buergeriana roots gainst CCL_4-induced toxicity: Action mechanism and structure activity realtionship. Plant Med 68:407–411

12. Kobata K, Tate H, Iwasaki Y, Tanaka Y, Ohtsu K, Yazawa S, Watanabe T (2008) Isolation of coniferyl esters from *Capsicum baccatum* L., and their enzymatic preparation and agonist activity for TRPV1. Phytochemistry 69: 1179–1184

13. Ikeya Y, Taguchi H, Mitsuhasi H, Takeda S, Kase Y, Aburada M (1988) A lignan from Schizandra Chinensis. Phytochemistry 27(2):569–573

14. Fukuyama Y, Mizuta K, Nakagawa K, Wenjuan Q, Xiue W (1986) A new lignan, a prostaglandin I2 inducer from the leaves of *Zizypus jujuba*. Planta Med 52:501–502

15. Takuya M, Chihiro I, Masataka I, Tadashi O, Hiroshi F (2007) Anti-inflammatory activity of phenylpropanoids and phytoquinoids from *Illicium* species in RBL-2H3 cells. Planta Med 73(7):662–665

16. Sahpaz S, Garbacki N, Tits M, Bailleuli F (2002) Isolation and Pharmacological acticity of phenylpropanoid esters from *Marrubium vulgare*. J Ethnopharmacol 79(3):389–392

17. Pisano M, Pagnan G, Loi M, Tilocca MEMG, Palmieri G, Fabbri D, Delogu MAG, Ponzoni M, Rozzo C (2007) Antiproliferative and pro-apoptotic activity of eugenol biphenyls on malignant melanoma cells Mol. Cancer 6(8):1–14

18. Hosoya T, Yun YS, Kunugi A (2008) Antioxidant phenylpropanoids glycosides from the leaves of *Wasabia japonica*. Phytochemistry 69:827–832

19. Tang W, Hioki H, Harada K, Kubo M, Fukuyama Y (2007) Antioxidant phenylpropanoids-substituted epicatechins from *Trichilia catgua*. J Nat Prod 70(12):2010–2013

20. David J, Barreiros A, David J (2004) Antioxidant phenylpropanoid esters of triterpenes from *Dioclea lasiophylla*. Pharm Biol 42(1):36

21. Daels-Rakotoarison DA, Seidel V, Gressier B, Brunet C, Tillequin F, Bailleulf F, Luyckx M, Dine T, Cazin M, Cazin JC (2000) Neurosedative and antioxidant activities of phenylpropanoids from Ballota nigra. ArzneimittelForschung 50:16–23

22. Choudhary MI, Begum A, Abbaskhan A, Ajaz A, Shafique-ur-Rehman, Atta-ur-Rahman (2005) Phenylpropanoids from *Lindelofia stylosa*. Chem Pharm Bull 53(11):1469–1471

23. Choudhary MI, Begum A, Abbaskhan A, Shafique-ur-Rehman, Atta-ur-Rahman (2006) Cinnamate derivatives of fructo-oligosaccharides from *Lidelofia stylosa*. Carbohydrate Res 341:2398–2405

24. Zhang L, Liao Chia-Ching, Huang Hui-Chi, Shen Ya-Ching, Yang Li-Ming, Kuo Yao-Haur (2008) Antioxidant phenylpropanoids glycosides from *Smilax bracteata*. Phytochemistry 69:1398–1404

25. Taira J, Ikemoto T, Yoneya T, Hagi A, Murakami A, Makino K (1992) Essential oil phenylpropanoids. Useful as.OH scavenger. Free Radic Res Commun 16(3):197–204

26. Foti MC, Daquino C, Geraci C (2004) Electron transfer reaction of cinnamic acid and their methyl esters with the DPPH radical in alcoholic solution. J Org Chem 69: 2309–2314

27. Lee SK, Zakaria H, Chung H, Luvengi L, Gamez EJC, Mehta RJ, Kinghorn D, Pezzuto JM (1998) Evaluation of antioxidant potential of natural products. Comb Chem High Throughput Screen 1:35

28. Decker EA, Welch B (1990) Role of ferritinas as lipid oxidation catalysts in muscles. J Agric Food Chem 38:674

29. Choudhary MI, Begum A, Abbaskhan A, Musharraf SG, Ejaz A, Atta-ur-Rahman (2008) Two new antioxidant phenylpropanoids from *Lindelofia stylosa*. Chem Biodivers 5:2676–2683

Part III

Gene Expression

Chapter 27

Generation of Antioxidant Adenovirus Gene Transfer Vectors Encoding CuZnSOD, MnSOD, and Catalase

Aoife M. Duffy, Timothy O'Brien, and Jillian M. McMahon

Abstract

Replication-deficient adenovirus gene transfer vectors are very useful for the experimental delivery of genes into cells and are widely used both in vitro and in vivo to determine the effects of transgene expression. Having a broad cell tropism, these vectors allow efficient transduction of many cell types and permit transfer of large amounts of DNA with resulting high expression levels within the target cell. Manganese superoxide dismutase (MnSOD), copper zinc superoxide dismutase (CuZnSOD) and catalase are all known antioxidants whose over-expression can result in amelioration of pathology brought about by an excess of reactive oxygen species within a cell. Their use has been suggested as therapies for many conditions, including cardiovascular disease, arthritis, diabetes, cancer, and damage to central nervous system cells. This chapter describes the methodology commonly used for production of replication-deficient adenovirus vectors encoding MnSOD, CuZnSOD, and catalase.

Key words: Oxidative stress, Adenovirus vector, Gene transfer, Gateway® system, Caesium chloride purification, Titration, Replication-competent adenovirus

1. Introduction

One of the main barriers to progress in clinical gene therapy is the absence of a safe and efficient vector system, having regulatable gene expression and tissue-specific characteristics. However, of the three major groups of viral vectors currently in use (adenovirus, retrovirus and adeno-associated virus) adenovirus vectors have many desirable characteristics making them ideal for gene transfer both in vitro and in vivo. These vectors allow a high level of expression of transgene, are capable of transducing both replicating and non-replicating cells, are easily prepared to high titres, can accommodate a large amount of foreign DNA and have a broad tissue tropism, allowing successful transduction of a wide variety of cells. The most widely used adenovirus vectors are, to

D. Armstrong (ed.), *Advanced Protocols in Oxidative Stress II*, Methods in Molecular Biology, vol. 594
DOI 10.1007/978-1-60761-411-1_27, © Humana Press, a part of Springer Science+Business Media, LLC 2010

date, the first generation E1 and/or E3-deleted vectors whereby sections of the viral genome, necessary for growth, replication, and for counteracting host-cell defence mechanisms, are replaced with the gene of interest under control of a suitable promoter.

Oxidative stress has been implicated in the pathophysiology of many disorders, including atherosclerosis (1), diabetes mellitus (2), amyotrophic lateral sclerosis, Alzheimer's disease, and Parkinson's disease (3). When it occurs, due to an imbalance between reactive oxygen species (ROS) production and the means by which the cell can deal with these reactive intermediates, oxidative damage can occur to proteins, membranes, and nucleic acids. Free-radical scavengers, such as the enzymes superoxide dismutase (SOD) and catalase, can help protect cells against oxidative damage and prevent cell death, as has been seen in a rabbit diabetic model (4). SOD is an important anti-oxidant and three isoforms are known in mammals: copper–zinc SOD (CuZnSOD) in the cytoplasm, the manganese-containing SOD (MnSOD) in the mitochondria, and the extracellular SOD isoform (EcSOD). Superoxide dismutase facilitates the dismutation of free oxygen radicals to hydrogen peroxide (H_2O_2). Catalase performs a similar anti-oxidant function by enabling the transformation of hydrogen peroxide to water and oxygen. Therefore, these enzymes perform a breakdown of the free oxygen radicals, which can be produced by oxidative stress in many disease states.

$$O_2^- \xrightarrow{\text{SOD}} H_2O_2 \xrightarrow{\text{Catalase}} O_2 + H_2O$$

Thus, over-expression of these enzymes (SOD and catalase) would allow reversal of oxidative stress in disease states and potentially prevent the cell death associated with the presence of excess ROS. Use of adenoviral gene transfer vectors, expressing these antioxidants, has a strong therapeutic potential and a wide variety of applications within the context of human disease. The following sections describe the methodology involved in producing replication-deficient adenoviruses encoding CuZnSOD, MnSOD and Catalase.

2. Materials

2.1. Equipment

1. PCR Machine.
2. DNA Gel Electrophoresis system.
3. Water Bath (37°C).
4. Vortex.
5. Benchtop microcentrifuge.

6. Heating block.

7. Gilson pipettes.

8. Biological Safety Cabinet Class II.

9. 37°C/5% CO_2 Incubator.

10. High performance centrifuge and rotor.

11. Ultra centrifuge and rotors, e.g. Beckman SW41 and SW55 rotors.

12. Retort Stand.

13. Spectrophotometer and Quartz cuvettes.

14. SDS-PAGE Electrophoresis system (e.g. BioRad).

2.2. Reagents

2.2.1. Generation of the Recombinant Expression Vector Using the Gateway® System

1. PCR primers (depending on gene of interest).

2. Qiagen PCR Cloning Kit (Qiagen, Catalogue number: 231122).

3. *Bam*HI (New England Biolabs, Catalogue Number: R0136L) and *Not*I (New England Biolabs, Catalogue Number: R0189L).

4. pENTR™1A (Invitrogen: 11813-011).

5. T4 ligase.

6. One-shot TOP10 chemically-competent *E. coli* (Invitrogen C4040-03).

7. Gateway LR Clonase™ II enzyme mix (Invitrogen 11791-20).

8. pAd/CMV/V-5-DEST™ (Invitrogen. V493-20).

9. Proteinase K.

10. pUC19 control DNA.

11. pENTR-gus (positive control vector) (Invitrogen. 11813-011).

2.2.2. Transfection of Recombinant Genome into Transcomplementing Cell Line

1. HEK-293 cells (ATCC CRL 1573).

2. *Infection medium*: Dulbeccos Modified Eagle Medium (DMEM) w/4.5 g/l glucose, 0.5 g/l l-Glutamine (Sigma: D 6429), 2% Fetal Bovine Serum, penicillin G (100 U/ml), and streptomycin sulphate (100 µg/ml).

3. *Production medium*: Dulbeccos Modified Eagle Medium (DMEM) w/4.5 g/l glucose, 0.5 g/l l-Glutamine (Sigma: D 6429), 10% Fetal Bovine Serum, penicillin G (100 U/ml), and streptomycin sulphate (100 µg/ml).

4. *Pac I* restriction enzyme (New England Biolabs: R0547S).

5. 2.5 M Calcium chloride (filter-sterilised and stored at −20°C).

6. 2× Hank's buffered saline (280-mM NaCl (16 g/l), 10 mM KCl (0.74 g/l), 1.5 mM Na2HPO4·2H$_2$O (0.27 g/l), 12 mM dextrose (2 g/l), 50 mM HEPES (10g/l), pH to 7.05, and filter (0.22 µm) and store at −20°C).

7. Cold absolute ethanol and 70% ethanol.

8. ddH$_2$O.

9. Dulbeccos Phosphate Buffered Saline, with Calcium and Magnesium (Sigma: D8662).

10. Agarose gel: 0.7% in TBE (90 mM Trizma base, 90 mM boric acid, 2.5 mM EDTA).

2.2.3. Adenovirus Titration

1. Infection medium (as in Subheading 27.2.2).

2. 2% (w/v) Sea Plaque Agar in ddH$_2$O.

3. 1.4% (w/v) Neutral Red solution in ddH$_2$O (autoclaved prior to use).

4. 20 mM sodium phosphate, 0.5% SDS, pH 7.2.

2.2.4. Propagation of Virus Vector Stock

1. Adenovirus working high titre stock.

2. HEK-293 cells (ATCC number: CRC-1573) (20× T175 flasks).

3. Infection medium (as in Subheading 27.2.2).

4. Production medium (as in Subheading 27.2.2).

2.2.5. Purification of Adenovirus

1. 1.43 g/ml CsCl$_2$.

2. 1.34 g/ml CsCl$_2$.

3. Disposable PD-10 desalting columns (GE Healthcare: 17-0851-01).

4. Dulbecco's Phosphate Buffered Saline (dPBS) without calcium and magnesium (Sigma: D 8537).

2.2.6. Virus Extraction and Wild-Type Virus Detection by PCR

1. PBS (as in Subheading 27.2.2).

2. DNase (Invitrogen: 18068015).

3. 500 mM EDTA.

4. Proteinase K.

5. Phenol:Chloroform:Isoamyl alcohol (50:50:1).

6. 10% (w/v) SDS.

7. Ammonium acetate (7.5 M).

8. Isopropanol.

9. PCR primers (5)(200 ng/µl):
E1A-1: 5′-ATT ACC GAA GAA ATG GCC GC-3′
E1A-2: 5′-CCC ATT TAA CAC GCC ATG CA-3′
E2B-1: 5′-TCG TTT CTC AGC AGC TGT TG-3′
E2B-2: 5′-CAT CTG AAC TCA AAG CGT GG-3′

10. 2 mM dNTP

11. Taq DNA polymerase

12. Sterile ddH$_2$O

2.2.7. SDS-PAGE and Immunoblotting for Transgene Expression

1. 12% acrylamide gel

2. Adenoviral vector

3. BCA protein assay kit (Pierce: Catalogue Number: 23225)

4. 0.2-mm nitrocellulose membrane (BioRad: Catalogue Number: 162-0112)

5. PBS (Make up PBS using tablets and H$_2$O. Sigma Catalogue Number: P4417))

6. 5% (w/v) non-fat milk in PBS

7. Polyclonal rabbit anti-human MnSOD antibody (Stressgen, Catalogue No: SOD-110D), anti-CuZnSOD (Stressgen, Catalogue No: SOD-100E), anti-catalase (Abcam, Catalogue Number: ab6572)

8. Horseradish peroxidase (HRP)-conjugated goat anti-rabbit immunoglobulin (Stressgen, Cat No: SAB-300J)

9. PBST (0.05% (v/v) Tween 20 in PBS)

10. ECL substrate (Amersham Life Sciences, Catalogue No: RPN2132)

3. Methods

3.1. Virus Vector Production

Human CuZnSOD, MnSOD and Catalase cDNAs were prepared and amplified as previously described (6–8).

3.1.1. Generation of the Recombinant Expression Vector Using the Gateway® System

3.1.2. Generating an Entry Clone

The PCR product is digested and cloned into the pDrive Cloning Vector (Qiagen). The fragment is released from the pDrive vector using restriction enzymes (e.g. *BamHI* and *Not I*) and then ligated into the pENTR™ Entry Vector (Invitrogen). Creating an expressing pENTR™ Entry Vector is carried out according to manufacturer's instructions.

3.1.3. LR reaction and Transformation

For the LR reaction, pENTR™-gus is used as a positive control and pUC19 plasmid as positive control for the transformation. pAd/ CMV/ V5-DEST™ is a destination vector adapted for use with the Gateway™ Technology, and allows a high- level, transient expression of recombinant fusion protein. The LR Clonase™

enzyme mix promotes in vitro recombination between an entry clone and pAd/ CMV/ V5-DEST™ vector to generate *att*B-containing expression clones. Please refer to manufacturer's instructions (see pAd/ CMV/ V5-DEST™ Instruction Manual and LR Clonase™ Instruction Manual).

3.1.4. Plasmid Digest and Purification

1. Adenovirus genome is removed from the plasmid backbone at the inverted terminal repeats using *Pac*I restriction digest.

2. Standard phenol:chloroform extraction is performed to purify the plasmid and DNA concentration determined by spectrophotometry.

3.1.5. Adenovirus Vector Production

Adenovirus (serotype 5) is first propagated in HEK293 cell line, as these cells contain the full E1 region of Ad5, making these cells suitable for the generation and growth of helper-independent recombinant adenovirus (9).

3.1.6. Transfection of Recombinant Genome into Transcomplementing Cells

1. Seed one 60-mm plate with 1×10^6 HEK-293 cells/per plate and grow for 24 h in production medium.

2. Replace medium on cells and transfect plasmid (*see* Note 1) using freshly prepared 2.5 M CaCl$_2$ and 2× HBSS solutions as follows:
   - Mix 25.6 µl 2.5 M CaCl$_2$ with plasmid and make up to 250 µl with ddH$_2$O.
   - Carefully add 250 µl of 2× HBSS into CaCl$_2$-plasmid solution (250 µl) while bubbling air into solution and allow solution to stand for 15 min.
   - This transfection solution is then added drop-wise into tissue culture while gently agitating the 60-mm plates.

3. Return plates to incubator for 24 h and replace medium.

4. Maintain cells in incubator until a cytopathic effect (CPE) is observed microscopically.

5. Harvest the in-solution cells and scrape any remaining cells, pellet and resuspend in 10% glycerol in PBS.

6. Carry out three freeze-thaw cycles on the harvested cells (using a dry-ice/ethanol bath and 37°C water bath).

7. Spin down at $500 \times g$ for 5 min.

8. Collect transfected cell suspension and store virus stock in aliquots at –80°C.

3.1.7. Production of Working High Titre Stock

1. Seed one 100-mm plate with 1.5×10^6 HEK-293 cells/per plate and grow for 24 h in production medium.

2. Infect cells with 200 µl of transfected cell suspension per plate and rock every 15 min for 1 h in the incubator.

3. Add fresh production medium to the cells and leave in incubator until cytopathic effect observed.

4. Harvest any floating cells and scrape any remaining cells and media, spin at $500 \times g$ for 5 min, discard the supernatant, and resuspend in 2 ml of infection medium.

5. Carry out three freeze-thaw cycles on the harvested cells (using a dry-ice/ethanol bath and 37°C water bath).

6. Spin down at $500 \times g$ for 5 min and collect the supernatant.

7. Use 2 ml of supernatant to infect a sub confluent T175 flask of HEK-293, seeded at 1×10^7 HEK-293 cells/ flask, rock every 15 min for 1 h in the incubator.

8. Add fresh production medium to the cells and leave again until cytopathic effect observed.

9. Harvest in-solution cells and scrape any remaining cells and medium, spin at $500 \times g$ of 5 min, discard supernatant and resuspend in 8 ml of infection medium.

10. Carry out three freeze-thaw cycles on the harvested cells (using a dry-ice/ethanol bath and 37°C water bath).

11. Spin down at $500 \times g$ for 5 min and collect the supernatant. This is now a working high titre stock

3.1.8. Virus Titration by Plaque Assay

1. Seed ten 60-mm tissue culture plates with 1×10^6 HEK-293 cells and leave in incubator for 24 h.

2. Make the following dilutions of high titre stock in infection medium:
10^2: 10 μl virus in 990 μl infection medium
10^4: 50 μl 10^2 dilution in 4.95-ml infection medium
10^6: 50 μl 10^4 dilution in 4.95-ml infection medium
10^7: 500 μl 10^6 dilution in 4.5-ml infection medium
10^8: 500 μl 10^7 dilution in 4.5-ml infection medium
10^9: 500 μl 10^8 dilution in 4.5-ml infection medium
10^{10}: 500 μl 10^9 dilution in 4.5-ml infection medium
10^{11}: 500 μl 10^{10} dilution in 4.5-ml infection medium

3. Remove medium from 60-mm plates and add 1 ml of each of the 10^9–10^{11}virus dilutions (in triplicate) gently. Add 1 ml of the infection medium to the final plate as a control.

4. Incubate plates at 37°C/5% CO_2 for 2 h. Rock gently every 15–30 min to increase vector–cell contact.

5. Melt 2% sea plaque agarose in microwave oven and place in 37°C for 15–30 min to allow the temperature to equilibrate.

6. Mix agarose and infection medium (1:1) immediately prior to use.

7. Aspirate the medium and replace with premixed DMEM/ agarose solution at 37°C, adding 2 ml per well.

8. Leave the plates to cool at room temp for 15–30 min.

9. Return to 37°C/5% CO_2 incubator.

10. Add 2 ml of melted agarose/ DMEM at 37°C, every 2–3 days and begin checking for plaques after 5 days.

11. On day 11, overlay each dish with 2–3 ml of agar/media solution containing 500 µl 1.4% Neutral Red. Once solidified, return to the incubator for one more day.

12. Count plaques at day 12 (if possible) and determine titre (Plaque Forming Units/ ml) as follows:

PFU/ml = average plaque count of the tree plates × dilution factor.

3.1.9. Propagation of Virus Vector Stock

1. Seed 20 T175 tissue culture flasks with HEK-293 cells (*see* Note 2).

2. Grow cells in 37°C/5% CO_2 incubator until they are approximately 70–80% confluent (*see* Note 3).

3. In a sterile bottle, mix 41-ml infection medium, heated to 37°C with viral stock according to the protocol below:

   (a) Low titre virus (1×10^9–1×10^{10}): add 50 µl

   (b) Average titre virus (1×10^{10}–7×10^{10}): add 25 µl

   (c) Titre $>7 \times 10^{10}$: add 10–15 µl

4. Aspirate media from T175 flasks in sets of 10. Add 2 ml of infection media to each flask.

5. Place the flasks into 37°C/5% CO_2 incubator for 1–2 h, rocking every 15 min.

6. After this time, add 20 ml of production medium (warmed to 37°C) to each flask and return flasks to 37°C/5% CO_2 incubator.

7. After 36–48 h, the cells are ready to be harvested. The cells will round up, float, and form a "bunch of grapes"-like structure (cytopathic effect). Loosen all remaining adhered cells by tapping the side of the flask.

8. Collect cells and medium (using cell scraper), spin at $500 \times g$ for 30 min at 4°C.

9. Remove supernatant from pellets, keeping 50-ml aside.

10. Resuspend pellets in the retained 50 ml of supernatant.

11. Carry out three freeze-thaw cycles on the harvested cells (using a dry-ice/ethanol bath and 37°C water bath). Vortex cells after each thaw.

12. Centrifuge cells at $500 \times g$ for 15 min at 4°C.

13. Collect the supernatant which now contains the adenovirus.

3.1.10. Adenoviral Purification

The standard method for purification of adenoviral vectors is based on using a caesium chloride density gradient combined with ultracentrifugation. Two rounds of centrifugation are performed on the virus, and the purified virus then extracted.

Purification of Vector Stock by Density Centrifugation Using Caesium Chloride

1. Place 2.5-ml CsCl 1.43 into six SW41 Beckman ultracentrifuge tubes.

2. Very carefully, overlay with 2 ml of CsCl 1.34. At this point, make sure there is no mixing of gradients as it will affect the purification procedure.

3. Overlay the gradients with 8 ml of virus supernatant, place 1 ml of paraffin oil on top of the supernatant and spin at 30,000 rpm for 18–20 h at 4°C.

4. Gently remove the tubes and place in testtube stand for band extraction.

5. Insert an 18 in. gauge needle through the tube wall around 5 mm below the purified band and pull off each band into a 4-ml syringe. The band representing defective particles can also be collected as they will be eliminated in the next round of purification.

6. Pool all six extracted bands.

7. Place 2 ml of 1.34 CsCl into a SW41 tube and gently overlay with the pooled viral bands. Overlay tube with 1 ml paraffin oil.

8. Spin at 35,000 rpm for18–20 h at 4°C.

9. Insert 18-in. gauge needle 5 mm below the viral band and extract the viral fraction.

10. Set up a PD-10 desalting column on a retort stand.

11. Cut the bottom off the column and wash six times with 5 ml of PBS.

12. Add the purified viral band to the column, along with 0.5 ml PBS.

13. Collect 0.5-ml eluted fraction (fraction 1).

14. Add 0.5 ml PBS to column and collect eluted fraction (fraction 2).

15. Repeat step 14 ten times (collecting a total of 12 fractions).

16. Examine all fractions for presence of virus, which will be evident by the appearance of opaque solution. This generally occurs in fractions 4–9, depending on the initial volume.

17. Pool all opaque fractions and determine total pooled volume.

18. Add 10% (in final volume) glycerol, aliquot and store at −80°C.

3.1.11. Determination of Adenovirus Titre

The total physical quantity of viral particles is determined by measuring spectrophotometry (at a wavelength of A260 nm).

Physical Titration

1. Make 1/10 and 1/20 dilutions of the purified vector stock in 20 mM sodium phosphate, 0.5% SDS, pH 7.2.

2. Centrifuge diluted samples at 10,000 rpm for 1 min, transfer the supernatant to a quartz cuvette and measure OD 260. 1 OD is equal to 1.1×10^{12} particle/ml (10).

Virus Titration by Plaque Assay

See Subheading 27.3.1.8

Using both physical titration and plaque assay titration allows determination of a VP: PFU ratio, which is important in Quality Control and Biosafety.

3.1.12. Detection of Wild Type Virus

An issue that can occur in adenoviral production is the occurrence of replication competent adenovirus (RCA) in a population of replication-deficient adenovirus. RCA can emerge as a result of recombination events that restore replication competency or from carry over of a replication-competent intermediate vector construct or contamination during batch processing (11). A sensitive PCR-based method of detecting wild-type virus contamination has been developed (5), based on the detection of adenoviral E1 DNA. This assay uses two pairs of primers in the same reaction to detect adenoviral E1 DNA with co-amplification of E2B DNA as an internal control.

3.1.13. Extraction of DNA from Virus

DNA is extracted using standard DNase I, Proteinase K and PCI extraction method.

3.1.14. Detection of RCA by PCR

1. Viral DNA is amplified by PCR (see Subheading 27.2.2.6 for primer sequences)

2. The following PCR conditions are used (a) 2 min at 94°C, (b) 30 s at 94°C, (c). 30 s at 56°C, (d) 1 min at 72°C, (e) indefinitely at 4°C.
   Steps b to d are repeated for 30 cycles
   Amplified DNA is electrophoresed on a 1.8% agarose gel.

3.1.15. SDS-PAGE and Immunoblotting for Transgene Expression

1. Coronary Aortic Smooth Muscle Cells (CASMCs) are transduced in 100-mm plates with relevant adenovirus vector at a Multiplicity of Infection (MOI) 100 and incubated for 72 h at 37°C.

2. Whole protein is extracted using standard techniques and total protein concentration determined by BCA protein assay (Pierce).

3. A total of 15 μg of protein is electrophoresed on a 12% SDS-PAGE gel and resolved proteins transferred to a 0.2-mm nitrocellulose membrane (Bio-Rad) using a semi-dry electrophoretic transfer system (Bio-Rad).

4. Following overnight blocking in 5% non-fat milk, the membrane is incubated with primary antibody (e.g. 1:2,000, polyclonal rabbit anti-human MnSOD) at 37°C for 30 min.

5. Secondary antibody (e.g. HRP-conjugated goat anti-rabbit, 1:4,000) incubation is performed for 1 h at RT.

6. HRP-bound protein was detected using enhanced chemiluminescence (ECL, Amersham Life Sciences).

All washes, carried out in between antibody incubations, were performed with PBS-0.05% Tween-20.

3.2. Results

In this chapter, we describe the protocols used to propagate, purify, quantify, assess the safety, and analyse transgene expression of serotype 5 adenovirus gene therapy vectors. Production and purification of adenovirus involves many different steps and takes approximately 8–10 weeks from start to finish. Using the Gateway® system, the gene of interest is ligated into an entry vector, which then combines with the destination adenoviral vector (pAd/CMV/V-5-DEST™; E1/ E3 deleted) by using a LR Clonase® enzyme. The vector is then *Pac-1* digested, PCI extracted, and transfected into HEK293 cell line, as 293 cells contain the full E1 region of Ad5. Adenovirus vector is further propagated in the cells and, following cell lysis, purification of adenovirus is usually undertaken by caesium chloride gradient purification (*see* Note 3), a procedure that takes 1.5 days. The yield of virus may be assessed by plaque assay (determining plaque forming units) and optical density (OD, determining number of viral particles) with average yields varying from $1 \times e^{10}$ to $1 \times e^{12}$ PFU/ml. This also allows a viral particle (VP) to plaque forming units (PFU) ratio to be determined.

While these vectors are an efficient means by which specific genes can be delivered to a broad range of cells in a disease setting, there are many factors that need to be considered during their preparation and propagation. Firstly, during their manufacture, replication-deficient adenovirus may theoretically recombine with wild-type virus to result in a replication-competent virus and, as the wild-type virus is common, this possibility must be considered when assigning biosafety levels for work with these vectors. Biosafety Level 2 containment is generally required for work with adenovirus. Also, it is imperative to determine if such a recombination event has occurred and a sensitive PCR-based method of detecting wild-type virus contamination has been developed (5), whereby the adenoviral E1 DNA is detected.

3.2.1. PCR

Recombinant adenovirus (RCA) detection in by PCR should be negative in all samples tested. Figure 27.1 shows the results of testing of a MnSOD-encoding adenoviral vector, including positive and negative controls.

Secondly, if vectors are to be used for clinical trials, the gene therapy products need to be produced under conditions of current good manufacturing practice (cGMP), where the environment

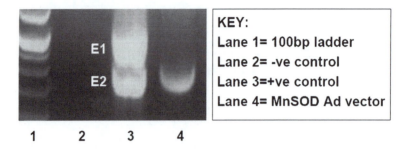

Fig. 27.1. PCR Detection of Wild-type Adenovirus Sequences. Using E1 and E2 region primers, amplification of viral DNA (extracted by DNase and phenol/chloroform) was performed over 30 cycles.

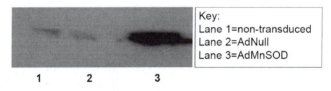

Fig. 27.2. Western blot for MnSOD expression in CASMC. Protein from 6× 100-mm plates of CASMC infected at MOI 100 for 72 h was extracted by cell lysis and 20 μg of protein from each preparation was electrophoresed, blotted, and probed for human MnSOD expression.

and procedures are controlled to produce a pharmaceutical grade product. It is well known and accepted that the VP: PFU ratio is a necessary component for the quality control of adenoviral vectors. This can be used as an indicator of the potential for generation of an immune response. Under FDA regulations, the recommended ratio is less than 30:1 (12).

Finally, the expression of transgene from the newly-manufactured vector should be determined using SDS-PAGE and western blotting (Fig. 27.2) to confirm that the vector can efficiently transduce target cells and allow production of the correct protein.

3.2.2. SDS-PAGE and Western Blotting

MnSOD expression can be detected in coronary artery smooth muscle cells (CASMC) by western blot. In the example shown below, the level of expression obtained was much higher than endogenous MnSOD (non-transduced cells and cells transduced with AdNull vector).

Transduction of cells with adenovirus encoding the SOD isoforms and catalase can facilitate the increased intracellular expression of these enzymes and, in turn, this will generate EcSOD. Such a strategy should help tackle the imbalance ROS, which results in oxidative stress and subsequent cell death and may have a broad therapeutic potential in the context of many human diseases.

4. Notes

1. When carrying out transfection of adenovirus plasmids (Subheading 27.3.1.6), it is recommended that 10 µg of plasmid is used to transfect each 60-mm plate.

2. When propagating virus stock (Subheading 27.3.1.9), the number of flasks used can be scaled up to 40 flasks for a larger prep.

3. When propagating virus stock (Subheading 27.3.1.9), HEK-293 cells should not be used if the cells are of a higher confluence than 80%.

4. HPLC can also be used for Adenoviral purification (13).

Acknowledgements

We would like to thank Martina Harte for her scientific and technical expertise.

References

1. Heistad DD (2006) Oxidative stress and vascular disease: 2005 Duff lecture. Arterioscler Thromb Vasc Biol 26:689–695

2. Baynes RW (1991) Role of oxidative stress in development of complications in diabetes. Diabetes 40:405–412

3. Calabrese V, Guaglino E, Sapienza M, Mancuso C, Butterfield DA, Stella AM (2006) Redox regulation of cellular stress in neurodegenerative disorders. Ital J Biochem 55:263–282

4. Zanetti M, Sato J, Kutusic Z, O'Brien, T (2001) Gene transfer of superoxide dismutase isoforms reverses endothelial dysfunction in diabetic rabbit aorta. Am J Physiol Heart Circ Physiol 280(6):H2516–H2523

5. Zhang WW, Koch PE, Roth JA (1995) Detection of wild-type contamination in a recombinant adenoviral preparation by PCR. Biotechniques 18:444–447

6. Greenberger JS, Epperly MW, Jahroudi N et al (1996) Role of bone marrow stromal cells in irradiation leukemogenesis. Acta Haematol 96:1–15

7. Zwacka RM, Dudus L, Epperley MW, Greenberger JS, Engelhardt JF (1998) Redox gene therapy protects human IB-3 lung epithelial cells against ionizing radiation-induced apoptosis. Hum Gene Ther 9:1381–1386

8. Brown MR, Miller FJ, Li W-G et al (1999) Overexpression of human catalase inhibits proliferation and promotes apoptosis in vascular smooth muscle cells. Circulation 85:524–533

9. Graham FL, Smiley J, Russel WC, Narin R (1997) Characteristics of a human cell line transformed by DNA from human adenovirus type 5. J Gen Virol 36:59–72

10. Maizel J V, White DO, Scharff MD (1968) The polypeptides of adenovirus I. Evidence for multiple protein components in the virion and a comparison of types 2, 7A and 12. 1968a. Virology 36:115–125

11. Zhu J, Grace M, Casale J et al. (1999) Characterization of replication-competent adenovirus isolates from large-scale production of a recombinant adenoviral vector. Hum Gene Ther 10:113–121

12. ARMWG (2001) ARMWG. BRMAC Meeting 30#: adenovirus titer measurements and RCA levels. FDA

13. Eglon M, McGrath B, O'Brien T (2009) HPLC purification of adenoviral vectors. In: Advanced protocols in oxidative stress Mn-SOD, and Catalase

Chapter 28

HPLC Purification of Adenoviral Vectors

Marc Eglon, Barry McGrath, and Timothy O'Brien

Abstract

Adenoviruses are attractive vectors for gene therapy where short-term transgene expression is required. In order to meet the clinical requirements of adenovirus for use beyond the laboratory, advanced methods are required for the purification and quantitation of recombinant adenoviral vectors (rAd). Chromatographic systems offer the advantages of linear scalability and reproducibility, and this method describes a laboratory-scale process based on liquid chromatography, which can be technically transferred and readily scaled-up according to the demands of the laboratory or clinic in which it will be used.

Key words: Adenovirus, Gene therapy vectors, Chromatography, Anion exchange, Gel filtration

1. Introduction

Owing to their transient but highly efficient transduction of a wide range of cell types both in vitro and in vivo, recombinant adenoviruses (rAd) are frequently used in both the modeling of diseases and gene therapy. They may serve as useful tools to enhance the biological and therapeutic effects of modulating oxidative stress.

Most recombinant adenoviral vectors being developed for clinical/gene therapy purposes are based on replication deficient rAd vectors, which have some critical functions deleted. The purpose of this attenuation is twofold: (1) deletion of viral genes enables packaging of additional foreign (therapeutic) genes and (2) the deletions render the viruses replication-incompetent, so that they may not proliferate outside of the laboratory (1). Adenoviral attenuation is achieved by the deletion of a combination of the four critical regions involved in early transcription, namely E1–E4, allowing the insertion of a therapeutic gene. Because these sequences are absent, they are provided *in trans* by a

D. Armstrong (ed.), *Advanced Protocols in Oxidative Stress II*, Methods in Molecular Biology, vol. 594
DOI 10.1007/978-1-60761-411-1_28, © Humana Press, a part of Springer Science+Business Media, LLC 2010

complimentary cell line, most commonly HEK-293 (human embryonic kidney) cells, to facilitate production.

In the first generation vectors, the E1 sequence is excised rendering the virus replication deficient, often in combination with the E3 region (as E3 functions are not required for the Ad life cycle in vitro), while the second generation vectors additionally have the E2 region deleted. The third generation adenovirus vectors are deficient in both early and late genes and contain only the viral inverted terminal repeats (ITRs) and packaging signals; for production, these "gutless" vectors require coinfection with a helper virus, which must later be separated from the therapeutic virus. In addition, gutless vectors possess a significantly greater cloning capacity than conventional adenoviral vectors, making the transfer of large cDNAs, multiple transgenes, and longer tissue-specific, or regulable promoters possible (2, 3). This protocol focuses on the preparation and purification of first generation vectors, produced in HEK-293 cells.

Traditionally, adenoviruses were purified using density caesium chloride (CsCl) density-gradient centrifugation but, while the product is pure and potent, the process is labor-intensive, slow and – most importantly – cannot be readily scaled up for Good Manufacturing Practice (GMP) production. More recently, techniques based on High Performance liquid Chromatography (HPLC) have emerged. This simple rapid protocol describes the use of a two-step process using anion exchange for product capture, followed by a size exclusion polishing step.

The process can be applied to the purification of adenoviral starting material produced as described. Since the upstream process can vary between laboratories, the process should be verified and adapted appropriately to ensure compatibility with the starting material. In our laboratory, all runs are performed at room temperature and have not been validated for cold-room conditions. If the procedure is to be performed under refrigerated conditions, the process should be verified to ensure that the binding conditions are consistent.

2. Materials

2.1. Equipment

1. Class II biological safety cabinet.
2. Humidified CO_2 incubator.
3. Light microscope.
4. Low speed centrifuge.
5. Waterbath.
6. Balance.

7. pH meter.

8. Vacuum pump.

9. –80°C Freezer.

10. Äkta Explorer 100 Air chromatography system (or similar instrument capable of measuring pH, conductivity and absorbance).

11. Microtitre plate reader.

12. Electrophoresis equipment – gel rigs and power packs.

2.2. Reagents and Supplies

2.2.1. Culture Media

1. *Complete Culture media*: Dulbecco's Modified Eagle's Medium (DMEM) containing L-glutamine and 4.5 g/l glucose (Sigma Catalogue number D6429), 10% Foetal Bovine Serum (FBS), supplemented with 100 units/ml of penicillin and 100 µg/ml of streptomycin.

2. *Infection Media*: DMEM containing 4.5 g/l glucose (Sigma D6429), 2% FBS, supplemented with 100 units/ml of penicillin and 100 µg/ml of streptomycin.

2.2.2. Solutions

1. Buffer A: 50 mM Tris, 2 mM $MgCl_2$, 5% (v/v) glycerol, pH 8.0 (*see* Notes 1–4).

2. Buffer B: 50 mM Tris, 2 mM $MgCl_2$, 5% (v/v) glycerol, 1 M NaCl, pH 8.0 (*see* Notes 1–4).

3. 1 M NaOH (*see* Note 3).

4. 20% Ethanol (*see* Note 3).

5. 18.2 MΩ dH_2O (*see* Note 3).

6. Tricorn 10/300 and Tricorn 5/150 HPLC Columns (GE Healthcare).

7. Q Sepharose XL (Virus licensed) chromatography media (GE Healthcare).

8. Sepharose 4 Fast-flow chromatography media (GE Healthcare).

9. 10-layer CellStacks™ (Corning) for adherent cell culture.

10. Sterile cell culture plasticware – serological pipettes, centrifuge tubes, etc.

11. General laboratory plasticware – beakers, measuring cylinders, weighing boats, etc.

3. Methods

3.1. Preparation of Adenovirus in HEK-293 Cells

The standardization and characterization of starting material for the downstream process is critical to ensure reproducible results and consistent quality between batches. This protocol describes an adherent culture system that exploits the natural events that occur during adenoviral infection. After 48–72 h, infected cells become cytopathic, taking on a rounded morphology and detaching into suspension. After around 96–120 h, cells enter a lytic phase, releasing free virions into the culture suspension. By monitoring the level of cytopathic effect (CPE) in the culture, the cell-bound virions are harvested by low-speed centrifugation, and liberated by freeze-thawing.

1. Grow a batch of HEK-293 cells to ~70% confluence in a 10-layer CellStack™ (Corning) before infecting with the adenovirus seed stock at a multiplicity of infection (MOI) of 10–50 in 150 ml of infection medium.

2. After 2–4 h, add 850 ml complete medium to the infected culture and incubate at 37°C, 5% CO_2. Observe the cells for CPE and harvest the cells after 48–72 h, when 100% CPE is apparent.

3. To detach the cells from the vessel surface, vigorously rock the vessel and transfer the cell suspension to sterile 500 ml centrifuge tubes. Pellet the cells at $3,000 \times g$ for 10 min at 4°C, and resuspend in 12.5 ml Buffer A in a 15 ml centrifuge tube.

4. Vortex the suspension and mechanically lyse the cells by three rounds of freeze-thawing. Freeze the suspension in a dry-ice/ethanol bath, ensuring that the suspension is completely frozen to allow the ice crystals to fully disrupt the cell membrane. Thaw the suspension in a waterbath at 37°C, vortex the suspension, and repeat the snap-freeze for a total of three cycles.

5. The method described here is capable of purifying 1 ml of crude lysate and is the first step used to verify the process. If desired, the clarified supernatant can be aliquoted and stored until use. This should be performed prior to the third freezing step and the aliquots stored at –80°C.

6. Benzonase™ is added to enzymatically hydrolyse free genomic DNA and unpackaged viral DNA, producing short oligomers of 3–8 base pairs. The digestion results in a significant reduction in sample viscosity and, moreover, curtails the deterioration in column performance that results from strong interactions between the phosphate backbone of the nucleic acid and the anion exchange media. After the final thaw, refer to the batch certificate for the Benzonase™ to determine the potency, and add 1,200 U to the crude cell lysate. Vortex the suspension for 5 s, and incubate at 37°C for 1 h in a stirred waterbath to allow the reaction to proceed. Vortex every 15 min during the incubation (*see* Note 5).

7. To remove the cellular debris from the crude lysate, centrifuge the suspension at $5,000 \times g$ for 10 min and discard the pellet. The supernatant (clarified lysate) contains the adenovirus.

8. To prevent the sample from reducing the conductivity of the mobile phase in the anion exchange column, the conductivity must be increased by mixing the sample with Buffer B to raise the NaCl concentration. Measure the volume of the supernatant and add 0.54 volumes of Buffer B to bring the final NaCl concentration to 350 mM. This will preclude nonspecific binding to the column and increase the elution of weak anionic contaminants in the flow-through.

9. Filter the clarified lysate using a prewetted 0.45 μm syringe filter. Change the filter if there is excessive resistance indicating that the pores are becoming blocked; depending on the viscosity of the clarified lysate, 2–3 filters may be required.

3.2. HPLC

In our laboratory, the Äkta Explorer 100 Air (GE Healthcare) is used for all HPLC operations – the unit is able to monitor pH, conductivity, absorbance, and pressure. Mixing is performed online, but if this is not possible, buffer compositions should be modified so that the Equilibration Buffer contains 400 mM NaCl and Elution Buffer contains 600 mM NaCl. A Stripping Buffer containing 1–2 M NaCl should also be prepared for cleaning the column.

3.3. HPLC: Column Packing

Columns must be thoroughly cleaned before packing; immediately prior to use, rinse the column with 18.2 MΩ dH$_2$O.

1. To pack the column, attach a packing reservoir and support the entire assembly vertically in a retort stand (use a spirit level to ensure optimal packing).

2. Wet a frit in 20% ethanol, insert it into the bottom of the column, and secure using the bottom adapter. Pour 1 ml of 20% ethanol on top of the frit.

3. Pump 20% ethanol at 5 ml/min through the bottom column port using a drop-to-drop connection to dispel any air from the column. Stop the pump.

4. To pack the Tricorn 5/150 QXL column, add 7.5 ml of 70% slurry of Q Sepharose XL media (QXL) to the top of the packing reservoir and top up with 18.2 MΩ dH$_2$O. Attach the top adapter, avoiding the introduction of any air to the column.

5. Using a drop-to-drop connection, attach the inlet tubing and pack the column at 2 ml/min (~600 cm/h) until the bed height has stabilized and the pressure is constant (~0.2 MPa). Stop the pump.

6. Remove the packing reservoir, and add the (prewetted) top frit to the column. Attach the top adapter, and reconnect the inlet tubing using a drop-to-drop connection. Continue pumping at 2 ml/min for at least 1 column volume, until the bed height is stabilized. Mark the top of the bed.

7. Stop the pump and adjust the top adapter so that the top of the frit is at the level marked in step 6.

8. Equilibrate the column in Buffer A (mixed with 10% Buffer B) for column volume at 1 ml/min.

9. A simple method to test the column packing quality is to measure the asymmetry of the elution peak of an inert marker such as 1% acetone after injection onto the column. Inject 100 μl of 1% acetone in Buffer B and elute at 1 ml/min. To calculate the asymmetry value, refer to Fig. 28.1; a value of 1.0 ± 0.1 is acceptable for separation. A low value indicates that the column is over-packed and the column should be repacked at a lower flow rate.

10. Equilibrate the column in 3 column volumes of 400 mM NaCl (formed by mixing Buffer A with 40% Buffer B) until conductivity is constant (35.1 ms/cm).

11. To pack the Tricorn 10/300 column, follow steps 1–10, substituting the relevant parameters listed in Table 28.1.

3.4. HPLC: Anion Exchange (Capture Step)

Adenovirus readily binds to the quaternary ammonium ligand at pH 8 and is eluted in the conductivity range 36–50 ms/cm. For the separation to proceed effectively, it is important to ensure that HPLC gradients are formed correctly and the desired conductivity is maintained for each step. In this protocol, the filtered starting material is loaded at around 35 ms/cm to allow complete

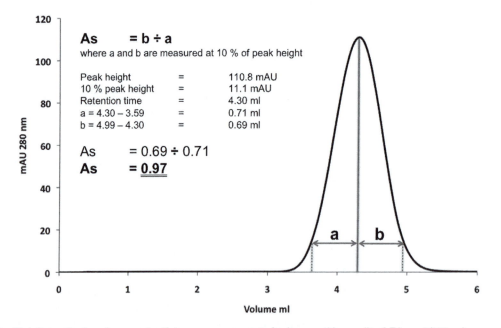

Fig. 28.1. Determination of asymmetry (As) as a measurement of column packing quality. A Tricorn 5/150 column was packed at 2 ml/min, and 100 μl of 1% acetone in Buffer B was used to measure the elution characteristics of the column (refer to **steps 1–9** above).

Table 28.1
Flow rates and packing parameters for Tricorn columns

	Tricorn 5/150	Tricorn 10/300
	1 ml/min≅300 cm/h	1 ml/min≅128 cm/h
Volume of 70% slurry	7 ml	45
Flow rate for pack 1 (with reservoir)	2 ml/min (for ~20 ml)	5 ml/min (for ~100 ml)
Flow rate for pack 2 (no reservoir)	2 ml/min (for ~20 ml)	5 ml/min (for ~100 ml)
Flow rate for pack 3 (top adapter adjusted)	2 ml/min (for ~20 ml)	5 ml/min (for ~100 ml)

Table 28.2a
Running parameters for the anion exchange step using a Tricorn 5/150 column

Step	Buffer	Volume	Flow rate
1. Equilibration	0.4 M NaCl (target 35.2 ms/cm)	2–3 cv	1 ml/min
2. Sample loading	Sample	~1.5 ml	0.35 ml/min
3. Wash	0.4 M NaCl (target 35.2 ms/cm)	2 cv	0.35 ml/min
4. Elution	0.6 M NaCl (target 50.0 ms/cm)	3 cv	0.35 ml/min

cv, column volume(s)

retention of the target virus while minimizing the binding of competitive anions. The virus is eluted in a single step gradient to 600 mM NaCl.

1. When switching buffers, particularly for elution, bypass the column and mix the buffers to ensure that all lines are fully primed prior to entering the column. If the mixing of buffers is allowed to proceed in the column, a linear gradient is formed, resulting in a broad dilute elution peak.

2. The running parameters for the anion exchange step are summarized in Table 28.2a.

3. Pool the eluted fractions containing the adenovirus peak, as shown in Fig. 28.2. It is important to collect only the most concentrated fractions to avoid dilution of the virus.

4. Following elution, the column should be "cleaned in place" (CIP) using the regime detailed in Table 28.2b. To expedite the anion exchange step, the CIP step can be performed following gel filtration.

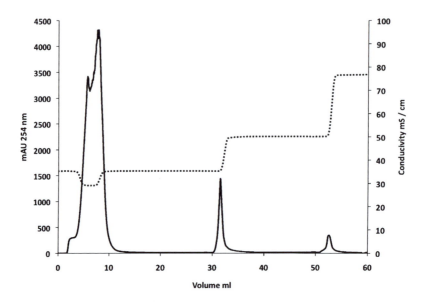

FIG. 28.2. ANION EXCHANGE CHROMATOGRAM. *SOLID LINE* INDICATES ABSORBANCE; *DOTTED LINE* INDICATES CONDUCTIVITY (SECONDARY AXIS); THE *ARROW* SHOWS THE PEAK CONTAINING THE TARGET ADENOVIRUS.

Table 28.2b
Running parameters for cleaning and regenerating the Tricorn 5/150 column

Step	Buffer	Volume	Flow rate
1. Strip	1 M NaCl	2 cv	0.35 ml/min
2. CIP	18.2 MΩ dH$_2$O	1 cv	1 (up-flow)
3. CIP	1 M NaOH	1.5 cv	0.2 (up-flow)
4. Charging	Buffer B	1 cv	1 (up-flow)

3.5. HPLC: Gel Filtration (Polishing Step)

Gel filtration exploits the relatively large macrostructure of the adenovirion, which is unable to penetrate the intra-bead pores and is eluted in the void fraction. Smaller contaminants – including NaCl – are eluted later, polishing the adenovirus and restoring the optimum NaCl concentration to maintain virus integrity during cryopreservation.

1. Equilibrate the column with 2.5 column volumes of buffer containing 150 mM NaCl (15% Buffer B) at 2 ml/min until conductivity is constant.

2. Reduce the flow rate to 0.78 ml/min (60 cm/h) and load the sample.

3. Continue to run the pump at 0.78 ml/min for 1.5 column volumes, under isocratic conditions of 150 mM NaCl, pH 8.

4. The virus will elute in the void fraction at around 0.3 column volumes, as indicated in Fig. 28.3. Pool the fractions containing the adenovirus, and remove samples for titration and quality control (QC) testing. The remaining purified adenovirus can be stored for up to 2 weeks at 4°C. If longer term storage is required, –80°C is preferred and the product should be aliquoted as desired to avoid freeze-thawing of the bulk, which may result in reduced integrity of infectious virions.

3.6. Online Measurement

To maintain consistency between batches and permit scaling up, it is important to monitor and control the critical operational parameters of the purification process, particularly with respect to conductivity in the anion exchange step. Where possible, the following parameters should be monitored throughout the process: absorbance at 280 nm (for protein) and 260 nm (for DNA), pH, and conductivity. It is possible to use the real-time absorbance values as an indication of the relative purity of the viral material by calculating the ratio of the mAU 260 nm to mAU 280 nm, whereby pure adenovirus should have a value in the range 1.23 ± 0.8 (4). This value should be verified by loading a purified sample as a control and by gel electrophoresis.

3.7. Quality Control

Several assays should be routinely performed to ensure satisfactory levels of potency and purity. In addition, freedom of replication-competent adenovirus (RCA) must be verified. RCA results when the E1 genes are recombined into the attenuated virus, restoring the ability to replicate in nonpermissive cells and tissues. For scale

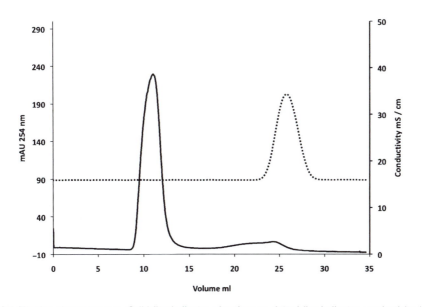

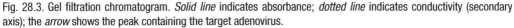

Fig. 28.3. Gel filtration chromatogram. *Solid line* indicates absorbance; *dotted line* indicates conductivity (secondary axis); the *arrow* shows the peak containing the target adenovirus.

up and translation into GMP production, a commercial enzyme-linked immunosorbent assay kit is available to demonstrate the removal of Benzonase™ in the finished product (Benzonase® ELISA-kit II, Merck catalogue number 1.01681.0001).

3.7.1. Infectious Titre

Infectious virus titration requires a limiting dilution of the sample, followed by the infection of a suitable cell line. Following a period of incubation ranging from 1 to 10 days, the degree of infectivity is quantified by measuring either cell death (CPE) or the expression of a suitable reporter gene. Several such methods are available such as the plaque assay (5), fluorescent focus assay (6), and end-point dilution (7). Owing to the inherent variability of these assays, several replicates are required at three or more dilutions to ensure the generation of reliable results. Consequently, the measurement of multiple samples is not possible, and these assays are generally applied only to the final product.

Where a suitable reporter gene such as green fluorescent protein (GFP) is included in the vector's genome, rapid assays utilizing flow cytometry have been developed to determine the infectious titre in less than 24 h (8–10). At a suitably high dilution where MOI <0.3, a linear relationship exists between the number of fluorescent cells and the dilution factor, such that each GFP positive cell corresponds to a single infectious viral unit (11).

3.7.2. Total Particle Titre

For clinical-grade adenoviral vectors, the potency of the viral preparation must be ≥3.3%; i.e. 1 in every 30 viral particles must be infectious (12). To measure total particle number, various methods exist such as DNA quantification by optical absorbance (13) or by PicoGreen staining (14) and anion exchange chromatography (15, 16).

3.7.3. RCA-PCR

To verify freedom form wild-type Ad as a result of recombination during production in HEK-293 cells, a simple PCR assay should be performed (17, 18).

3.7.4. Purity Analysis: SDS–PAGE and Silver Staining

Sodium dodecyl sulphate polyacrylamide gel electrophoresis (SDS–PAGE) enables the rapid assessment of adenovirus purity compared to a standardized control such as a preparation purified CsCl density-gradient ultracentrifugation. All steps can be performed at room temperature.

1. Denature the purified viral material in a microcentrifuge tube by adding 2.5 µl of sample to 2.5 µl of 1% SDS in Buffer A.

2. Incubate at room temperature for 20 min.

3. Quantify the protein content using an appropriate assay such as the detergent compatible (DC) protein assay (Biorad).

4. Dilute fresh samples in Final Formulation Buffer (Buffer A containing 15% Buffer B) to a protein concentration of ~2–4 µg/µl and add Laemmli Buffer to a final concentration of 1×.

5. Cap the tube securely, vortex the mixture, and incubate at 95°C for 5 min.

6. Invert the tube to mix the suspension and vortex the suspension for 5 s.

7. To resolve the proteins, add 20 µg of protein to each lane of a suitable polyacrylamide gel (e.g. 10% bis-Tris) and run at 100 V for ~1 h.

8. Stain the gel using the SilverQuest™ Silver Stain kit (Invitrogen). More than 15 bands may be visible in the stained gel, depending on the resolution of the gel and the sensitivity of the silver staining (19), but a characteristic banding pattern consisting of 5 or 6 major bands should always be visible, as shown in Fig. 28.4a.

3.8. Western Blot

To verify the identity of the bands, Western blots can be performed using primary antibodies to detect adenoviral capsid monomers or residual host cell proteins. This protocol describes the positive identification of viral components using a rabbit polyclonal antibody to Ad5 (Abcam, catalogue number ab6982). All procedures can be performed at room temperature (*see* Note 6).

1. Transfer the proteins to a nitrocellulose membrane using semi-dry blotting apparatus at 15 V for 40 min.

2. Rinsing, blocking, and antibody incubation should be performed on a rotational rocker platform. Wash the membrane in Phosphate Buffered Saline (PBS) for 5 min. Repeat a total of five times.

3. Block the membrane in PBS containing 5% milk and 0.2% Tween 20 for at least 1 h. The membrane can be blocked overnight if necessary.

4. Rinse the membrane three to five times in PBS-Tween before incubation with the primary antibody.

5. Dilute the primary antibody 1/6,000 in 30 ml of blocking solution (PBS containing 5% milk and 0.2% Tween 20). Apply the solution to the membrane and incubate for 1 h.

6. Decant the antibody solution and rinse the membrane as described at step 4.

7. Prepare the secondary antibody (ECL™ Anti Rabbit IgG HPO linked, raised in donkey. Amersham NA934V) at 1/10,000 in 30 ml blocking solution (*see* step 5). Apply the solution to the membrane in incubate for 1 h.

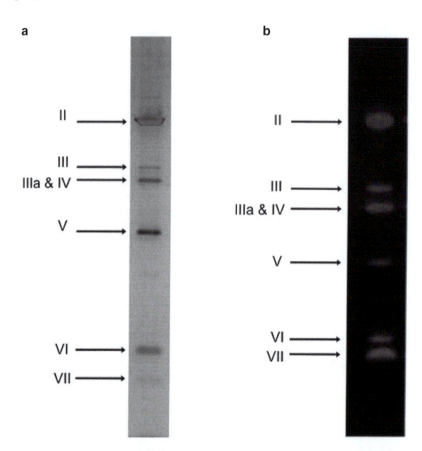

Fig. 28.4. Electrophoretograms for purified adenovirus. The bands are annotated with their protein identities based on published data (19–21). (a) Silver Stain – proteins were resolved using a Novex 4–12% gradient bis – Tris gel and stained using the Silverquest silver stain kit (both Invitrogen). (b) Western Blot for Adenoviral Capsid Components – viral bands were separated using a Novex 10% bis – Tris gel (Invitrogen) and detected using a rabbit polyclonal primary antibody to Ad5 (Abcam) followed by a donkey anti Rabbit IgG HPO-linked, secondary antibody (GE Healthcare).

8. Rinse the membrane for 5–10 min in PBS and decant. Repeat three times.

9. Using the ECL detection kit (GE Healthcare), dilute the reaction substrate 1/40 as directed and add 2 ml to the membrane.

10. Allow the reaction to proceed for 1 min at room temperature, and use a chemiluminescent imager such as the Alpha Innotech Fluorchem system to document the positive bands as shown in Fig. 28.4b.

3.9. Benzonase® Endonuclease Quantitation

When validating the HPLC steps, use the Benzonase® Elisa-Kit II (Merck KGaA) enzyme ELISA to detect residual Benzonase in the final product and intermediate samples, as required. Prepare the adenovirus samples in a microtitre plate as follows:

1. Dilute the samples by adding 55 µl to 55 µl of buffer 1 (provided in the Benzonase® Elisa-Kit II).

2. Prepare a standard curve in a microtitre plate using a fivefold dilution series by diluting the standard in buffer 1.

3. Pipette 100 µl of diluted test sample or standard into the polystyrene microtitre plates coated with the primary antibody (provided in the Benzonase® Elisa-Kit II) and proceed with the protocol as described by the manufacturer.

4. Notes

1. If online mixing of buffers is not possible with your chromatography system, prepare ready-made buffers as above, but adjust the NaCl concentration as follows: Equilibration Buffer – 400 mM; Elution Buffer – 600 mM; Stripping Buffer 1.5 M.

2. Prepare all solutions from concentrated stocks using 18.2 MΩ dH$_2$O and adjust to pH 8.0 by the addition of concentrated hydrochloric acid or 10 M sodium hydroxide (NaOH). Prepare the Tris Buffer at the target pH by combining Trizma Hydrochloride and Trizma base (Sigma-Aldrich) in the ratios specified by the manufacturer.

3. Filter and degas the buffers using a vacuum-driven 0.2 µm filter unit. Provided the buffers are stored in sterile containers, they can be stored unopened for up to 6 months.

4. To compensate for batch-to-batch variability when preparing buffers and ambient temperature deviations, buffers should be allowed to equilibrate to the running temperature and the conductivity verified prior to use. Using an automated mixing system, a conductivity curve for each pair of buffers can be calibrated to achieve the optimum conductivity.

5. Benzonase® (synonym:Benzon Nuclease) (Merck).

6. To detect Host-cell proteins [HCP] follow the protocol as described in section 3.8, Western blot, substituting the following antibodies: (1) Primary antibody – Goat polyclonal to HEK-293 lysate, *Fitzgerald catalogue number HK293/11119a* (dilute 1/3,000); (2) Secondary antibody – donkey anti-goat IgG – HRP, *Santa Cruz Biotech SC-2020* (dilute 1/5,000).

References

1. Lusky M (2005) Good manufacturing practice production of adenoviral vectors for clinical trials. Hum Gene Ther 16(3):281–291

2. Segura MM, Alba R, Bosch A, Chillon M (2008) Advances in helper-dependent adenoviral vector research. Curr Gene Ther 8(4):222–235

3. Volpers C, Kochanek S (2004) Adenoviral vectors for gene transfer and therapy. J Gene Med 6(Suppl 1):S164–S171

4. Huyghe BG, Liu X, Sutjipto S, Sugarman BJ, Horn MT, Shepard HM et al (1995) Purification of a type 5 recombinant adenovirus encoding human p53 by column chromatography. Hum Gene Ther 6(11):1403–1416

5. Cooper PD (1961) The plaque assay of animal viruses. Adv Virus Res 8:319–378

6. Philipson L (1961) Adenovirus assay by the fluorescent cell-counting procedure. Virology 15:263–268

7. Nielsen LK, Smyth GK, Greenfield PF (1992) Accuracy of the endpoint assay for virus titration. Cytotechnology 8(3):231–236

8. Gueret V, Negrete-Virgen JA, Lyddiatt A, Al-Rubeai M (2002) Rapid titration of adenoviral infectivity by flow cytometry in batch culture of infected HEK293 cells. Cytotechnology 38(1–2):87–97

9. Park MT, Lee GM (2000) Rapid titer assay of adenovirus containing green fluorescent protein gene using flow cytometric analysis. Bioprocess Eng 22(5):403–406

10. Sandhu KS, Al-Rubeai M (2008) Monitoring of the adenovirus production process by flow cytometry. Biotechnol Prog 24(1):250–261

11. Nadeau I, Jacob D, Perrier M, Kamen A (2000) 293SF metabolic flux analysis during cell growth and infection with an adenoviral vector. Biotechnol Prog 16(5):872–884

12. ARMWG (2001) BRMAC meeting #30: Adenovirus titer measurements and RCA levels. FDA

13. Maizel JV Jr, White DO, Scharff MD (1968) The polypeptides of adenovirus. I. Evidence for multiple protein components in the virion and a comparison of types 2, 7A, and 12. Virology 36(1):115–125

14. Murakami P, McCaman MT (1999) Quantitation of adenovirus DNA and virus particles with the PicoGreen fluorescent dye. Anal Biochem 274(2):283–288

15. Kuhn I, Larsen B, Gross C, Hermiston T (2007) High-performance liquid chromatography method for rapid assessment of viral particle number in crude adenoviral lysates of mixed serotype. Gene Ther 14(2):180–184

16. Shabram PW, Giroux DD, Goudreau AM, Gregory RJ, Horn MT, Huyghe BG et al (1997) Analytical anion-exchange HPLC of recombinant type-5 adenoviral particles. Hum Gene Ther 8(4):453–465

17. Duffy AM, McMahon JM, O'Brien T. (2009) Generation of Anti-Oxidant Adenovirus Gene Transfer Vectors Encoding CuZnSOD, MnSOD and Catalase. Meth Mol Biol, Humana Press, NY, in press

18. Zhang WW, Koch PE, Roth JA (1995) Detection of wild-type contamination in a recombinant adenoviral preparation by PCR. Biotechniques 18(3):444–446

19. Rexroad J, Wiethoff CM, Green AP, Kierstead TD, Scott MO, Middaugh CR (2003) Structural stability of adenovirus type 5. J Pharm Sci 92(3):665–678

20. Blanche F, Monegier B, Faucher D, Duchesne M, Audhuy F, Barbot A et al (2001) Polypeptide composition of an adenovirus type 5 used in cancer gene therapy. J Chromatogr A 921(1):39–48

21. Rux JJ, Kuser PR, Burnett RM (2003) Structural and phylogenetic analysis of adenovirus hexons by use of high-resolution X-ray crystallographic, molecular modeling, and sequence-based methods. J Virol 77(17):9553–9566

Chapter 29

Mapping of Oxidative Stress Response Elements of the Caveolin-1 Promoter

Janine N. Bartholomew and Ferruccio Galbiati

Summary

According to the "free radical theory" of aging, normal aging occurs as the result of tissue damages inflicted by reactive oxygen species (ROS). ROS are known to induce cellular senescence, and senescent cells are believed to contribute to organismal aging. The molecular mechanisms that mediate the cellular response to oxidants remain to be fully identified. We have shown that oxidative stress induces cellular senescence through activation of the caveolin-1 promoter and upregulation of caveolin-1 protein expression. Here, we describe how reactive oxygen species activate the caveolin-1 promoter and how the signaling may be assayed. These approaches provide insight into the functional role of caveolin-1 and potentially allow the identification of novel ROS-regulated genes that are part of the signaling machinery regulating cellular senescence/aging.

Key words: Caveolae, Caveolin-1, Cellular senescence, Free radicals, Tumor suppressor

1. Introduction

Several theories have been proposed in the past to explain why and how living organisms cannot escape aging. One of these is the "free radical theory" of aging, proposed by Denham Harman in the 1950s. According to this theory, normal aging occurs as the result of tissue damages inflicted by reactive oxygen species (ROS). In support of this theory, increased oxidative damage of DNA, proteins, and lipids has been reported in aged animals (1). Thus, endogenous and exogenous stimuli may significantly increase oxidant levels within the cell and induce a series of cellular damages.

Most cells cannot divide indefinitely due to a process termed cellular senescence (2–8). Growth arrest is associated with well-defined biochemical alterations. These include cell cycle arrest,

D. Armstrong (ed.), *Advanced Protocols in Oxidative Stress II,* Methods in Molecular Biology, vol. 594
DOI 10.1007/978-1-60761-411-1_29, © Humana Press, a part of Springer Science+Business Media, LLC 2010

increased p53 activity, increased $p21^{Wafl/Cip1}$ and p16 protein expression, and hypo-phosphorylation of pRb (2–6). Interestingly, oxidative stress has been shown to induce premature senescence in fibroblasts in culture (9–11). Because a number of molecular changes that are observed in senescent cells occurs in somatic cells during the aging process, investigating the molecular mechanisms underlying oxidative stress induced premature senescence will allow us to better understand the more complicated aging process.

Caveolae are vesicular invaginations of the plasma membrane. Caveolin-1 is the structural component of caveolae. It has been proposed that caveolin-1 participates in vesicular trafficking events and signal transduction processes (12–14) by acting as a scaffolding protein (15) to organize and concentrate specific lipids (cholesterol and glyco-sphingolipids (16, 17)) and lipid-modified signaling molecules (Src-like kinases, H-Ras, eNOS, components of the p42/44 MAP kinase pathway, G-proteins, EGF-R, Neu, protein kinase A, and protein kinase C) within caveolar membranes (18–26). In addition to concentrating these signaling molecules within a specific region of the plasma membrane, caveolin-1 binding functionally inhibits the activity of caveolae-associated molecules.

We have shown that overexpression of caveolin-1 in fibroblasts is sufficient to arrest the cells in the G_0/G_1 phase of the cell cycle and induce premature senescence. Consistent with these data, senescent human diploid fibroblasts have been shown to express higher levels of caveolin-1, as compared to younger human diploid fibroblasts (27). In addition, caveolin-1 has been shown to play an important role in senescence-associated morphological changes by regulating focal adhesion kinase activity and actin stress fiber formation in senescent cells (28). Finally, Park and colleagues have shown reentry of replicative senescent cells into cell cycle upon EGF stimulation after downregulation of caveolin-1 (29).

We have also demonstrated that oxidative stress induces cellular senescence in fibroblasts by stimulating caveolin-1 gene transcription through p38 MAPK/Sp1-mediated activation of two GC-rich promoter elements and upregulation of caveolin-1 protein expression (30). Interestingly, quercetin and vitamin E, two antioxidant agents, successfully prevent the premature senescent phenotype and the upregulation of caveolin-1 induced by hydrogen peroxide. Moreover, downregulation of caveolin-1 expression using an antisense-based approach inhibits oxidative stress-induced cellular senescence (31). Thus, caveolin-1 appears to play a major role in the signaling events linking oxidative stress to cellular senescence. Because oxidative stress induces cellular senescence and senescent cells are believed to contribute to organismal aging, studying ROS-mediated gene regulation will allow us to gain mechanistic insight into in vivo aging.

The assays described here were used to determine the transcription factor that mediates the oxidant-induced activation of the caveolin-1 promoter and include Luciferase-based Reporter assays, Electrophoretic Mobility Shift Assay (EMSA), and Chromatic Immunoprecipitation (ChIP) analysis. More precisely, luciferase-based reporter assays were used to define the sequence within the caveolin-1 promoter, which is responsive to free radical stimulation. Electrophoretic mobility shift assays were employed to determine whether the oxidant-responsive caveolin-1 promoter sequence identified with luciferase-based reporter assays formed a nucleoprotein complex after oxidative stress, which is indicative of binding of a transcription factor to the DNA. Finally, ChIP analysis was performed to pinpoint the transcription factor involved in the oxidant-induced activation of the caveolin-1 promoter.

2. Materials

2.1. Oxidative Stress

1. Cell culture: NIH 3 T3 Fibroblasts.
2. Cellular media: Dulbecco's Modified Essential Medium supplemented with 10% Donor Bovine Calf Serum, Glutamine, and Antibiotics (Penicillin and Streptomycin).
3. 150 µM Hydrogen peroxide (H_2O_2) diluted in cellular media.
4. Phosphate buffered saline (PBS).

2.2. Luciferase-Based Reporter Assay

1. GME Buffer (25 mM glycylglycine, 15 mM $MgSO_4$, 4 mM EGTA, dissolved in H_2O). GME Buffer must be stored at 4°C.
2. Z Buffer (100 mM NaH_2PO_4, 10 mM KCl, 1 mM $MgSO_4$, pH 7.0). Right before using the solution, 50 mM β-mercaptoethanol should be added.
3. Extraction buffer (1% w/v Trition-X-100, 1 mM DTT in GME Buffer). Extraction buffer must be stored at 4°C. Additionally, 500 µL will be needed for each sample.
4. ATP mix (17 mM K Phosphate, 10 mM DTT, 2 mM ATP in GME Buffer). Mix must be kept at room temperature. Each sample needs 300 µL.
5. Luciferin solution (Add 1 ml of 1 mM luciferin and 50 µl of 1 M DTT to 4 ml of GME buffer). Do not add luciferin until right before using the solution. Solution can be kept at room temperature (*see* Note 1).

6. Calcium phosphate transfection reagents ($CaCl_2$ and HeBs).

7. Caveolin-1 promoter luciferase reporter construct, luciferase reporter plasmid pTA-luc, and β-galactosidase-expressing construct.

8. PBS.

9. 1 M Na_2CO_3.

10. 4 mg/ml chlorophenol red-β-Dgalactopyranoside (CPRG) in ddH_2O.

11. Luminometer reading at 562 nm and spectrometer reading at 574 nm.

2.3. Electrophoretic Mobility Shift Assay

1. NIH 3 T3 fibroblasts cultured in 10 cm dishes.

2. Cellular media: Dulbecco's Modified Essential Medium supplemented with 10% Donor Bovine Calf Serum, Glutamine, and Antibiotics (Penicillin and Streptomycin).

3. Nuclear Extraction Buffer A (10 mM HEPES pH 7.9, 1.5 mM $MgCl_2$, 10 mM KCl, 0.5 mM DTT, 1 mM EDTA, and protease inhibitor tablet).

4. Nuclear Extraction Buffer B (20 mM HEPES pH 7.9, 25% glycerol, 0.43 M NaCl, 1.5 mM $MgCl_2$, 0.2 mM EDTA, 0.5 mM DTT, and protease inhibitor tablet).

5. Phosphate buffered saline (PBS).

6. 3′ end biotin-labeled double-stranded oligonucleotides containing a GC-rich box (in bold) (sequences are listed 5′ to 3′): Cav-1 (-244/-222): ggcact**ccccgccc**tctgctgcc; Cav-1 (-124/-101): cagcca**ccgccccccgcc**agcgc.

7. Annealing Buffer (10 mM Tris-Hcl, 0.5 mM EDTA, 0.5 mM trisodium phosphate, and 1 mM NaCl in sterile H_2O).

8. 10× Binding Buffer (100 mM Tris–HCl pH 8.0, 50% glycerol, 10 mM EDTA, 10 mM DTT, and 500 μg/mL) poly (Deoxyinosinic-deoxycytidylic acid).

9. 5% nondenaturing polyacrylamide gel in 1× TBE along with appropriate running and gel transfer apparatus.

10. 10× TBE Buffer (108 g Tris–base, 55 g Boric Acid, and 20 ml 0.5 M EDTA in 1 l of H_2O; pH 8.0)

11. Positively charged Biodyne B nylon membrane.

12. Chemiluminescent Nucleic Acid Detection Module (Pierce Biotechnology, Illinois).

13. Film and developing cassettes.

2.4. Chromatin Immunoprecipitation Analysis

1. Cellular media: Dulbecco's Modified Essential Medium supplemented with 10% Donor Bovine Calf Serum, Glutamine, and Antibiotics (Penicillin and Streptomycin).

2. Formaldehyde.

3. 1.4 M glycine.

4. Chromatin IP buffer (50 mM HEPES KOH pH 8.0, 1 mM EDTA pH 8.0, 0.5 mM EGTA pH 8.0, 140 mM NaCl, 10% glycerol, 0.5% IGEPAL, 0.25% Triton X-100, and protease inhibitor tablet).

5. Wash Buffer (10 mM Tris–HCl pH 8.0, 1 mM EDTA pH 8.0, 0.5 mM EGTA pH 8.0, 200 mM NaCl, and protease inhibitor tablet).

6. RIPA buffer (10 mM Tris–HCl pH 8.0, 1 mM EDTA pH 8.0, 0.5 mM EGTA pH 8.0, 140 mM NaCl, 1% Trition X-100, 0.1% Na-deoxycholate, 0.1% SDS, and protease inhibitor tablet).

7. Protein A Sepharose beads conjugated to salmon sperm DNA.

8. ChIP Dilution Buffer (0.01% SDS, 1.1% Triton X-100, 1.2 mM EDTA, 16.7 mM Tris–HCl pH 8.1, and 167 mM NaCl).

9. Antibody of interest. For this protocol, Sp1 antibody was used.

10. LiCl Buffer (10 mM Tris–HCl pH 8.0, 1 mM EDTA pH 8.0, 0.5 mM EGTA pH 8.0, 250 mM LiCl, 1% Triton X-100, 1% Na-deoxycholate, and protease inhibitor tablet).

11. Elution Buffer (1% SDS and 0.1 M NaHCO$_3$).

12. Proteinase K.

13. Qiagen PCR Purification Kit.

14. Polymerase chain reaction (PCR) primers for the Caveolin-1 gene promoter: Sense strand (5′ to 3′): caggctctcagctc-cccgcgc; antisense strand (5′ to 3′): gtatagagggggggaaaggcgc

15. PCR reagents (DNA template, primers, dNTPs, 10× reaction buffer, Taq enzyme, and H$_2$O).

16. 1.2% agarose DNA gel (with ethidium bromide) and TAE 6× DNA loading dye.

17. UV light gel documentation system.

3. Methods

3.1. Oxidative Stress

This section describes how to subject cells to oxidative stress using hydrogen peroxide. Hydrogen peroxide has been widely used as a source of free radicals and is shown by a number of groups to cause senescence (10, 11). Additionally, it is known to trigger the upregulation of caveolin-1 (30, 31).

1. H_2O_2 is diluted in cellular media to a concentration of 150 µM.

2. Media is removed from cell culture dishes, and H_2O_2 media is placed on cells (appropriate volume for dish size).

3. Cells are incubated for 2 h at 37°C.

4. Cells are washed twice in PBS to remove all traces of H_2O_2 media and are replated with fresh media (*see* Note 2).

3.2. Luciferase-Based Reporter Assay

This technique is commonly known as reporter gene assay. The firefly luciferase gene used in this assay is cloned downstream of the DNA promoter sequence under investigation in a promoter-less DNA vector. The luciferase enzyme is synthesized after transient transfection of the DNA vector in cells only if the upstream DNA promoter sequence drives transcription of the luciferase gene. The ability of the cloned DNA promoter sequence to be activated by a certain stimulus is proportional to the amount of light produced by the oxidation of luciferin by luciferase in the presence of ATP in an in vitro reaction performed using cell exctracts. Cotransfection with a β-galactosidase expressing vector is commonly used to compensate for variations in transfection efficiency/sample manipulation. By generating a series of deletion mutants of the caveolin-1 promoter fused to the luciferase gene, we have identified two independent sequences within the caveolin-1 promoter that responded to oxidative stress.

1. Cells are seeded in 6-well plates with 270,000 cells per well. The following day, cells are transiently transfected with 2 µg of the caveolin-1 promoter-luciferase reporter constructs or the luciferase reporter construct alone and a β-galactosidase expressing vector by calcium phosphate precipitation method (*see* Note 3).

2. Twenty-four hours posttransfection, cells are rinsed with PBS, subjected to oxidative stress (see Subheading 29.3.1), and recovered in complete medium for 48 h.

3. Cells are washed twice with ice cold PBS and lysed in 500 µL of Extraction Buffer on a rocker at 4°C for 30 min.

4. During this incubation, place 300 µL of ATP Mix into cuvettes for luciferase assay and 600 µL of Z buffer into eppendorf tubes for the β-galactosidase assay.

5. Pipet 200 µL of sample into each cuvette for luciferase assay and 150 µL of sample into each eppendorf tube for the β-galactosidase assay. Vortex briefly to mix.

6. Luciferase activity is then measured by assessing light production at 562 nm using a luminometer, which inject 100 µl of luciferin solution into each sample.

7. For β-galactosidase activity, add 50 µl of CPRG to the Z buffer/sample mix, and let the incubation proceed for 30 min at 37°C. The reaction is stopped by adding 200 µL of 1 M Na_2CO_3. Light production is then measured at 574 nm using a spectrometer.

8. Variations in transfection efficiency/sample manipulation among the experimental points are compensated for by adjusting the luminometer readings for β-galactosidase readings (β-galactosidase activity is considered a nonvariable factor among the experimental groups so that variations of β-galactosidase readings reflect variations in transfection efficiency/sample manipulation).

9. Figure 29.1a shows that while Cav-1 (-1296/-1), Cav-1 (-800/-1), and Cav-1 (-372/-1) were activated by oxidative stress by ~15-fold, Cav-1 (-222/-1), and Cav-1 (-150/-1) showed instead only a fourfold induction upon hydrogen peroxide treatment (Fig. 29.1a). H_2O_2 did not activate the first 91 nucleotides of the caveolin-1 promoter (Fig. 29.1a). We concluded from these data that each of the nucleotides-372/-222 and -150/-91 of the caveolin-1 promoter contains an oxidative stress responsive element. Interestingly, both regions of the caveolin-1 promoter contain GC-rich boxes, which represent putative binding sites for the Sp1 transcription factor.

10. In Fig. 29.1b, we demonstrate that the two caveolin-1 promoter sequences containing GC-rich boxes (-244/-222 and -124/-101) were indeed oxidant-responsive elements. In fact, when these two sequences were fused to the luciferase gene, they were able to respond to hydrogen peroxide.

3.3. Electrophoretic Mobility Shift Assay

Electromobility Shift Assay, or EMSA, allows for detection of sequence-specific DNA-transcription factor interactions. It is based on the premise that free DNA will run quicker than DNA bound to a protein when resolved on a polyacrylamide gel. When an antibody against the protein of interest is introduced into the mix, it further adds to the retardation of the complex in the gel matrix and allows the identification of the transcription factor, which binds the DNA sequence under investigation. This is deemed a supershift. Incubation with excess unlabeled double-stranded oligonucleotides representing the consensus sequence for the investigated transcription factor prevents the formation of the labeled DNA-protein complex and is often used to corroborate supershift data. Although ^{32}P-labeled DNA has been the cornerstone for EMSA studies, there are alternative ways to label DNA, including the biotin-based method used in this study. Here, by using EMSA analysis, we found that oxidative stress promotes binding of nuclear proteins to Sp1 consensus elements within the Cav-1 promoter.

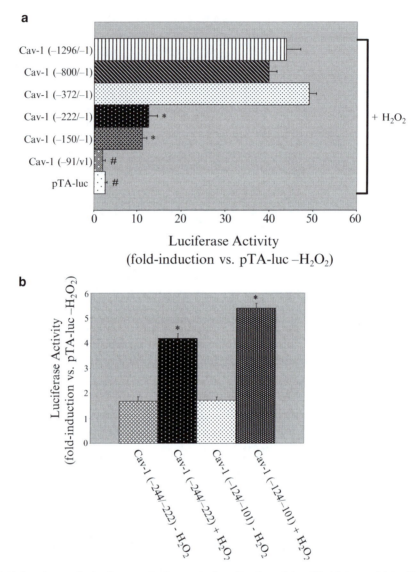

Fig. 29.1. Oxidative stress activates the caveolin-1 promoter by acting through two GC-rich boxes. (a) *Luciferase assay.* Caveolin-1 promoter deletion mutant constructs were transiently transfected in NIH 3 T3 cells. pTA-luc alone was used as a control. Twenty-four hours after transfection, cells were treated with or without 150 μM H_2O_2 for 2 h. Cells were collected 48 h after oxidative stress and luciferase activity measured. Values represent means ±SEM. */#$P < 0.001$. (b) *Luciferase assay.* The caveolin-1 promoter -244/-222 and -124/-101 regions, both containing a GC-rich box, were cloned upstream of the luciferase gene in the pTA-luc vector. These constructs were transiently transfected in NIH 3 T3 cells. pTA-luc alone was used as a control. Twenty-four hours after transfection, cells were treated with or without 150 μM H_2O_2 for 2 h. Cells were collected 48 h after oxidative stress and luciferase activity measured. Values represent means ±SEM. *$P < 0.001$. Figure adapted from ref. 30.

1. NIH 3 T3 cells are cultured in 10 cm dishes. They are treated with oxidative stress as described in Subheading 29.3.1. Control plates that are not treated are split to be the same confluency as the treated dishes at the time of the experiment.

2. Forty-eight hours after recovery from oxidative stress, nuclear extracts are prepared as follows: cells are washed twice with PBS and then collected in PBS. Cells are centrifuged at $1,500 \times g$ at 4°C for 10 min. The pellet is resuspended in 400 µL ice cold Buffer A, and incubated on ice for 10 min. Samples are vortexed on high for 10 s and centrifuged again for 10 s at $12,000 \times g$. The supernatant is saved (*see* Note 4). Nuclear pellet is resuspended by pipetting up and down in 100 µL ice cold Buffer B and incubated on ice for 20 min at 4°C. Samples are centrifuged for 5 min at 4°C at $16,000 \times g$ and the supernatant containing nuclear proteins is used for the experiment.

3. 3′ end biotin-labeled double-stranded oligonucleotides containing a GC-rich box (see Subheading 2.3, item 6 for sequence) are resuspended in an eppendorf tube with Annealing Buffer. The oligonucleotides are serially diluted to a final concentration of 50 fmol/µl. Equal volumes (2 µl) of complementary single stranded oligonucleotides are incubated in an eppendorf tube, and the tube is placed into a beaker with 500 ml of water at 95°C. The beaker is then allowed to cool down slowly to room temperature.

4. 5 µg of nuclear protein extracts from untreated and H_2O_2-treated cells are incubated with 100 fmol 3′ end biotin-labeled double-stranded oligonucleotides and 1× Binding buffer for 15 min at room temperature in a final volume of 20 µL.

5. *See* Note 5 for proper controls to run.

6. Samples are run on a 5% nondenaturing polyacrylamide gel in 1× TBE buffer. The gel is prerun at a constant voltage until the current no longer varies with time.

7. The gel is then transferred onto a positively charged Biodyne B nylon membrane at 380 mA for 60 min.

8. The DNA is then UV-light crosslinked to the membrane at 120 mJ/cm².

9. DNA is visualized using Chemiluminescent Nucleic Acid Detection Module (Pierce) according to manufacturer's instructions.

10. Membrane is exposed to film for 1–4 min depending on intensity of signal.

11. Figure 29.2a, b illustrate that oxidative stress promoted the formation of nucleoprotein complexes by the -244/-222 and -124/-101 caveolin-1 promoter sequences (Complex I and complex II, respectively). Because incubation with excess unlabeled double-stranded oligonucleotides representing the Sp1 consensus sequence totally prevented the formation of both complex I and complex II in Fig. 29.2a, b, these results

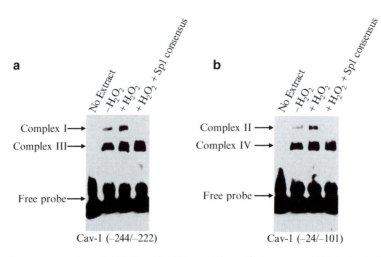

Fig. 29.2. Oxidative stress stimulates the binding of nuclear proteins to Sp1 consensus elements within the caveolin-1 promoter. (a) and (b) *EMSA studies*. Electrophoretic mobility shift assays were performed with nuclear extracts from untreated and H_2O_2-treated (150 μM for 2 h) NIH 3 T3 cells 48 h after oxidative stress. Nuclear extracts were incubated with either Cav-1 (-244/-222) (a) or Cav-1 (-124/-101) (b) biotin-labeled oligonucleotides. Lack of nuclear extract was used as a negative control. Note that two nucleoprotein complexes were identified in (a) (Complex I and III) and two in (b) (Complex II and IV). Incubation with excess unlabeled Sp1 consensus oligonucleotides was performed to show specificity of complex I and II. Figure adapted from ref. 30.

suggest that Sp1 may indeed represent the transcription factor that mediates the response of the caveolin-1 promoter to oxidative stress.

3.4. Chromatin Immunoprecipitation Assay

Chromatin immunoprecipitation (ChIP) assays can determine what transcription factors interact in vivo with a specific DNA sequence under certain conditions. Cells are exposed to a stimulus, and then, proteins that interact with the DNA are cross-linked to the DNA by formaldehyde. The DNA is broken into fragments by sonication, and antibodies precipitate the particular protein of interest attached to DNA. Reverse cross-linking and PCR amplification allow resolve whether the hypothesized protein-interaction takes place under the given conditions. This method is very efficient for determining transcription factors that bind to the Caveolin-1 promoter under oxidative stress. Using ChIP analysis, we identified that the transcription factor Sp1 binds to GC-rich elements within the Cav-1 promoter upon oxidative stress (30). Other groups have used ChIP assays to map transcription factor binding and epigenetic changes to the caveolin-1 promoter (32, 33).

1. Cells are subjected to oxidative stress as described in Subheading 3.1. Plates are washed twice with PBS and then incubated with regular media for 48 h. Untreated cells are used as controls.

2. Cells are crosslinked with 1% formaldehyde (final concentration), which is added to the cell plate containing 8 ml of growth media for 10 min at 37°C. The crosslinking is quenched by the addition of 0.8 ml of 1.4 M glycine to each plate for 5 min at 4°C. Media is promptly removed, and cells are washed twice with 4 ml of cold PBS, and are then scraped twice in 4 ml of PBS and placed into a 50 ml tube. An aliquot should be taken at this time to count the number of cells and the volume normalized for each sample so that there are approximately 5 million cells per time point for the assay.

3. Cells are spun down at $600 \times g$ for 10 min at 4°C, and the pellet is resuspended in 500 µl of cold ChromatinIP buffer. Cells are rocked at 4°C for 10 min and then spun down at $600 \times g$ at 4°C for 10 min.

4. The pellet is resuspended in 500 µl of cold Wash Buffer. Samples are rocked at 4°C for 10 min and then spun down at $600 \times g$ at 4°C for 10 min.

5. The pellet is resuspended in 500 µl cold RIPA buffer. Samples are rocked at 4°C for 10 min.

6. Samples are then sonicated in polypropylene tubes at amplitude of 21%, 2 pulses on 1 pulse off, for 10 s with three cycles (so that the DNA is fragmented to ~250–1,000 bp sizes) (*see* Note 6).

7. Samples are transferred to eppendorf tubes and spun down at max speed for 15 min. The supernatant is transferred to a new eppendorf tube and the final volume brought to 500 µl in RIPA buffer + protease inhibitors.

8. ChIP lysate is precleared with Protein A Sepharose beads conjugated to salmon sperm DNA for 30 min at 4°C. Afterward, samples are spun down and the supernant saved. About 10% of the sample (50 µl) is taken as input. The inputs are reverse crosslinked immediately by adding 950 µl ChIP Dilution Buffer with 40 µl 5 M NaCl and incubated at 65°C for 4 h. Inputs are saved at –20°C.

9. ChIP lysate is combined with 70 µl of Protein A Sepharose beads conjugated to salmon sperm DNA in addition to 1–3 µg of antibody and rotated over night at 4°C.

10. Beads are centrifuged at $600 \times g$ and washed twice with cold RIPA buffer for 5 min on rotation at 4°C, twice with cold RIPA buffer + 500 mM EGTA pH 8.0, and once with LiCl Buffer.

11. Complexes are eluted twice in 250 µl of fresh Elution Buffer.

12. Samples are reverse crosslinked by adding 20 µl of 5 M NaCl and incubating at 65°C for 4 h. To both ChIP samples and

thawed inputs, 10 μl of 0.5 M EDTA, 20 μl of 1 M Tris–HCl pH 6.5, and 2 μl of 10 mg/ml Proteinase K are added, and samples are incubated at 45°C for 1 h.

13. Inputs and sample DNA are recovered by using Qiagen PCR purification kit according to manufacturer's instructions and resuspended in 30 μl of H_2O (*see* Note 7).

14. PCR was conducted using the following sequences for primers: sense (5′ to 3′) CAG GCT CTC AGC TCC CCG CCG. The antisense strand was (5′ to 3′) GTA TAG AGG GGG GAA AGG CGC. For both ChIP samples and inputs, PCR reactions were performed using 5 μl of sample DNA, 0.3 μM of sense and antisense primers, 0.3 mM dNTPs, 1× PCR Buffer, 1 unit of Taq enzyme, and H_2O to a final volume of 50 μl. The PCR was run at (1) 94°C 5 min (2) 94°C 30 s (3) 58°C 30 s (4) 72°C 1 min (steps 2–4 were repeated 29 times) (5) 72°C 7 min and (6) 4°C 16 h.

15. 8 μl of TAE 6× DNA loading dye is added to 40 μl of ChIP PCR, and run out on a 1.2% agarose gel with ethidium bromide. PRC products are visualized by UV light and documented.

16. In Fig. 29.3, a ChIP analysis was performed on chromatin from untreated and hydrogen peroxide-treated cells using an antibody probe specific for Sp1. We found that oxidative stress increased binding of Sp1 to the caveolin-1 promoter region containing the two GC-rich boxes. This result shows that free radicals stimulate direct binding of Sp1 to the GC-rich boxes of the caveolin-1 promoter in vivo.

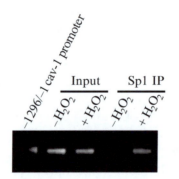

Fig. 29.3. Oxidative stress promotes the binding of Sp1 to GC-rich elements within the caveolin-1 promoter. *ChIP assay.* Chromatin immunoprecipitation assay was done on chromatin derived from untreated or hydrogen peroxide-treated (150 μM for 2 h) NIH 3 T3 cells 48 h after oxidative stress using an antibody probe specific for Sp1. PCR was performed using primers surrounding the region of the caveolin-1 promoter containing the two GC-rich boxes. Amplification of input DNA from both untreated and H_2O_2-treated cells was performed before immunoprecipitation. A vector containing the entire caveolin-1 promoter sequence was used as a positive control for PCR. Figure adapted from ref. (30).

4. Notes

1. For the Luciferase assay, GME buffer and Z buffer (lacking β-mercaptoethanol) can be stored as stocks. Extraction buffer, ATP mix, and Luciferin solution should be made fresh on the day of the experiment. Luciferin is light sensitive and should be protected from light by wrapping container in foil.

2. Oxidative stress generated by hydrogen peroxide has been used on a number of cell lines. There can be slight differences in the concentrations needed to activate the caveolin-1 promoter and upregulate caveolin-1 protein expression, depending on the cell type used. Therefore, it is recommended to do a concentration dose curve to determine which dosage is needed for a particular cell line. For example, incubation of NIH 3 T3 murine fibroblasts with $150 \mu M$ H_2O_2 for 2 h is sufficient to upregulate caveolin-1 protein expression after 48–72 h of recovery from oxidative stress (31).

3. We have used the calcium phosphate precipitation method to transfect our cells. However, other transfection methods will work well, and should be used based on the efficiency with a particular cell line. The amount of DNA should be adjusted according to the particular transfection method used.

4. Supernatant can be run on a protein gel as an internal control.

5. Controls for EMSA assay. Run DNA only without nuclear extract. This will have biotin-labeled oligonucleotides without protein bound to it. This control will establish the nonshifted DNA band. Additionally, a competition assay should be done. Excess unlabeled double-stranded oligonucleotides containing an Sp1 consensus site should be added (in 200-fold molar excess) to the labeled oligonucleotides with nuclear extract. There should be competitive binding of the unlabeled oligonucleotides and a decrease in the bandshift intensity of the labeled oligo-nucleoprotein complex.

6. It is of utmost importance to have the DNA sheared to 200–1,000 base pairs in length. This must be worked out before the full experiment is conducted. The degree of sonication necessary to shear to this size will differ between cell lines. It is recommended to treat cells as would be called for by the experimental design, and fix them in 1% formaldehyde at 37°C for 10 min. Quench with glycine as described in Subheading 29.3.4, step 2. Wash and collect cells in PBS. Count cells and normalize to approximately 5 million for each experimental point. Pellet cells for 5 min at $600 \times g$ at 4°C. Resuspend pellet in $200 \mu L$ of SDS Lysis Buffer and incubate

10 min on ice. Each experimental point can now be used to test different degrees of sonication. Add 8 µL of 5 M NaCl to lysate and incubate at 65°C to reverse crosslinks. Purify DNA by phenol/chloroform extraction, and run samples on an agarose gel with DNA ladder to determine fragmentation size.

7. Although we have had success with a Qiagen kit, there are other commercially available kits as well and should be used according to manufacturer's instructions. Purification is needed to remove salts from the mixture.

Acknowledgments

Ferruccio Galbiati and this work were supported by Grant AG022548 from the National Institutes of Health (NIH). Janine N. Bartholomew was supported by a National Institute of Health Predoctoral Training Grant in Pharmacological Sciences (T32GM008424).

References

1. Chen QM (2000) Replicative senescence and oxidant-induced premature senescence. Beyond the control of cell cycle checkpoints. Ann N Y Acad Sci 908:111–125

2. Lundberg AS, Hahn WC, Gupta P, Weinberg RA (2000) Genes involved in senescence and immortalization. Curr Opin Cell Biol 12:705–709

3. Dimri GP, Lee X, Basile G, Acosta M, Scott G, Roskelley C, Medrano EE, Linskens M, Rubelj I, Pereira-Smith O et al (1995) A biomarker that identifies senescent human cells in culture and in aging skin in vivo. Proc Natl Acad Sci U S A 92:9363–9367

4. Black EJ, Clark W, Gillespie DA (2000) Transient deactivation of ERK signalling is sufficient for stable entry into G0 in primary avian fibroblasts. Curr Biol 10:1119–1122

5. Sherr CJ, DePinho RA (2000) Cellular senescence: mitotic clock or culture shock? Cell 102:407–410

6. Wynford-Thomas D (1999) Cellular senescence and cancer. J Pathol 187:100–111

7. Kim NW, Piatyszek MA, Prowse KR, Harley CB, West MD, Ho PL, Coviello GM, Wright WE, Weinrich SL, Shay JW (1994) Specific association of human telomerase activity with immortal cells and cancer. Science 266:2011–2015

8. Lee SW, Reimer CL, Oh P, Campbel lDB, Schnitzer JE (1998) Tumor cell growth inhibition by caveolin re-expression in human breast cancer cells. Oncogene 16:1391–1397

9. Chen Q, Ames BN (1994) Senescence-like growth arrest induced by hydrogen peroxide in human diploid fibroblast F65 cells. Proc Natl Acad Sci U S A 91:4130–4134

10. Frippiat C, Chen QM, Zdanov S, Magalhaes JP, Remacle J, Toussaint O (2001) Subcytotoxic H2O2 stress triggers a release of transforming growth factor-beta 1, which induces biomarkers of cellular senescence of human diploid fibroblasts. J Biol Chem 276:2531–2537

11. Chen QM, Bartholomew JC, Campisi J, Acosta M, Reagan JD, Ames BN (1998) Molecular analysis of H2O2-induced senescent-like growth arrest in normal human fibroblasts: p53 and Rb control G1 arrest but not cell replication. Biochem J 332(Pt 1):43–50

12. Lisanti MP, Scherer P, Tang Z-L, Sargiacomo M (1994) Caveolae, caveolin and caveolin-rich membrane domains: A signalling hypothesis. Trends In Cell Biology 4:231–235

13. Couet J, Li S, Okamoto T, Scherer PS, Lisanti MP (1997) Molecular and cellular biology of caveolae: Paradoxes and Plasticities. Trends Cardiovasc Med 7:103–110

14. Okamoto T, Schlegel A, Scherer PE, Lisanti MP (1998) Caveolins, a family of scaffolding proteins for organizing "pre-assembled signaling complexes" at the plasma membrane. J Biol Chem (Mini-review) 273:5419–5422

15. Sargiacomo M, Scherer PE, Tang Z-L, Kubler E, Song KS, Sanders MC, Lisanti MP (1995) Oligomeric structure of caveolin: Implications for caveolae membrane organization. Proc Natl Acad Sci U S A 92:9407–9411

16. Li S, Song KS, Lisanti MP (1996) Expression and characterization of recombinant caveolin: Purification by poly-histidine tagging and cholesterol-dependent incorporation into defined lipid membranes. J Biol Chem 271:568–573

17. Murata M, Peranen J, Schreiner R, Weiland F, Kurzchalia T, Simons K (1995) VIP21/caveolin is a cholesterol-binding protein. Proc Natl Acad Sci USA 92:10339–10343

18. Song KS, Li S, Okamoto T, Quilliam L, Sargiacomo M, Lisanti MP (1996) Copurification and direct interaction of Ras with caveolin, an integral membrane protein of caveolae microdomains. Detergent free purification of caveolae membranes. J Biol Chem 271:9690–9697

19. Garcia-Cardena G, Oh P, Liu J, Schnitzer JE, Sessa WC (1996) Targeting of nitric oxide synthase to endothelilal cell caveolae via palmitoylation: implications for caveolae localization. Proc Natl Acad Sci USA 93:6448–6453

20. Scherer PE, Lisanti MP, Baldini G, Sargiacomo M, Corley-Mastick C, Lodish HF (1994) Induction of caveolin during adipogenesis and association of GLUT4 with caveolin-rich vesicles. J Cell Biol 127:1233–1243

21. Mineo C, James GL, Smart EJ, Anderson RGW (1996) Localization of EGF-stimulated Ras/Raf-1 interaction to caveolae membrane. J Biol Chem 271:11930–11935

22. Liu P, Ying Y, Anderson RG (1997) Platelet-derived growth factor activates mitogen-activated protein kinase in isolated caveolae. Proc Natl Acad Sci U S A 94:13666–13670

23. Smart EJ, Foster D, Ying Y-S, Kamen BA, Anderson RGW (1993) Protein kinase C activators inhibit receptor-mediated potocytosis by preventing internalization of caveolae. J Cell Biol 124:307–313

24. Smart EJ, Ying Y-S, Anderson RGW (1995) Hormonal regulation of caveolae internalization. J Cell Biol 131:929–938

25. Schnitzer JE, Liu J, Oh P (1995) Endothelial caveolae have the molecular transport machinery for vesicle budding, docking, and fusion including VAMP, NSF, SNAP, annexins, and GTPases. J Biol Chem 270:14399–14404

26. Feron O, Belhassen L, Kobzik L, Smith TW, Kelly RA, Michel T (1996) Endothelial nitric oxide synthase targeting to caveolae. Specific interactions with caveolin isoforms in cardiac myocytes and endothelial cells. J Biol Chem 271:22810–22814

27. Park WY, Park JS, Cho KA, Kim DI, Ko YG, Seo JS, Park SC (2000) Up-regulation of caveolin attenuates epidermal growth factor signaling in senescent cells. J Biol Chem 275:20847–20852

28. Cho KA, Ryu SJ, Oh YS, Park JH, Lee JW, Kim HP, Kim KT, Jang IS, Park SC (2004) Morphological adjustment of senescent cells by modulating caveolin-1 status. J Biol Chem 279:42270–42278

29. Cho KA, Ryu SJ, Park JS, Jang IS, Ahn JS, Kim KT, Park SC (2003) Senescent phenotype can be reversed by reduction of caveolin status. J Biol Chem 278:27789–27795

30. Dasari A, Bartholomew JN, Volonte D, Galbiati F (2006) Oxidative stress induces premature senescence by stimulating caveolin-1 gene transcription through p38 mitogen-activated protein kinase/Sp1-mediated activation of two GC-rich promoter elements. Cancer Res 66:10805–10814

31. Volonte D, Zhang K, Lisanti MP, Galbiati F (2002) Expression of caveolin-1 induces premature cellular senescence in primary cultures of murine fibroblasts. Mol Biol Cell 13:2502–2517

32. Kathuria H, Cao Y, Hinds A, Ramirez MI, Williams MC (2007) ERM is expressed by alveolar epithelial cells in adult mouse lung and regulates caveolin-1 transcription in mouse lung epithelial cell lines. J Cell Biochem 102:13–27

33. van den Heuvel AP, Schulze A, Burgering BM (2005) Direct control of caveolin-1 expression by FOXO transcription factors. Biochem J 385:795–802

Part IV

Biostatistics

Chapter 30

Meta-Analysis: Drawing Conclusions When Study Results Vary

Leslie Rosenthal and Enrique Schisterman

Summary

Low-dose aspirin has been suggested to positively impact a number of clinical outcomes associated with oxidative stress; however, results of clinical trials surrounding its effect on a woman's ability to achieve and sustain pregnancy have been inconclusive. A meta-analysis is an advantageous tool in this situation. Meta-analyses allow researchers to formally and systematically pool together all relevant research in order to clarify findings and form conclusions based on all currently available information. The purpose of this chapter is to describe how to perform a meta-analysis, clarify the impact of model selection, and provide examples of implementation.

Key words: Meta-analysis, Low dose aspirin, Oxidative stress, Fixed-effects model, Random-effects model

1. Introduction

Because of its low cost, wide availability, and apparent safety, low-dose aspirin has become an increasingly popular therapy to consider for a number of conditions requiring a treatment with anti-inflammatory, vasodilatory, and platelet aggregation inhibition properties. Its efficacy has been and is currently being investigated in relation to improving health issues ranging from cardiovascular disease (CVD) (1) to reproduction and sub fertility (2–10). Elevated levels of oxidative stress (OS) are associated with both of these conditions, as well as their respective risk factors. Low-dose aspirin has been suggested to have a cytoprotective effect against OS. As a result, improving understanding of how low-dose aspirin use affects oxidative stress may translate to improved understanding of its impact on corresponding clinical outcomes.

D. Armstrong (ed.), *Advanced Protocols in Oxidative Stress II,* Methods in Molecular Biology, vol. 594
DOI 10.1007/978-1-60761-411-1_30, © Humana Press, a part of Springer Science+Business Media, LLC 2010

While there has been a lot of research concerning the effects of low-dose aspirin use on both individuals at risk for CVD and individuals demonstrating signs of sub fertility who desire to become pregnant, the results of the research surrounding its effect on a woman's ability to achieve and sustain pregnancy have been inconsistent. These discrepant findings and/or findings occurring in trials that lack significant power have made it difficult to draw conclusions concerning the hypothesized benefits and, in turn, form inferences concerning the effects on oxidative stress.

Under circumstances like this where results of different studies are conflicting or inconclusive, it can be advantageous to perform a meta-analysis. Meta-analyses allow researchers to formally and systematically pool together all relevant research in order to clarify findings and form conclusions based on all currently available information. Essentially, this procedure allows researchers to address: "Was an effect seen in the trials included in the analysis?" or "Will an effect be seen in a given trial in the future?" (11). The purpose of this chapter is to describe how to perform a meta-analysis, clarify the impact of model selection, and provide examples of implementation.

2. Materials

Software for data extraction can be accessed through The Cochrane Collaboration website: http://www.cochrane.org (12). As noted in the procedure section of this chapter, data extraction should be performed with the use of structured forms, which are made available through this website.

3. Methods

3.1. Procedure

A meta-analysis is comprised of four main steps: identifying pertinent studies/ trials, determining criteria for inclusion and exclusion, data extraction, and data analysis (11).

1. To begin, one must first seek out all information concerning studies or trials relevant to the topic of interest. In order to ensure the scientific integrity of a meta-analysis, however, this must be done in a systematic and explicit manor to allow future researchers who follow the same procedure to achieve the same results. This generally begins with a thorough search of multiple electronic databases, for example, MEDLINE. All articles are then read to determine if they are applicable to the topic at hand. Next, the references of selected studies are

searched manually to ensure recovery of any articles related to studies not identified in the computerized search. For completeness, this step is repeated for any new studies identified. Moreover, the list is often verified through submission to an expert in the field, or by having a second independent researcher perform the search.

2. The second step is clearly defining the criteria for inclusion and exclusion of studies from the meta-analysis. Explicitly stating these criteria helps to guarantee reproducibility as stated in step one, as well as avoid unnecessary sources of bias (*see* Note 1). Such criteria include, but are not limited to, factors surrounding the design of a study or trial, the presentation of the study or trial, and treatments undergone in a study or trial.

3. Data from the identified studies (or trials) are then extracted for both determining eligibility and analyzing. All data extraction should be done with the use of structured forms. Details on data combination are described elsewhere (13).

4. Data from those studies (or trials) included in the meta-analysis are compiled and analyzed statistically. Analysis is performed with the use of either the fixed-effects (Mantel–Haenszel (14)) or the random-effects (DerSimonian and Laird (15)), or in rare cases the Peto (16) model to determine both a test statistic and confidence interval (CI), as well as any other relevant information (17). Table 30.1 provides formulas for these methods (*see* Notes 2 and 3).

The fixed-effects model weights the studies by the inverse of the variance of estimates and produces a CI that takes into account the random variation within each trial. This model aims to understand: "Was an effect seen in the trials included in the analysis?" Fixed-effects model weighting thus constitutes a direct synthesis of the literature (11, 18).

Conversely, the random-effects model includes a between-study component of variance. It is sometimes thought to be a more conservative approach, because the CIs produced are wider than those produced from the fixed-effects model due to the contribution of the inter-study variability to the total variability. 'Random-effects models use study power to estimate inter-study covariance, and the weighting of each study in the model is determined by both intra-study and inter-study variance. Accordingly, the random-effects model attempts to answer the question: "Will an effect be seen in a given trial in the future?"' (11, 18).

Thoughtful consideration should be used in determining which model is most appropriate for statistical analysis. While the random-effects model can sometimes be considered more conservative, its use is accompanied by a loss of precision. Moreover, the strong assumptions it requires may not be valid in practice.

Table 30.1
Formulas for calculating odds ratios and their corresponding 95% confidence intervals

	Model		
	Mantel–Haenszel	**Peto**	**Random-effects**
OR_M	$\dfrac{\sum(\text{weight}_i \times OR_i)}{\sum \text{weight}_i}$	$e^{\sum(O_i - E_i) / \sum \text{variance}_i}$	$e^{\sum(w_i^* \times \ln OR_i) / \sum(w_i^*)}$
	$OR_i = \dfrac{(a_i \times d_i)}{(b_i \times c_i)}$	$\ln OR_P = \dfrac{\sum(O_i - E_i)}{\sum \text{variance}_i}$	$\ln OR_{RE} = \dfrac{\sum(w_i^* \times \ln OR_i)}{\sum(w_i^*)}$
	$\text{weight}_i = \dfrac{1}{\text{variance}_i}$	$E_i = \dfrac{(e_i \times g_i)}{n_i}$	$w_i^* = \dfrac{1}{[D + 1/w_i]}$, $\quad w_i = \dfrac{1}{\text{variance}_i}$
variance_i	$\dfrac{n_i}{b_i \times c_i}$	$\dfrac{(E_i \times f_i \times h_i)}{(n_i \times (n_i - 1))}$	$\dfrac{n_i}{b_i \times c_i}$
95% Confidence Interval	$e^{\ln OR_{mh} \pm 1.96\sqrt{\text{variance } OR_{mh}}}$	$e^{\ln OR_P \,\backslash pm\pm 1.96\sqrt{\text{variance}_i}}$	$e^{\ln OR_{dl} + 1.96\backslash pm \times \sqrt{\text{variance}_s^*}}$, $\text{variance}_s^* = \sum \text{weight}_i^*$

"Petitti suggests that the random-effects model should be used only when the absence of inter-study heterogeneity can be assumed, because any significant between-study heterogeneity dominates the weights assigned to the studies (11). As a consequence, in the presence of substantial between-study heterogeneity, small and large trials become weighted the same, and the summary statistic is greatly affected by the inclusion of small trials in the analysis. The fixed-effects model, however, takes into account only between-study variances and thus weights studies according to [their] sample size." (18)

3.2. Applications and Results

1. In response to inconsistent evidence, Gelbaya et al. conducted a meta-analysis to examine the impact of low-dose aspirin use on women undergoing *in vitro* fertilization (IVF). They began by "[searching] four electronic databases – MEDLINE, EMBASE, Cochrane Controlled Trials Register (CENTRAL), and The UK National Research Register of ongoing and completed research projects undertaken in or for the UK National Health Service – from January 1980 to March 2006 using the key words '(aspirin or acetylsalicylic acid) and (IVF or ICSI).'" (19). To follow, they performed manual searches consistent with the above listed recommendations, and finally verified the findings through an independent search by a second investigator.

They outlined the inclusion and exclusion criteria, such as they would include "studies investigating the effect of low-dose aspirin alone or in conjunction with heparin or glucocorticoids on IVF or ICSI outcome" (19) and there would not be any language restrictions. As a result, they had six trials to be included in the meta-analysis. After extracting the data, they determined that due "to significant heterogeneity among trials, [they would use] the random-effects model (15) to derive the summary estimates of the effect of treatment" (19).

Statistical analysis was conducted to examine the rate of clinical pregnancy (CP) per embryo transfer (ET) between individuals ingesting daily low-dose aspirin and those who received a placebo, or, no treatment. Their results indicated no statistically significant "[difference] between aspirin and no treatment groups (RR 1.09, 95% CI 0.92–1.29)" (19). Furthermore, they had similar findings when they examined other outcomes of interest.

2. To assess and evaluate the conclusions mentioned above, Ruopp et al. conducted a *reanalysis* of the work described above (18). We began in a similar fashion, having two reviewers conduct independent searches of MEDLINE, Web of Science, EMBASE, TOXLINE, DART, and the Cochrane Database of Systematic Reviews and following through with a manual search of the references for each selected trial/ study.

Inclusion criteria included prospective trials using low-dose aspirin (<150 mg) during IVF randomized or matched controlled trials. Unlike Gelbaya et al., abstracts from conferences, subgroups of infertile patients, and frozen ET were included; however, Roupp et al. only included those written in English.

After determining which studies were to be included, we extracted and analyzed the data using 2×2 structured forms and Cochrane Review Manager Software (version 4.1; Update Software, Oxford, England), respectively. We calculated odds ratios, 95% confidence intervals, and heterogeneity. As opposed to Gelbaya et al., Ruopp et al. assumed fixed effects in utilizing tthe Mantel–Haenszel method to calculate heterogeneity. This allowed us to pool data from all of the studies and assign weights on the basis of sample size.

From analysis of the ten studies that met the selection criteria as well as a subgroup analysis evaluating only the studies included in Gelbaya et al., Ruopp et al. arrived at somewhat different conclusions. The results of this analysis are illustrated in Table 30.2, from Ruopp et al. When comparing clinical pregnancy per embryo transfer rates between the low-dose aspirin and no-treatment groups, Ruopp et al. found a significant risk ratio of 1.15 with data from all ten studies, and a risk ratio of 1.12 in the subgroup of studies selected by Gelbaya et al. (both greater than the value estimated by Gelbaya et al.). This and other similar results led

Table 30.2
Summary statistics and CIs using fixed-and random-effects models for full and subgroup analysis

	Model	
Outcome	Fixed effects, risk ratio (95% CI)	Random effects, risk ratio (95% CI)
Full analysis: all studies		
Pregnancy rate	1.15 (1.03, 1.27)	1.14 (0.95, 1.35)
Implantation rate	1.08 (0.69, 1.71)	1.00 (0.34, 2.93)
Miscarriage rate	1.19 (0.86, 1.65)	1.18 (0.85, 1.64)
Subgroup analysis: studies including only fresh ET		
Pregnancy rate	1.16 (1.04, 1.29)	1.15 (0.98, 1.36)
Implantation rate	1.32 (0.81, 2.16)	1.49 (0.50, 4.46)
Miscarriage rate	1.19 (0.86, 1.65)	1.18 (0.85, 1.64)
Subgroup analysis: studies included by Gelbaya et al.		
Pregnancy rate	1.12 (1.00, 1.25)	1.09 (0.92, 1.29)
Implantation rate	0.89 (0.48, 1.66)	0.89 (0.48, 1.66)
Miscarriage rate	1.17 (0.84, 1.63)	1.17 (0.84, 1.63)

Ruopp et al. to conclude that the effect of low-dose aspirin on women undergoing *in vitro* fertilization is still uncertain (18).

Variation between the two meta-analyses described above can be attributed to differences in statistical modeling, study selection, and statistical inference. Table 30.2 highlights these differences demonstrating how model selection can lead to different results in some cases. And, while the variation is not necessarily substantial, in some cases this can determine whether or not findings are considered significant.

4. Notes

1. "The systematic, explicit nature of the procedures for study identification distinguishes meta-analysis from [a] qualitative literature review" (11).

2. "The decision whether to use a fixed- or random-effects model is frequently misunderstood...[Although] both models produce

an estimate for the effect of treatment, the interpretation of the regression coefficient is subtly different" (18). When the aim is to understand if an effect was seen among data from the included trials, a fixed-effect model should be used. Conversely, the random-effects model should be used to address predictions for future trials.

The pregnancy rates compared in Table 30.2 demonstrate that while at times differences may be minimal, discrepancies between estimates calculated with the two models could affect conclusions concerning statistical significance. As a result, model selection may substantially impact inferences drawn.

3. "The Peto method performs well with sparse data and is then the best choice, but when events are common there is usually no preference to use it over the other methods. It is *not* a good idea to use the Peto method when the treatment effect is very large, as the result may be misleading. This method is also unsuitable if there are large imbalances in the size of groups within trials" (17).

References

1. Berger JS, Brown DL, Becker RC (2008) Low-dose aspirin in patients with stable cardiovascular disease: A meta-analysis. Am J Med 121:43–49

2. James AH, Brancazio LR, Price T (2008) Aspirin and reproductive outcomes. Obstet Gynecol Surv 63(1):49–57

3. Bordes A, Bied Dmaon VA, Hadj S, Nicollet B, Chomier M, Salle B (2003) Does aspirin improve IVF results? Hum Reprod 18(Suppl 1):119

4. Lentini GM, Falcone P, Guidetti R, Mencaglia L (2003) Effects of low-dose aspirin on oocyte quality, fertilization rate, implantation and pregnancy rates in unselected patients undergoing IVF. Hum Reprod 18(Suppl 1):40

5. Pakkila M, Rasanen J, Heinonen S, Tinkanen H, Tuomivaara L, Makikallio K et al (2005) Low-dose aspirin does not improve ovarian responsiveness or pregnancy rate in IVF and ICSI patients: a randomized, placebo-controlled double-blind study. Hum Reprod 20:2211–2214

6. Rubinstein M, Marazzi A, Polak de Fried E (1999) Low-dose aspirin treatment improves ovarian responsiveness, uterine and ovarian blood flow velocity, implantation, and pregnancy rates in patients undergoing in vitro fertilization: a prospective, randomized, double-blind placebo-controlled assay. Fertil Steril 71:825–829

7. Wada I, Hsu CC, Williams G, Macnamee MC, Brinsden PR (1994) The benefits of low-dose aspirin therapy in women with impaired uterine perfusion during assisted conception. Hum Reprod 9:1954–1957

8. Lok IH, Yip S, Cheung LP (2004) Adjuvant low-dose aspirin therapy in poor responders undergoing in vitro fertilization: a prospective, randomized, double-blind, placebo-controlled trial. Fertil Steril 81:556–561

9. Van Dooren IM, Schoot BC, Dargel E, Maas P (2004) Low-dose aspirin demonstrates no positive effect on clinical results in the first in vitro fertilization (IVF) cycle. Fertil Steril 82(Suppl 2):18

10. Waldenstrom U, Hellberg D, Nilsson S (2006) Low-dose aspirin in a short regimen as standard treatment in in vitro fertilization: a randomized, prospective study. Fertil Steril 81:1560–1564

11. Petitti DB (1994) Meta-analysis decision analysis and cost-effectiveness analysis. Methods for quantitative synthesis in medicine. Oxford University Press, New York

12. The Cochrane Collaboration (2002) Collecting data from relevant studies. The Cochrane Collaboration Open Learning Material. http://www.cochrane-net.org/openlearning/HTML/mod7-2.htm. Accessed 13 June 2008

13. Radhakrishna S (1965) Combination of results from several 2×2 contingency tables. Biometrics 21:86–98

14. Mantel N, Haenszel W (1959) Statistical aspects of the analysis of data from retrospective studies of disease. J Natl Cancer Inst 22:719–748

15. DerSimonian R, Laird N (1986) Meta-analysis in clinical trials. Control Clin Trials 7:177–188

16. Peto R (1987) Why do we need systematic overviews of randomized trials? Stat Med 6:233–240

17. The Cochrane Collaboration (2002) Combining studies. The Cochrane Collaboration Open Learning Material. http://www.cochrane-net. org/openlearning/html/mod12-4.htm. Accessed 13 June 2008

18. Roupp M, Collins T, Whitcomb BW, Schisterman EF (2008) Evidence of absence or absence of evidence? A re-analysis of the effects of low-dose aspirin in IVF. Fertil Steril 90:71–76

19. Gelbaya TA, Kyrgiou M, Li TC, Stern C, Nardo LG (2007) Low-dose aspirin for in vitro fertilization: a systematic review and meta-analysis. Hum Reprod Update 13:357–364

INDEX